The Successful Internship

PERSONAL, PROFESSIONAL, AND CIVIC DEVELOPMENT

The Successful Internship

PERSONAL, PROFESSIONAL, AND CIVIC DEVELOPMENT

THIRD EDITION

H. Frederick Sweitzer
University of Hartford

Mary A. King
Fitchburg State College

BROOKS/COLE
CENGAGE Learning™

Australia • Brazil • Japan • Korea • Mexico • Singapore • Spain • United Kingdom • United States

BROOKS/COLE
CENGAGE Learning™

The Successful Internship: Personal, Professional, and Civic Development, Third Edition

H. Frederick Sweitzer and Mary A. King

Publisher/Executive Editor: Marcus Boggs

Acquisitions Editor: Marquita Flemming

Assistant Editor: Christina Ganim

Editorial Assistant: Ashley Cronin

Technology Project Manager: Andrew Keay

Marketing Manager: Karin Sandberg

Marketing Assistant: Ting Jian Yap

Marketing Communications Manager: Shemika Britt

Project Manager, Editorial Production: Abigail Greshik

Creative Director: Rob Hugel

Art Director: Vernon Boes

Print Buyer: Rebecca Cross

Permissions Coordinator: Scott Bragg

Production Service: Pre-Press PMG

Copy Editor: Pamela Rockwell

Cover Designer: Andy Norris

Cover Artist: Britt Howe

Compositor: Pre-Press PMG

For product information and technology assistance, contact us at **Cengage Learning Academic Resource Center, 1-800-423-0563**

For permission to use material from this text or product, submit all requests online at **www.cengage.com/permissions**
Further permissions questions can be e-mailed to **permissionrequest@cengage.com**

Library of Congress Control Number: 2007943402

Student Edition:

ISBN-13: 978-0-495-38500-4

ISBN-10: 0-495-38500-X

Brooks/Cole
10 Davis Drive
Belmont, CA 94002-3098
USA

Cengage Learning is a leading provider of customized learning solutions with office locations around the globe, including Singapore, the United Kingdom, Australia, Mexico, Brazil, and Japan. Locate your local office at: **international.cengage.com/region**

Cengage Learning products are represented in Canada by Nelson Education, Ltd.

For your course and learning solutions, visit **academic.cengage.com**.

Purchase any of our products at your local college store or at our preferred online store **www.ichapters.com**.

Printed in the United States of America
1 2 3 4 5 6 7 12 11 10 09 08

This book is dedicated, with love

From Fred

To my sister, Sally Sweitzer, whose light and love shine so brightly on all of us and whose dedication, passion, and integrity are an inspiration to me.

And to my wife, Martha Sandefer. This time, more than ever, I couldn't have done it without you.

From Mary

To my son, Patrick, a natural teacher, my very best, full of heart, soul, and indomitable spirit.

And to my husband, Peter, whose patience, guidance, and understanding make it all possible.

The Koru

The Koru is a symbol used by the Maori culture
in New Zealand to represent new beginnings, growth,
and harmony. This symbol also represents a fern
slowly unfolding toward the light.

 *Contents*

Foreword   xvii

Preface   xix

Section One   Beginning the Journey   1

Chapter 1   Surveying the Landscape   3

Welcome to Your Internship   3
  A Few Basic Terms   4

What Can You Learn from an Internship?   5
  Personal Development   5
  Professional Development   5
  Civic Development   6
  So Why Do You Need a Book?   8

The Concepts Underlying This Book   10
  Experiential Education   10
  Predictable Stages   12
  Self-Understanding   13

The Internship Seminar   13

Overview of the Text   16
  Chapter Organization   16
  Chapter Exercises   17
For Contemplation   17
For Further Exploration   18
References   20

Chapter 2   Essentials for the Journey   22

Essential Personal Resources   22
  Your Learning Style   23
  Your Life Context   24
  Your Support Systems   26

Essential Attitudes   28
  Being Open-Minded   28
  Being Flexible   28
  Being Receptive   28
  Being Open to Diversity   29

Essential Skills 29
  The Skills and Habit of Reflection 30
  Communication Skills 36

Essential Knowledge 37
  Information About the Site 37
  Knowing What Is Expected 38
  On-Site Resources 39
  Campus Resources 39
  Liability Insurance 39
  Professional, Ethical, and Legal Issues 39
  Information About Personal Safety 41
  The Developmental Stages of an Internship 42

Essentials for Empowerment 42
  The Power of Positive Expectations 42
  The Power of Perspective 43
  The Power of Discussion 43
  The Power of Humor 44
  The Power of Mindfulness 44

Summary 45
For Contemplation 45
For Further Exploration 47
References 47

**Chapter 3 Framing the Experience:
The Developmental Stages
of an Internship 49**

The Developmental Stage Model 50
  Stage 1: Anticipation 51
  Stage 2: Disillusionment 52
  Stage 3: Confrontation 54
  Stage 4: Competence 56
  Stage 5: Culmination 57

Summary 58
For Contemplation 58
For Further Exploration 59
References 61

Chapter 4 Understanding Yourself 62

Introduction 62
  Meaning Making 62
  Dealing with Difference 63
  Recognizing the Stages 64
  *The Special Relevance of Self-Understanding
    for the Helping Professions* 65

Components of Self-Understanding   66
   Knowing Your Values   66
   Recognizing Reaction Patterns   68
   Appreciating Your Learning Style   69
   Recognizing Family Patterns   73
   Giving Thought to Your Cultural Identity   74
   For Helping and Service Professionals:
      Grasping Your Psychosocial Identity Issues   78
   Remembering What Motivates You   81
   Considering Unresolved Issues   82

Summary   83
For Contemplation   84
For Further Exploration   85
References   87

Section Two   Discovering the Field   89

Chapter 5 Experiencing the "What Ifs": The Anticipation Stage   91

The Tasks at Hand   93
   Examine and Critique   93
   Develop Relationships   93
   Acknowledge Concerns   94
   Clarify Role and Purpose   95

The Learning Contract   97
   The Importance of a Learning Contract   97
   Get Involved in Your Contract!   98
   The Timing of the Contract   98
   Fundamentals of the Learning Contract   99
   The Goals of the Learning Contract   99
   Choosing Activities   104
   Assessing Your Progress   105

Summary   107
For Contemplation   107
For Further Exploration   107
References   108

Chapter 6 Getting to Know Your Colleagues   110

Getting the Most from Supervision   111
   A Focus on Learning   112
   Developing Realistic Expectations of Your Supervisors   112
   Your Supervisor as a Person and a Professional   114

The Match Between You and Your Supervisors 114
Approaches to Supervision and Evaluation 118
Your Reaction to Supervision 121
Managing the Triadic Relationship 122

Co-Workers 123
Expectations 123
Acceptance Issues 124
For Helping and Service Professionals: Russo's Patterns of Adjustment 126

Summary 127
For Contemplation 128
For Further Exploration 129
References 130

Chapter 7 Getting to Know the Placement Site 132

Don't Skip This Chapter! 132

Lenses on Your Placement Site 133
Systems Concepts 133
Organizational Concepts 133

Background Information 134
History 134
Mission 134
Goals and Objectives 135
Values 135
Funding 137

Organizational Structure 139
Division of Responsibilities and Tasks 139
Coordination and Control of the Work 141

Human Resources 145
Communication Skills 145
Norms 145
Informal Roles 147
Cliques 147
Management Style 148
Staff Development 148

Organizational Politics 148
Power and Influence in an Organization 149

Organizations as Cultures 150
For Helping and Service Professionals: The Importance of Language 151

The External Environment 151
The Task Environment 151
The Sociopolitical Environment 152

The Organization and Your Civic Development   152
  For Helping and Service Professionals: The Civic Dimension   153
Summary   154
For Contemplation   154
For Further Exploration   156
References   156

Chapter 8 Getting to Know the Community   158

Introduction   158
The Community Context and the Civic Professional   159
  *For Helping and Service Professionals: But I Don't Want to Work
    in the Community!   159*
What Is a Community?   161
A Community Inventory   161
  Basic Information About Communities   162
  Assets and Needs   162
  Structural Considerations   163
  Human Resources   164
  Community Symbols   164
  Political Considerations   164
How Do I Find All This Out?   166
Summary   166
For Contemplation   167
For Further Exploration   167
References   168

Chapter 9 Getting to Know the Clients:
A Chapter of Special Relevance for
Helping and Service Professionals   169

Recognizing the Traps: Assumptions and Stereotypes   170
  Uncovering the Roots   172
  Getting Beyond the Traps   172
  Rethinking Client Success   172
Acceptance—The First Step   173
  Being Accepted by Clients   173
  Learning to Accept Clients   176
  Managing Value Differences   178
  Specific Client Issues   179
Personal Safety of the Professional   181
  Assessing and Minimizing Your Levels of Risk   182
  Facing the Fears   184
Summary   185

For Contemplation   185
For Further Exploration   186
References   187

Section Three   Facing New Frontiers   189

Chapter 10 Taking Stock and Facing Reality: The Disillusionment Stage   191

Thinking About Growth   192
The Human Side of the Internship   192
Experiencing Change   193

Taking Stock of Your Progress   194
Keeping the Contract Alive   195

Considering the Issues: Predictable Challenges   197
Hitting the First Bump   198
Rising Expectations   198
Anticipation and Actuality   198
Reactions and Responses   199

Considering the Issues: Variable Challenges   200
Issues with the Work   200
Issues with the People   201
For Those in Direct Service: Concerns with Clientele   205
Issues with the Site: Values, Systems, and Philosophy   210
Issues in Civic Development   211
Issues with Yourself   212

What Happened to My Internship?   215
Encountering the Unexpected   216
Realizing the Changes   216
Managing the Feelings   217
Engaging the Seminar Class and the Supervisors   218
Understanding the Crisis   219

Summary   221
For Contemplation   222
For Further Exploration   223
References   224

Chapter 11 Breaking Through Barriers: The Confrontation Stage   227

The Tasks You Face   227

Leading with Your Heart   228
The Power of Belief   229
The Power of Will   229
The Power of Effort   231

A Metamodel for Breaking Through Barriers:
    Eight Steps to Creating Change   231
  One Step at a Time   233
Summary   242
For Contemplation   243
For Further Exploration   243
References   245

Section Four Going the Distance   247

Chapter 12 Riding High: The Competence Stage   249
The Tasks at Hand   249
Taking Charge of the Journey   250
  The Concept of Competence   251
  A Time of Transformation and Empowerment   251
Enjoying the Ride   252
  Redefining Supervisory Relationships   253
  Becoming the New Me   254
Bracing for the Bumps   256
  Recognizing the Bumps   256
  Freezing the Moment   257
  Feeling Success on the Ride   257
  Transition Issues and New Realities   261
Preparing for the Profession   265
  Leaving Your Footprint   265
  Moving Beyond the Textbooks   265
Developing as a Civic Professional   266
  The Public Relevance of the Work   267
  Issues in Civic Development   267
  Civic Participation   268
Summary   269
For Contemplation   269
For Further Exploration   270
References   270

Chapter 13 Considering the Issues: Professional, Ethical, and Legal   272
Internship Issues   273
  Internship Role Issues   274
  Practice Issues   275
  Integrity Issues   275
  Intervention Issues   277

Professional Issues: A World of Responsibilities
    and Relationships   277
  Questioning the Professional Conduct of Others   278
  Questioning Your Own Professional Conduct   279

Ethical Issues: A World of Principles and Decisions   282
  Talking the Talk   283
  Rules of the Trade   283
  Ethical Principles and Ethical Values   285

Legal Issues: A World of Laws and Interpretations   288
  Talking the Talk   288
  Rules of the Trade   290
  Relevant Legal Matters   291

Grappling with Dilemmas   293
  Recognizing Dilemmas   294
  Walking the Walk   295
  Reasoned Steps to Resolving Dilemmas   295

Managing a Professional Crisis   298
  Have Resources in Place   298
  Expect to Learn from the Crisis   299
  Lay Out a Crisis Response Plan   299
  Practice Self-Awareness   300

Summary   300
For Contemplation   300
For Further Exploration   301
References   302

Chapter 14 Traveling the Last Mile: The Culmination Stage   306

Making Sense of Endings   306
  A Myriad of Feelings   307
  Changes and More Changes   307

Seizing the Opportunities   308
  Thinking About Endings   309
  Considering Your Competence   310
  The Tasks at Hand   310

Finishing the Work   311

Closure with Supervisors   312
  The Final Evaluation   313
  Preparing for the Final Conference   313
  The Final Conference   315
  Feedback for the Supervisor   315
  Ending the Supervisory Relationship   316

For Helping and Service Professionals:
      Saying Goodbye to Clients   317
  Timing and Style   318
  Unfinished Business   318
  Overarching Feelings   319
  The Client's Experience   319
  The Intern's Experience   320
  The Future   320
Bidding Farewell to the Placement Site   323
  Rituals and Remembrances   323
A Self-Evaluation: The Learning Contract and Beyond   325
  Preparing a Portfolio   325
Looking Ahead and Moving On   326
  A Fond Farewell . . .   327
For Contemplation   328
For Further Exploration   329
References   329

Index   331

Foreword

Benjamin Franklin wrote in his will, "I have considered that, among artisans, good apprentices are most likely to make good citizens. . . ." This third edition of *The Successful Internship* masterfully links the apprenticeship traditions of internships and the civic dimensions of experiential learning, such as service-learning, in a way that has been much needed in higher education. After years of educators trying to define and distinguish internships, service-learning, cooperative education, and so on, Sweitzer and King have achieved a synthesis of both the career dimensions and the civic dimensions of internships. Each chapter includes the relevant element of the civic dimension of the internship experience, such as civic professionalism, issues in civic development, and the civic context of the internship placement's environment.

As a director of internships and field study for twenty years, I appreciate the two earlier editions of *The Successful Internship* because it is a text on the EXPERIENCE of being an intern that weaves the theoretical, the personal, the professional, and now the civic aspects of holistic learning that is possible with carefully constructed and reflectively instructed internships.

The addition of "Essentials" as Chapter 2 addresses the practical issues, concerns, and anxieties that students (and faculty) have at the beginning of the internship experience. With these essentials addressed, students are then guided in subsequent chapters on their journey through the five stages of internships with practical tools, theoretical insights, and exercises to draw out the meanings of each stage. I have found that my students were able to learn better from their internships when they could name, understand, and analyze the stages that are the core of the internship experience.

One of the major enhancements of this edition is the authors' goal of producing a text for a wider audience than just the helping professions for which the original edition was intended. This goal is realized in a number of ways throughout the text. Examples are drawn from diverse fields such as business, humanities, government, education, and nonprofit agencies. The carefully designed format includes the text as well as boxed features, such as *Focus on Theory*, *Focus on Skill*, *Think About It*, and *In Their Own Words*; these provide a flexible menu that can be adapted to a variety of settings.

I am often asked how to teach using the experiences of students in internship placements and whether there are any textbooks. While there are only a few such texts, the previous editions of this one have stood out as unique texts that blend the personal, professional, and academic dimensions of internship learning. This third edition promises to be the most widely applicable internship text available in higher education. When paired with discipline-specific readings, this book forms the foundation of a sound curriculum for experiential education. As a pedagogy, experiential education has become much more sophisticated over the past four decades; this text not only reflects that development but also the promise of furthering our understanding of how to enhance the learning potential that is inherent in the successful internship.

Dwight E. Giles Jr., Ph.D.
Professor of Higher Education
Senior Associate, New England Resource Center
   for Higher Education
University of Massachusetts, Boston
October 2007

 *Preface*

In this edition of *The Successful Internship*, as in the previous one, we chose the Koru as the symbolic representation of the internship. This symbol, taken from Maori culture, is meaningful to us in several ways, one of which is its depiction of a fern unfolding toward the light. Like the fern, interns unfold as they learn through experience. The new beginnings an internship affords the student are replete with opportunities for growth, leading to a harmony of one's self in relation to one's work. Our thinking about internships also has unfolded over time and continues to do so. Thus, this latest edition is more than an update of research in the field; it reflects the evolution of our thinking about the roles of growth and development in learning through experience. William Sullivan writes that the heart of professionalism is the dynamic interplay of theory and practice; our thinking has been influenced by our own theory, as well as by other theories we continue to discover, and by the experience we and others have had in applying those theories across a wide variety of academic disciplines. It is our hope that we have made a contribution to the understanding of the internship as a *deep* learning experience—one that has the capacity to transform and empower the learner. At the same time, we are aware of how the experiences of interns and faculty members have transformed our conceptual understanding of internships and experiential learning.

Part of the evolution of our thinking is that a fourth thread has been added to the three that were interwoven in past editions. Those threads are: a phenomenological approach to understanding internships; the idea of predictable stages and sequences of concerns; the critical importance of self-examination and reflection to the learning process; and, new to this edition, the internship as a vehicle for civic development.

This book is not primarily about the terminology, skills, and knowledge that interns need in order to do particular work in a particular setting or with a particular population. Nor is it a manual or workbook preparing students for internships. Rather, it is about the *lived experience* of the internship, the evolving process of self-knowledge, and the confrontation and overcoming of barriers to growth. The ideas, theories, and exercises in the text are designed to help interns make sense of the *experience* of the internship and of the emotions and reactions they go through every day and every week. These, we have found, are useful and vital across many different contexts and settings.

We continue to believe that there are some issues and concerns that students encounter at certain stages in their experiences while in the field. We use the concept of stages to describe a level or period of time in the *process* of development. We find that interns tend to progress through these stages in a reasonably predictable order, though not at any predictable rate of speed. We also find as interns move forward through the stages, that in any one stage they often find themselves reexperiencing earlier stages when faced with new experiences. Organizing interns' experiences in this way allows them to demystify and normalize the inevitable emotional ups and downs that occur over the course of an internship.

We continue to believe that a commitment to self-knowledge and self-examination makes for better people, better citizens, and better practitioners. It also helps interns recognize and be more mindful of the stages because individual learners experience the stages in their own unique ways. It is careful, consistent, and consciously intended reflection that links experience to learning of all kinds, and it is that type of reflection that is emphasized throughout the book. Nowhere is this connection more vital than in the domain of self-understanding.

Our interest in the civic development aspect of internships has been stimulated by three related sources. The first is a growing body of literature in higher education that emphasizes citizenship, engagement, and social responsibility as important outcomes of a college education, regardless of academic major. The second is our deep interest in and involvement with service-learning. The theory and practice of service-learning has long held that deliberate civic outcomes must be part of such experiences. The third is the work of the Carnegie Foundation for the Advancement of Teaching on the education of professionals. William Sullivan in particular has written about the concept of Civic Professionalism, emphasizing that professional education must ground students in the public relevance of their work.

Civic engagement not only refers to the college or university's relationship with the community at the institutional level; it refers to that relationship at the student level. Volunteerism and community service are familiar forms of civic engagement on college and university campuses. So, too, are the academically based internships, course-based practica, project-based partnerships with local agencies, co-op education, international education programs, community-based research, and service-learning. For many of these approaches, experiential education is the pedagogy used to ensure that optimal learning occurs. The internship is often regarded primarily as a vehicle for professional or academic development; we believe that an internship can be a powerful vehicle for civic and personal development as well.

We have been delighted to learn that students and instructors bring great diversity of field experiences to the use of this book. The book has been adopted by a wide variety of helping professions (which were its original audience) and has also been adopted by other academic disciplines and professions. Because the internship also serves different roles across academic programs, student use of this book varies. For example, in some cases, the students are in a culminating internship, and most of the specific skill development will have been accomplished earlier in their programs, while in other cases, internships and field experiences are woven into the entire program, and skill development proceeds in tandem. Some students will use this book as part of an on-campus or on-line seminar that accompanies their experience; that is how we use it. But others will be using it as a self-guide or a resource, selecting chapters based on what is important at the moment and communicating with instructors individually, though not necessarily meeting in groups.

Because of this diversity, we have tried to give interns and instructors as much flexibility as possible in using the book. We have included concepts and examples from a range of professions. We have included resources for further exploration so that students and instructors can build on those areas that are of interest and relevance. We have also provided a wide range of reflective questions at the end of each chapter, anticipating that students and instructors will choose from among them. Some of those questions work best for individual reflection and some are designed for group work. We have also tried to create a book that can be augmented or supplemented with more discipline- and skill-specific assignments, readings, and instruction.

Our experience has been that students go through the stages in the order we discuss but often at different rates of speed and emotional intensity. In our seminar classes, we do cover the chapters in order and at a particular time in the internship, and that seems to work for the majority of the group. However, we also listen carefully to what individual students are saying in class, on-line, and in their journals, and we often suggest that a particular student reread a chapter or skip ahead to future ones as needed.

The chapters do not need to be read in the order in which they are presented, with some caveats. Chapters 1–4 are sometimes read before the internship or field experience begins or as part of a prerequisite seminar. Some instructors will want to have students read the chapter on legal and ethical issues much earlier. Others may want to have students read and think about clients before colleagues or organizations. Still others, depending on the context, may want to have students consider the community or organizational context before turning to consideration of either clients or colleagues. We do suggest, though, that the order of the stages be preserved. Chapters 5–9 pertain to the Anticipation stage, and we recommend reading them, in whatever order seems appropriate, before moving on. Chapters 10 and 11 should come next; they speak to the concerns of the Disillusionment and Confrontation stages. Chapters 12 and 13 pertain to the Competence stage. And Chapter 14 is designed to help students think about the ending of the internship. It should be read, though, in enough time for students to anticipate and plan for the concerns and issues of the Culmination stage.

THE THIRD EDITION

Each chapter of the book has been updated and augmented, but some particular changes are worth emphasizing:

- Most chapters include a section on civic development.
- We have found that although the full treatment of the stage-related concerns makes sense in the order in which they are presented, there are some critical attitudes, skills, personal resources, and pieces of knowledge that interns need as they begin. Therefore, Chapter 2 focuses on these essentials. Some students and instructors may find that they do not need to spend a lot of time on this chapter; for others, it will require a careful review and some focused practice.
- In former editions, there were two chapters on self-understanding; they have now been condensed into one chapter (Chapter 4).

- We have used a number of boxed features in the book to emphasize certain aspects of the internship. The main text of the chapters flows past these boxes so that instructors and students can choose the ones they want to read. These boxes include *Focus on Theory, Focus on Skills, Think About It,* and *In Their Own Words.*

- Because we have tried to make the book more inclusive of a wider variety of professions, we have also included separate sections called "For the Helping and Service Professions" so that interns in these professions can continue to get what they need and others may choose to skip these sections.

- We have reordered the chapters in the previous edition of the book. Because not all interns, even in the helping professions, are doing direct service with clients, we begin the second section with aspects of the internship that are more universal—colleagues (especially supervisors), the organization, and the community—and we end the section with a chapter on clients.

WITH SPECIAL APPRECIATION

This book is about the *lived* experience of an internship. Our lived experience of learning and writing about internships includes many colleagues and friends to whom we owe a great debt of thanks. Most importantly, though, we are indebted to the students, both graduate and undergraduate, across many disciplines, who inform our thinking and educate us each time we have the opportunity to be part of their journeys in learning.

Our professional lives are enriched by our contact with two different (although sometimes overlapping) groups of colleagues: those involved in human services education and those involved in civic engagement and the broader arena of experiential education, particularly with internships, co-ops, and service-learning. We recognize the contributions of national organizations such as the National Society for Experiential Learning (NSEE), the National Organization for Human Services (NOHS), the former American Association of Higher Education (AAHE), and the Association of American Colleges and Universities (AACU). They have been inspiring professional homes for us and have provided us with forums to learn as well as present and discuss our ideas. Through these organizations, we have been introduced to the work of a number of colleagues whose publications, presentations, or personal conversations have had an impact on our thinking. In the field of human services education, we especially acknowledge Cynthia Crosson-Tower, Georgianna Glose, Jackie Griswold, Mark Homan, Will Holton, Lynne Kellner, Pam Kiser, Heather Lagace, Ed Neukrug, Tricia McClam, Lynn McKinney, Diane McMillen, Trula Nicholas, Bill Oswald, Vicky Totten, and Marianne Woodside. In the fields of civic engagement and experiential education, we acknowledge Gene Alpert, Richard Battistoni, Al Cabral, Janet Eyler, Andy Furco, Michael Goldstein, Jeffrey Howard, Lynne Montrose, Gene Rice, Roseanna Ross, John Saltmarsh, Lee Shulman, William Sullivan, Nancy Thomas, Mike True, and Ed Zlotkowski. Very special appreciation and recognition go to Dwight Giles and Garry Hesser. Their friendships and support along with their contributions to this field have affected both our thinking and the direction of our work in important ways.

Many reviewers contributed their time and expertise to making this a better book: Eugene Alpert, The Washington Center for Internships and Academic Seminars; David Cessna, past president of the Cooperative Education and Internship Association (CEIA); Dwight Giles, University of Massachusetts, Boston; Iris Wilkinson, Washburn University; Laura Woliver, University of South Carolina-Columbia; Roberta Magarrell, Brigham Young University; Janet Hagen, University of Wisconsin–Osh Kosh; Jill Jurgens, Old Dominion University; Garry Hesser, Augsburg College; Linda Long, SUNY Corning Community College; and Heather Lagace, University of Hartford, provided thorough reviews, astute insights, and valuable suggestions at various stages in the writing process.

In addition, our heartfelt thanks go to Professor Steve Eisenstat at Suffolk University School of Law for giving generously of his time and expertise to the chapter on legal issues; to Janice Ouellette, librarian extraordinaire at Fitchburg State College, for her ongoing, commendable sleuth work; and to Maile Moore and Lindsey Kyte, technical assistants, who kept pace with us all the way.

Our appreciation goes to the staff at Cengage. Marquita Flemming stayed with us during times of transition for us and for the publishing industry. She showed real interest in us and our work and was as flexible and patient as anyone could hope for. As this book goes into production, Marquita has moved on to new professional challenges and we wish her the very best. Her assistant, Ashley Cronin, made the transition seamless for us. In addition, we thank Brenda Ginty, Production Director, and Tanya Nigh, Senior Content Project Manager. We would also like to thank Lisa Gebo, who has moved on to other professional frontiers but who has been an invaluable colleague, and Claire Verduin, who was relentless in getting the first edition of this book to the market.

Our families and friends give meaning and depth to our lives and inspiration to our work. Family members Charlie and Phyllis King, Skip and Betty Sweitzer, Sally Sweitzer and Britt Howe, T. Frederick Sweitzer-Howe (go Freddie!), and Phyllis and Dave Agurkis deserve special mention. So, too, do our dear friends Jeff and Judy Bauman, Steph Chiha, Judith and George Collison, Vicky Day, Mary Ann Hanley, Anita Hotchkiss, Margot Kempers, Regina Miller, Peter Oliver, Ken Pollak and Peter Gabriel, Sam Pukitis, Bev Roder, Kathy Rondeau, Gin Sgan, Nancy Thomas, and Mary Jean Zuttermeister. Our partners in marriage, Martha Sandefer and Peter Zimmermann, and Mary's son, Patrick Zimmermann, as well as Tikka, Max, Ravi, and Misty, have our enduring thanks and love. And, now, they have much more of our time.

ABOUT THE AUTHORS

H. Frederick Sweitzer is Professor of Human Services at the University of Hartford in Connecticut, where he also serves as Associate Dean of the College of Education, Nursing and Health Professions. Fred has over thirty years' experience in human services as a social worker, administrator, teacher, and consultant. He has placed and supervised undergraduate interns for twenty years and developed the internship seminar at the University of Hartford. Fred brings to his work a strong background in self-understanding, human development, experiential education, service-learning, civic engagement, professional education, and group dynamics. He is on the editorial boards for the journals *Human Service Education and Human Services Today* and has published widely in the field.

Mary A. King is Professor Emerita at Fitchburg State College in Massachusetts, where she instructs seminar classes for interns and colloquia in service-learning. She has held administrative positions in academic programs and has placed and supervised interns in a variety of academic majors. Her teaching specialties include applied ethics across disciplines in the behavioral sciences. She has published in the fields of human services and experiential education and holds several professional licenses. Mary brings to her academic work a background in public education, juvenile justice, clinical counseling, and consultation. She has served on national and regional boards in human service education and local social service boards and presently serves on the board of the National Society of Experiential Education, overseeing its Experiential Education Academy.

BEGINNING
THE JOURNEY

CHAPTER 1

Surveying the Landscape

Education is revelation that affects the individual.
GOTTHOLD EPHRAIM LESSING, 1780

(Internship) gives meaning to everything you have learned and makes practical sense of something you've known as theoretical.
STUDENT REFLECTION

WELCOME TO YOUR INTERNSHIP

You are beginning what is, for most students, the most exciting experience of your education program. Chances are you have looked forward to this experience for a long time. You've probably heard your share of stories—both good and bad—from other, more experienced students. And while you may be a minority on your campus, you join virtually thousands of other students all over the country. An intensive field experience is a critical component of many academic programs—including psychology, sociology, and criminal justice—and of almost every human service education program (Simon, 1989). Many human services students conduct their internships in social service settings; others choose corporate, government, or research settings. Field-based learning, such as internships, is also an option in many other liberal arts and professional studies programs, such as communication media, journalism, and business. There are many types of field-based learning experiences besides internship, including field work, field experience, co-op education, field education, and practicum; but there is little to no consensus on the meaning of all those terms. We will use *internship* to refer to those learning experiences that involve receiving academic credit for learning at an approved site, under supervision, for at least eight hours per week over the course of a semester.

3

A Few Basic Terms

Although internships exist at many colleges and universities, different language is often used to describe the various aspects of the experience and the people associated with it. For example, the term *supervisor* sometimes refers to a person employed by the placement site and sometimes to a faculty or staff member on campus. So, at the risk of boring those of you who are very clear about these terms, we take a moment now to be clear about what we mean by them.

- **Campus Supervisor(s) or Instructors**   These are the faculty or staff members on your campus who oversee your placement. These are the people who may have helped you find the placement, who may meet with you individually during the semester, visit you at the site, hold conferences with you and your supervisor, conduct a seminar class for you and your peers, evaluate your performance, or do all the above. It is possible for up to three different people to fill these roles. Even though they may go by different titles on various campuses (internship coordinator, seminar leader, facilitator, and so on), for simplicity's sake, we will use this one term to refer to all those roles. Please note that in using the term *instructor*, we are describing a role, not an academic rank. People filling these roles may be staff members. They may also be part-time or full-time faculty members and may hold faculty rank up to Full Professor.

- **Co-Workers**   This term refers to the other people who work at the placement, regardless of their title, status, or how much you interact with them. If there are other students at the site, from your school or some other school, they are functioning in the role of co-worker when you are at the placement site.

- **Intern**   This is the term that refers to you, the student who is working at the site, even though you may not be called an intern at your college or university.

- **Placement or Site**   This term refers to the place where you are working, and sites can vary quite a bit. It could be a social service agency, a corporate setting, a college or university office, a hospital, or a school. Through the process of finding a placement, you are probably aware of the incredible variety of opportunities that exist in the community.

- **Site Supervisor**   Your site supervisor is the person assigned by the placement site to meet regularly with you, answer your questions, and give you feedback on your progress. Most placements assign one site supervisor to one student, although in some cases, there may be more than one person fulfilling these functions. Some academic programs use the term *field instructor* to describe this person in order to emphasize the educational (as opposed to managerial) nature of the role.

- **Clients or Population**   This term refers to the people who are served by your placement site or with whom the site does business. Given the wide variety of internships, it is not possible to use one term that works in all settings. For example, in human or social service settings, the term *clients* is very common, but the people served are also called *customers, consumers, residents, students,* or *patients*, depending in part on the philosophy of the site and the nature of the work. Other organizations, such as advertising agencies or public relations firms, have clients as well, although of a different nature and with different needs.

Still other settings, such as business or retail, use the terms *customer* or *consumer* more commonly. We will use the terms *clients*, *population*, and sometimes *clientele* to refer to these individuals and groups.

WHAT CAN YOU LEARN FROM AN INTERNSHIP?

As is suggested by the title of this book, we view a successful internship as one that facilitates three significant aspects of your development: personal, professional, and civic. You enter the internship at different points in your development in these three categories, and with care and attention, you can (and we hope you will) grow in all three areas as well.

Personal Development

The internship is an opportunity for intellectual and emotional development that may be important for an internship but will also be important in your life, whatever path you choose. The ability to look critically at information, as well as to think creatively, and to look at issues from multiple viewpoints are essential abilities. So is the ability to communicate clearly both orally and in writing. Solving problems and working in teams are abilities that will serve you at home, at work, and in the community. Many of these abilities are traditional outcomes of a liberal education (Crutcher, 2007), but they are also critical components of any profession (Lemann, 2004).

If you give yourself a chance, you can learn a tremendous amount about yourself. The internship can be a powerful catalyst for personal growth, providing opportunities to have a sense of your potential through work under the supervision of experienced and qualified supervisors. There will be opportunities to accomplish tasks independently and test your creative capacities while doing so. You'll learn more in Chapter 12 about what makes an internship feel satisfying.

Professional Development

Some students enter an internship primarily for career exploration. They may be studying a traditional liberal arts discipline such as sociology or psychology and want to see some ways in which those disciplines are put into practice. For other students, the internship is the culminating academic experience in a highly structured and sequenced set of experiences and can be a chance to pull together and apply much of what they have learned. And, of course, there are internships whose purpose falls somewhere in between these two positions. For everyone, though, the internship is a chance to take the next step, to acquire more of the knowledge skills, attitudes, and values of a profession, and to explore how well it fits with personal interests and strengths.

The internship also affords you the opportunity to understand the world of work in a more complete way than you do now. Even if you have had full-time jobs, presumably your internship is taking you into an area in which you have little professional experience. It is also an opportunity to become socialized into the norms and values of a profession (Royse, Dhooper, & Rompf, 2007).

Many internship programs also emphasize academic learning: that is, the applied learning of a particular academic discipline. Internships are a wonderful opportunity for this sort of learning, and in some internships, it is the primary purpose. Even if your goal and primary purpose are to enter a profession, there is an academic component to your learning.

Civic Development

Civic development refers to the development of personal and professional knowledge, skills, and values for participation in a healthy democracy (Colby, Erlich, Beaumont, & Stephens, 2003; Howard, 2001). An internship can be a vehicle for civic development in two ways: The internship can help you develop knowledge, skills, and values that will make you become a more responsible and contributing member of your community and society regardless of where you live and what you choose to do for your life's work; and the internship can help you become what William Sullivan calls a "Civic Professional" (2005). Because this is an aspect of the internship that in our experience is the least well understood by most interns, let us spend some time elaborating on these two aspects of civic development.

CIVIC LEARNING

Regardless of what profession you enter, or even whether you enter a profession at all, the internship is an opportunity for you to learn some of the knowledge, skills, and values that will help you participate fully and productively in your community. Several authors have written about the domains of civic learning (Battistoni, 2006; Colby et al., 2003; Howard, 2001), and we will discuss them in more detail when we discuss your learning contract, but here are a few examples. Civic *knowledge* might mean learning not just about the challenges faced by the people your profession serves but about some of the historical and current social forces that bring about those challenges. It might mean, for example, learning that people who are hungry, poor, or underemployed are not necessarily lazy or unintelligent but that their condition results at least in part from social conditions over which they have no control (Godfrey, 2000). Civic *skills* might mean learning how to advocate successfully for change in a workplace, a neighborhood, or a community to make conditions there more equitable. And civic *values* might include the belief that understanding social issues is an obligation for everyone, not just those in politics or journalism.

CIVIC PROFESSIONALISM

In order to understand the term *civic professional*, you need to understand the word *professional* a bit more deeply than perhaps you do now and then consider where the civic component fits in.

Being a Professional Many students perceive the internship as a chance to (finally) learn to actually *do* something in their academic major or in an area of interest or curiosity. Although you may have had courses that developed particular skills, the internship is a chance to test and improve those skills and acquire lots of new ones as well. Skill development, though, is only one of the possible goals and outcomes of your experience. Being a professional, or learning how to be one, means far more than learning new skills. A profession is based on solid and evolving theory, referred to in some instances as "evidence based practice" (Baird, 2007); and, as we mentioned earlier, every profession has core attitudes and values. For those in the helping professions, empathy might be such a value. Other examples of professional values include fairness, a belief in change, and integrity.

Internships are often described as a time when theory is applied to real-life settings; we believe that the relationship between theory and practice is more complex than that. The internship is a chance to *develop* the relationship between theory and practice, for each should inform the other (Sgroi & Ryniker, 2002). According to Sullivan (2005), professionals excel at the art of what he refers to as Practical Reasoning, which literally means reasoning in and about practice. Professionals must move with fluidity between their understanding of theory and the real, human situations that they face (which do not always quite conform to how the professional understands that theory). This movement is not easy; theory is abstract and relatively objective, whereas the human context of your work is entirely subjective and concrete. Your experience will help you see where the theories do not quite apply or where you need to search for a new theoretical model to help you. Thus, theories are transformed through their application, and you will be actively involved in that process as an intern.

Being a Civic Professional   A *civic professional*, then, is one who embraces and always attempts to understand the human context of the work. For some professions, such as counseling, this context begins with the individual. For all professions, however, it includes a broad and complex social context of families, diverse cultures, communities, and political dynamics.

A civic professional is also someone who understands the *public relevance* of the profession. To quote Sullivan, "To neglect formation in the meaning of community, and the larger public purposes for which the profession stands, is to risk educating mere technicians for hire in place of genuine professionals" (2005, p. 254). Each profession has an implicit contract with society. Some professions exist only to serve society, and they are funded largely by society because of the public value placed on that service. Even those professions, however, must grapple with the nature of their social mission or contract. For example, there is a history of debate within the field of criminal justice about whether its primary purpose is to protect society in the short term by incarcerating, monitoring, and/or punishing those who have committed crimes or to try and rehabilitate those who have committed crimes so that they may become productive, contributing citizens in the long term. Regardless of what you believe about that issue, an internship in criminal justice is an opportunity to explore it.

All professions also have ethical and moral obligations to the society in which they function, and the work of each professional is by definition connected to a larger social purpose. Journalism should be about more than entertainment; a free press should be an anchor of a healthy democracy. Even the intensely private domain of business can be seen as a public good as well as a private benefit (Waddock & Post, 2000). Business educators have argued that those entering the business world need to understand that business is about more than maximizing profits and creating wealth; that corporations should contribute to community issues such as social justice and ecological stability (Godfrey, 2000). The internship, then, is a chance for you to learn about the public relevance and social obligations of a profession and about how those obligations are (or are not) carried out at your internship site.

One hallmark of professional education is the use of an apprenticeship model (Sullivan, 2005). If professionals are expected to apply their knowledge in a human context, then at least some of their preparation must take place in that context. Again, not just the skills but the knowledge and the values of the profession are learned by doing. This approach to learning has long been understood by the service-learning movement and used for a variety of learning outcomes (Eyler & Giles, 1999; Howard, 2001; Saltmarsh, 2005). The internship has the same potential; it is an intensive form of apprenticeship.

So Why Do You Need a Book?

The internship is a learning experience like no other. In a classroom learning experience, you learn through readings, lectures, discussions, and exercises. These are the raw materials you are given. You bring your ability to memorize, analyze, synthesize, and evaluate to these materials—these are your learning tools. The arena for learning is the classroom, and classrooms vary in the amount and quality of interaction between students and the instructor and among students. Some can be very interactive places, with mutual dialogue among students as well as between students and teachers. Other classrooms are interactive but only between teacher and student; it is almost as if there is a multitude of individual relationships being carried on in isolation. The internship is different in every way, and this book is designed to help you and your instructor with this new approach.

Experience, both intellectual and emotional, is the raw material of the internship. You will be learning mostly through experience, although you may engage in some traditional academic activities as well. However, one noted theorist in the field of experiential education, David Kolb, suggests that experience alone does not automatically lead to learning or growth. Rather, the experience must be processed and organized in some way (D. A. Kolb, 1984). More specifically, Kolb contends that you must think about your experience, sometimes in structured ways, and discuss it with others in reflective ways. Reflective dialogue with yourself and your peers is the primary tool for learning in the internship. This book is designed to help you structure that reflection and dialogue. It will invite you to think about your internship in a variety of ways, some of which may be new to you. It may also help you anticipate some of the challenges that await you and move successfully through them.

Furthermore, the internship is not just an intellectual experience. It is a *human* experience, full of all the wonderful and less-than-wonderful feelings that people bring to their interactions and struggles. This emotional, human side of the internship is more than a backdrop to the real work and the real learning; it is every bit as real and important.

Relationships are the medium of the internship; they are the context in which most of your learning and growth occurs. Many of you will work in direct service to others at your placement site, but even if you do not, you will be involved in relationships with a site supervisor, a campus supervisor, other interns on site and on campus, and co-workers at your placement. These relationships offer rich and varied opportunities for learning and growth. This book will help you think about these relationships, capitalize on the opportunities they present, and address any problems that may arise.

In addition to excitement and satisfaction, most interns also experience some real difficulties—moments when they question themselves, their career choices, their placements, or all the above. We like to think of these moments as crises but not in the way you are probably familiar with the term. We prefer to think of crises as the Chinese do. The symbol for crisis in Chinese is a combination of two symbols: danger and opportunity (Figure 1.1).

So, while there is some risk and certainly some discomfort in these challenges, there is also tremendous opportunity for growth. We hope this book will help you see both the dangers and the opportunities inherent in an internship and to grow from both. We will encourage you not to run from these crises but to meet them head on, with your mind and your heart open to the experience.

The truth is that some of you may not need a book; you will learn all you can and survive the challenges just fine. In other cases, you may want to use this book as a self-guiding tool; this is often the case when there is no seminar class to accompany the field experience (either in actuality or on the web). Some interns have used the book as a resource to draw upon as needed, especially if they are in the field for only one day a week, the class is one hour, and the focus of the class is the integration of what is being learned in the field with the theories of the academic major. In such a situation, there is little time to discuss the issues in this book in a meaningful way. We have tried to write the book so that you can do just that. However, we believe that for most students, this book—in combination with a skilled campus supervisor, supportive peers, and some of your time and energy—can add to your learning in meaningful and substantive ways.

We first conceived of this book for a human service internship audience, and there are an incredible variety of programs and placements that fall under that umbrella. Add to that the audiences in other professions, such as communications, nursing, and business, and the variety becomes even wider. So, too, is the amount of preparation

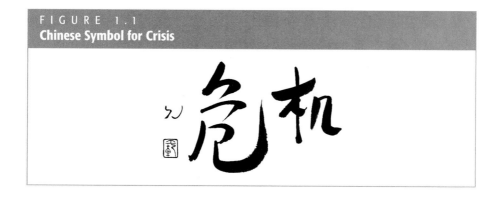

FIGURE 1.1
Chinese Symbol for Crisis

that students bring to these experiences. This book is meant as a guide to the phenomenological experience of the internship to help you anticipate and make sense of the emotional aspect of your work. But because of the varieties we just mentioned, it is very important that you consult your campus supervisor(s) early and often if needed, even if there is no seminar class that accompanies your experience. Decisions about what theories to explore and what to emphasize warrant at the least a consultation with the person overseeing your placement in the field.

THE CONCEPTS UNDERLYING THIS BOOK

Experiential Education

An internship, like other kinds of field instruction, is a form of experiential education. Although this approach to learning may not be well understood in many places on your campus, it comes out of a long theoretical and practical tradition, as discussed in the Focus on Theory box below. Experiential education is based on

---

Focus on THEORY

How Do You Learn As An Intern?

Experiential learning has philosophical roots dating back to the guild and apprenticeship systems of medieval times through the Industrial Revolution. Toward the end of the 19th century, professional schools required direct and practical experiences as integral components of the academic programs: for example, medical schools and hospital internships, law schools and moot courts and clerkships, normal schools and practice teaching, forestry/agriculture, and field work (Chickering, 1977). Experiential education is described by the National Society for Experiential Education (2006) as *learning activities that involve the learner in the process of active engagement with and critical reflection about phenomena being studied.*

Perhaps the best-known proponent of experiential education was the educational philosopher John Dewey (1916/1944, 1933, 1938, 1940). Dewey strongly believed that "an ounce of experience is better than a ton of theory simply because it is only in experience that any theory has vital and verifiable significance" (1916/44, p. 144). However, he was convinced that even though all real education comes through experience, not all experience is necessarily educative. This idea was reiterated by David Kolb (D. A. Kolb, 1984, 1985; D. A. Kolb & Fry, 1975), who emphasized along with Dewey the need for experience to be organized and processed in some way to facilitate learning. Dewey also felt strongly that the educational environment needs to actively stimulate the student's development, and it does so through genuine and resolvable problems or conflicts that the student must confront with active thinking in order to grow and learn through the experience.

the premise that for real learning to happen, students need to be active participants in the learning process rather than passive recipients of information given by a teacher. When learning is a passive process, teachers are the centers of energy who tell you the information that they think you need to know. But when learning is an active process, students are the centers of energy. The teacher's role is to guide or facilitate your learning by taking an interest in your work and coaching you through the experience (Garvin, 1991). As an active participant in the learning process, you play a central role in shaping the content, direction, and pace of your learning.

A key component to experiential education is reflection. Dwight Giles, who has written extensively about service-learning and internships, makes a point about service-learning that we think applies to all the experiences covered in this book. He says that reflection is what connects and integrates the service, or the work in the field, to the learning. Otherwise, whatever theory you study can be emphasized in your classes but not necessarily integrated with practical experience. At the other extreme, practical experience is left to stand on its own. Reflection is the connection, and it is a powerful key to your success, your growth, your learning, and your development (Eyler & Giles, 1999; Giles, 1990, 2002).

David Kolb (1984) originally set forth a cycle of four phases that people go through to benefit from experiential learning, as illustrated in Figure 1.2. The first phase is

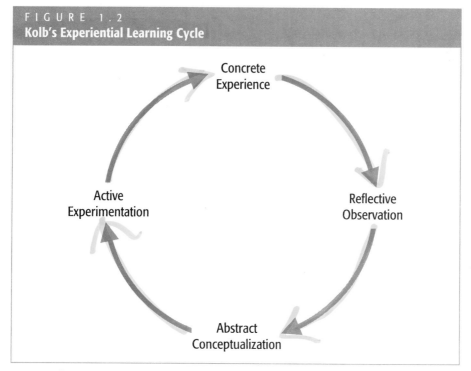

FIGURE 1.2
Kolb's Experiential Learning Cycle

Source: Kolb and Fry (1975)

concrete experience (CE); students have a specific experience in the classroom, at home, in a field placement, or in some other context. They then reflect on that experience from a variety of perspectives (reflective observation, or RO). During the abstract conceptualization (AC) phase, they try to form generalizations or principles based on their experience and reflection. Finally, they test that theory or idea in a new situation (active experimentation, or AE) and the cycle begins again, since this is another concrete experience. James Zull's research (2002) on the relationship between the brain, learning, and education suggests that Kolb has it right: Education can affect your brain maximally when you, the student intern, have a concrete experience in the field, reflect upon it, connect it to what you already know, and then create your own abstract hypotheses about what you've experienced and test them actively, which in turn will produce a new concrete experience for you.

You may recognize this cycle from your internship. For example, suppose you observe a customer arguing with one of your co-workers. You could then draw on several theories or ideas you have studied to try to understand what was happening or you might seek out some new information from staff. You then begin to form your own ideas about what happened and why, and you might use this knowledge to guide your own interactions with that or another customer. Once you do that, the interaction is itself a new concrete experience, and the cycle begins again.

Predictable Stages

This book, while grounded in experiential education theory, is based on two "big ideas." The first big idea is that interns go through predictable stages of development during the course of the internship. Over the years, as we have supervised interns and worked with students engaged in service-learning, listened to their concerns and read their journals, talked with their site supervisors, and discussed similar experiences with colleagues and students at other institutions, a predictable progression of concerns and challenges began to emerge. We have organized these concerns into stages, which are modeled after the work of Lacoursiere, Tuckman, and Schutz (Lacoursiere, 1980; Schutz, 1973; Sweitzer & King, 1994; Tuckman & Jensen, 1977). Understanding this progression of concerns will help you, your campus supervisor, and your site supervisor predict and make sense of some of the things that may happen during the course of your placement and think in advance about how to respond. It will also help you view many of your thoughts, feelings, and reactions as normal and even necessary. The experience then becomes a bit less mysterious, and for some people that makes it more comfortable.

For example, if you are feeling excited but also pretty anxious as you begin the placement, you may wonder whether that anxiety is a sign of trouble or where it may have come from. Knowing that it is a common and predictable experience will help you stop worrying about the fact that you are anxious and let you direct your energy toward moving through that anxiety. Here is what one of our students had to say: *Now that I know I am not the only one that is concerned with these feelings, I am better able to share them with others without feeling embarrassed.*

Self-Understanding

The other "big idea" is that to make sense of your internship, you need to understand more than a stage theory and the experiential nature of learning in the field; you need to understand yourself as well. No two students have the same experience, even if they are working at the same site. That is because any internship experience is the result of a complex interaction between the individuals and groups that comprise the placement site and each individual intern. You are a unique individual, and that uniqueness influences both how people react to you and how you react to people and situations. You view the world through a set of lenses that are yours alone. Therefore, each of you will go through these stages at your own pace and in your own way. Events that trouble you may not trouble your peers and vice versa. Some of you will be very visible and dramatic in both your trials and your tribulations. Others will experience changes more subtly and express them more quietly.

No one experiences an internship in a vacuum and you won't either. You have a life outside the placement (although it may not seem like it sometimes), and your network of family, friends, and academic and professional obligations will shape your experience in a powerful way. We want to help you think about yourself throughout your internship in ways that we believe will lead you to important insights and to a smoother journey on your path to personal, professional, and civic development.

In summary, appreciating the experiential nature of the learning that takes place will help you to understand how different learning by doing is from traditional classroom learning. The stages of an internship will help you understand internships in general and some of the experiences that are apt to happen during your internship. Understanding yourself will help you recognize the particular style in which you will experience the internship. Combining these pieces of knowledge will give you a powerful tool to understand what is happening to you, to meet and deal successfully with the challenges you face, and to take a proactive stance in making your internship the most rewarding experience it can be.

THE INTERNSHIP SEMINAR

Many of you will be meeting with an instructor and other interns on campus during the semester. We refer to these meetings as seminars. The word *seminar* comes from the Italian *seminare*, which means to sow or seed. The class sessions are a medium that is most helpful in the integration of intellectual and affective learning, encouraging new understanding and creative responses, and strengthening the effectiveness of interpersonal relationships (Williams, 1975). The seminar provides an opportunity for you to share your experiences and what you've learned in a mutually supportive place, discuss problems you are facing and concerns that you have, and seek guidance from your campus instructor for the journey you are taking (Sweitzer & King, 1995; Williams, 1975).

A seminar may be a bit different from other classes you have taken (see table below). For example, one basic assumption of a seminar is that each person has something to contribute (Royse et al., 2007), unlike many classroom experiences where the assumption often is that only the teacher has something to contribute. If everyone has something to contribute, everyone shares the responsibility for the success of the experience. You have additional responsibilities in this type of class, but then you also reap additional benefits.

The importance of the seminar to the quality of the internship cannot be overstated. A workshop given at the National Society for Experiential Education described the seminar as "a keystone to learning" (Hesser & King, 1995) because the quality of learning is enhanced when interns come together as a community of learners.

Traditional Model	Seminar Model
• The teacher has the most to contribute	• Everyone has something to contribute
• Many students, one teacher	• All are teachers; all are learners
• Students and teachers blame one another when things go wrong	• Students and teachers take collective responsibility
• The group is a collection of individuals	• The group is necessary for the accomplishment of learning goals

An effective seminar affords opportunities for reflective dialogue, support, the development of important relationships, and a variety of new learning experiences. A seminar class is one in which an exchange of ideas takes place, where information is shared, and mutual problems are discussed. It is also a forum for problem-centered learning. You will have the chance to hear about and perhaps learn new professional skills, strengthen analytic and problem-solving skills, and develop knowledge of other placement sites as well as the communities in which they are located.

Although many courses carry the title of "seminar," they are not all run the same way. Michael Kahn uses clever metaphors to distinguish different kinds of seminars: the Free for All, the Beauty Contest, the Distinguished House Tour, and the Barn Raising (Kahn, 2003). In the Free for All approach, students compete for the teacher's attention and approval. The group itself is not necessary; its only function is efficiency, since a group of twenty-five costs the same to run as a group of five. From the individual student's point of view, however, the other students are only seen as competitors. In the Beauty Contest approach, students show off their wonderful ideas and then spend their time thinking of the next wonderful idea rather than listening to anyone else's ideas. Once again, the size or even the presence of the group serves no educational purpose. In the Distinguished House Tours approach, each student might present a case, a piece of writing, or an idea, and the group *does* pay careful attention, perhaps with skilled guidance from the instructor. Then they turn their attention to the next student. In a Barn Raising approach, everyone has a role and the group is essential. The group must be of a certain size, and the task can only be accomplished through cooperation and collaboration.

As you think about your internship seminar and its goals, ask yourself where they fit? Is the group really necessary or is it only there for efficiency? We believe that the last two types, the Distinguished House Tour and the Barn Raising, are the most appropriate models for an internship seminar. Using the first of these two models, each intern can be given a chance to talk about his or her experience or perhaps present a case. The intern practices the skill of reflection by trying to interpret the story through various theoretical lenses. The group is there to give feedback and support and perhaps to assist in the reflection and analysis.

If we use the barn raising metaphor, then what is the barn? It is what the group is trying to accomplish together. One possibility is that it is trying to build a collective understanding of the profession using the experiences of different people at different sites. Another possibility is that it is trying to build a collective understanding of the community or communities, of their assets and challenges, and the role that each of the sites and professions might play in maximizing those assets and addressing those challenges.

A third and very common possibility is that the group is trying to create a community of support, where support is provided not simply by the professor but by everyone. It is important that interns have a place where they can talk about their experiences, their feelings and reactions, and their struggles and achievements. Although your friends and family can do some of this for you, it is often helpful to have this exchange with others who are undergoing a similar experience. Support groups exist for almost every purpose; perhaps you have participated in some. While the seminar is not a support group in the formal sense, one of its principal benefits is the quality of connections that you develop with your peers. Through these relationships, you give and receive support. In fact, you receive a double benefit—not only do you give and receive support but you also become more skilled at each of these functions. We will talk more about these skills in the next chapter.

In trying to build a community of support, it is also important to remember that the seminar class is a group, and it goes through developmental stages and group dynamics like any other group. It is not hard to describe the atmosphere that you would want in a group where people are sharing ideas, joys, and fears. But the atmosphere of trust, openness, safety, honesty, and feedback that characterize successful groups (Baird, 2007) does not just happen. It happens in stages, over time, with all members investing in it. Because of the sensitivity of some of the information that is shared in the seminar class (about clients or customers, supervisors, co-workers, the site, or yourself), it is very important to be clear early on about what is expected in terms of confidentiality, disclosure of information, and how interns are to conduct themselves in the class. The obligation to keep that information "in the room" is a common guideline for this course.

Please remember that this is not a therapy group. You are not therapists (nor are your instructors in therapist roles, although they may have those skills). There may be times when certain individuals encounter a challenge in the internship or an experience that evokes feelings that need the attention of a counselor or therapist. Your campus supervisor can help you recognize those instances and locate appropriate resources to deal with them.

Finally, if you are seeking to build a community of support, then it is important that there is time in the seminar to reflect upon how that endeavor is going. Early in the seminar, you may want to discuss issues such as listening, effective feedback, and the overall goal of learning to function as a supportive group. As the semester progresses, take time to celebrate your successes and growth in this endeavor, to discuss and try to solve problems, and to give one another feedback about the achievement of the goals.

OVERVIEW OF THE TEXT

Chapter Organization

This book is organized into four sections. In Section One, *Beginning the Journey* (Chapters 1–4), we present the conceptual framework that underlies the book and give you some basic tools to proceed. In Chapter 2, we introduce you to some critical knowledge, skills, and attitudes that we think you will need right away. Chapter 3 introduces you to the first big idea underlying this book—the developmental stages of an internship. Chapter 4 focuses on self-understanding; we ask you to look at and think about yourselves through a variety of lenses.

In Section Two, *Discovering the Field* (Chapters 5–9), we deal with the issues and concerns associated with getting started in an internship. We call this the *Anticipation* stage. Some of our colleagues and students have told us that they prefer to know and think about some of these issues and concerns before the actual placement begins, and this section certainly could also be used in that way. Chapter 5 discusses the Anticipation stage itself and emphasizes the careful construction of a learning contract. The remaining chapters can be read in any order, depending on the setting you are in. Chapters 6 and 7 help you become oriented to supervisors, co-workers, and the agency itself. These two chapters could easily be read out of order. Chapter 8 asks you to consider the community context of your work, and Chapter 9 focuses on clients and other populations served by your site.

Section Three, *Facing New Frontiers* (Chapters 10–11), looks at the challenges that await you once the initial concerns are resolved. In Chapter 10, we help you take stock of where you are and where you want to be. We also try to help you anticipate some of the more common challenges that interns encounter at some point in their placements. The final portion of Chapter 10 deals with the *Disillusionment* stage, which can be a disheartening stage for students and campus and site supervisors alike but which we view as a normal and necessary part of the experience. In Chapter 11, we give you the tools to move from Disillusionment to *Confrontation* as you attempt to identify and resolve issues that are standing in the way of your continued progress. We emphasize that this process is a learning experience in and of itself as well as one that paves the way for future learning. There is a problem-solving model presented in Chapter 11; a different model certainly could also be used, and your campus supervisor may provide you with options to choose from.

The final section of the book, *Going the Distance* (Chapters 12–14), examines issues and concerns that are common in the latter stages of the experience: *Competence* and *Culmination*. Chapter 12 deals with several professional issues that await, even though you are feeling and doing quite well in the internship. Chapter 13 deals with legal and ethical issues. Of course, ethical issues can arise throughout the internship, but it has been our experience that interns often do not *notice* them in a conscientious way, regardless of whether and how often they are covered in class, until some other concerns have been resolved. We suggest some common ethical issues that arise in an internship and also provide a decision-making model for dealing with ethical issues. Students who have had a course on ethical issues, or who have encountered this theme in many of their courses, may already have a model for thinking about and resolving ethical dilemmas. However, the issues raised in the internship are probably new to you or at least you may be encountering them for the first time in a professional context. The final chapter, Chapter 14, guides you to end the internship on a number of levels and in productive, meaningful ways.

Chapter Exercises

At the end of each chapter, we offer you several ways to extend and enhance your learning. There is a list of additional resources in case there are areas you want to explore further. There are also several exercises, and these fall into two categories. In the section "Personal Reflection," we pose questions for you to think about and perhaps to write about in your journal. You should select, in cooperation with your instructor, those questions that seem most relevant and meaningful to you. In the section "Seminar Springboards," we offer questions and exercises for group discussion or skill building or that can be undertaken individually.

We wish you good fortune with your journey. Before proceeding further, take some time to reflect upon the exercises that follow; you then will be ready to move on to the experience of the internship!

For Contemplation

PERSONAL REFLECTION

The beginning of your internship or service-learning experience is a good time to review your academic program's expectations of you, your supervisor, and your instructor during the internship. This is particularly important in terms of knowing what you can expect from others and what others, including the staff at the placement site, might expect from you. Take time now to locate any written documents from your program that specify these responsibilities. Make copies and keep them with your other internship paperwork.

SEMINAR SPRINGBOARDS

1. Internship students often use a language of their own. Your supervisor or co-workers may appear puzzled when you use certain terms, even though they are commonly understood on your campus. We call this language *fieldspeak*. There is also *agencyspeak*, which you may adopt without even thinking about it after a few days at

the placement, but it will puzzle your seminar classmates or even your instructor. Review your program's definition of terms and compare them to the ones in this chapter. Be ready to explain them to people at your placement site. In class, share the important terms and slang that make up your field site's *agencyspeak*.

2. Seminar class can become an important part of the internship experience. Now is a good time to think about what you want from that class. For example, what are your major objectives for the time spent in seminar? What works best for you as a way of teaching and learning? What role do you see yourself having in class? In class, share these ideas with your peers and instructor. Make a list of the major goals and objectives and learning preferences and discuss ways you can help one another.

For Further Exploration

Battistoni, R. (2006). Civic engagement: A broad perspective. In K. Kecskes (Ed.), *Engaging departments: Moving faculty culture from private to public, individual to collective focus for the common good.* Bolton, MA: Anker Publications. pp. 11–26.

An excellent summary of how academic majors can be vehicles for civic development.

Colby, A., Erlich, T., Beaumont, E., & Stephens, J. (2003). *Educating citizens: Preparing America's undergraduates for lives of moral and civic responsibility.* San Francisco: Jossey-Bass.

Thorough discussion of the theoretical and philosophical underpinnings of the education for citizenship movement as well as sections focusing on examples of best practice at a range of higher education settings.

Coles, R. (1993). *The call of service: A witness to idealism.* New York: Houghton Mifflin.

This book draws on the author's direct experience with service, as well as his experience with countless volunteers, to examine the individual urge toward idealistic action. Strategies for using literature to illuminate service concepts are also discussed.

Collison, G., Elbaum, B., Haavind, S., & Tinker, R. (2000). *Facilitating on-line learning: Effective strategies for moderators.* Madison, WI: Atwood Publishing.

Informative, useful, and classic guide for netcourse instructors for moderating a web-based course such as the internship seminar.

Eyler, J., & Giles, D. E. (1999). *Where's the learning in service-learning?* San Francisco: Jossey-Bass.

A thorough discussion of service-learning concepts and the results of the first large-scale, systematic study of the impact of service on learning.

Inkster, R. P., & Ross, R. G. (1995). *The internship as partnership: A handbook for campus-based coordinators and advisors.* National Society for Experiential Education. www.nsee.org.

Inkster, R. P., & Ross, R. G. (1998). *The internship as partnership: A handbook for businesses, nonprofits, and government agencies.* National Society for Experiential Education. www.nsee.org.

A series of two comprehensive and invaluable handbooks for internship programs and field sites. The first edition focuses on the responsibilities and challenges of the work of the field coordinator; the second edition does the same for the work of the site supervisors. Both editions have significant value for faculty and other campus supervisors in developing their effectiveness in working with placement sites.

Jackson, R. (1997). Alive in the world: The transformative power of experience. *National Society for Experiential Education Quarterly*, 22(3), 1: 24–26.

Excerpts from a keynote address in which the author explores the *experiences* needed to be considered an educated citizen.

McKenzie, R. H. (1996). Experiential education and civic learning. *National Society for Experiential Education Quarterly*, 22(2), 1: 20–23.

Based on a deliberative democracy seminar sponsored by the Kettering Foundation; participants spanned middle school through four-year colleges and included an international contingent, drawn from a network of those who use National Issues Forums in their classes. The role of deliberation in learning is underscored.

NSEE Foundations Document Committee (1998). Foundations of experiential education, December 1997. *National Society for Experiential Education Quarterly*, 23(3), 1: 18–22.

A living document that describes the common ground of the members of NSEE at this time, reflecting the thinking of the membership and operating assumptions about defining experiential education; intended to initiate discussion about experiential learning and education.

Parilla, P., & Hesser, G. (October 1998). Internships and the sociological perspective: Applying principles of experiential learning. *Teaching Sociology*, 26(4), 310–329.

The argument is made that internships do achieve educational goals by providing opportunities to apply sociological principles, improve analytical skills, and use Mills's concept of "sociological imagination."

Russo, F. X., & Willis, G. (1986). *Human services in America.* Upper Saddle River, NJ: Prentice Hall.

A good introduction to the six major categories of human service delivery systems as well as an introduction to public policy in human services.

Schon, D. A. (1987). *Educating the reflective practitioner.* San Francisco: Jossey-Bass.
Schon, D. A. (1995). *The reflective practitioner: How professionals think in action* (2nd ed.). Aldershott, UK: Ashgate Publishing.

Schon's work on reflectivity is a classic and forms a foundation for subsequent work on practical reasoning.

Sullivan, W. M. (2005). *Work and integrity: The crisis and promise of professionalism in America* (2nd ed.). San Francisco: Jossey-Bass.

Excellent discussion of civic professionalism.

References

Baird, B. N. (2007). *The internship, practicum and field placement handbook: A guide for the helping professions* (5th ed.). Upper Saddle River, NJ: Prentice Hall.

Battistoni, R. (2006). Civic engagement: A broad perspective. In K. Kecskes (Ed.), *Engaging departments: Moving faculty culture from private to public, individual to collective focus for the common good.* Bolton, MA: Anker Publications, 11–26.

Chickering, A. W. (1977). *Experience and learning: An introduction to experiential learning.* Rochelle, NY: Change Magazine Press.

Colby, A., Erlich, T., Beaumont, E., & Stephens, J. (2003). *Educating citizens: Preparing America's undergraduates for lives of moral and civic responsibility.* San Francisco: Jossey-Bass.

Crutcher, R. A., Corrigan, R., O'Brien, P., & Schneider, C. G. (2007). *College learning for the new global century: A report from the National Leadership Council for Liberal Education and America's Promise.* Washington, DC: AAC&U.

Dewey, J. (1916/1944). *Democracy and education.* New York: MacMillan Publishing Co.

Dewey, J. (1933). *How we think.* Lexington, MA: D.C. Heath and Co.

Dewey, J. (1938). *Experience and education.* New York: MacMillan Publishers.

Dewey, J. (1940). *Education today.* New York: Greenwood Press.

Eyler, J., & Giles, D. E. (1999). *Where's the learning in service-learning?* San Francisco: Jossey-Bass.

Garvin, D. A. (1991). Barriers and gateways to learning. In C. R. Christensen, D. A. Garvin, & A. Sweet (Eds.), *Education for judgment.* Boston: Harvard Business School Press, 3–13.

Giles, D. E. (1990). Dewey's theory of experience: Implications for service-learning. In J. C. K. Associates (Ed.), *Combining service and learning.* National Society for Internships and Experiential Education (Vol. I, pp. 257–260). www.nsee.org.

Giles, D. E. (2002). *Assessing service-learning.* Paper presented at the Annual Conference of the National Organization for Human Service Education.

Godfrey, P. C. (2000). A moral argument for service-learning in management education. In P. C. Godfrey & E. T. Grasso (Eds.), *Working for the common good: Concepts and models for service-learning in management.* Sterling, VA: Stylus Publications, 21–42.

Hesser, G., & King, M. A. (1995). *Internship seminar: A keystone to learning.* Paper presented at the National Society for Experiential Education, New Orleans, LA.

Howard, J. (Ed.). (2001). *Service-learning course design workbook.* Ann Arbor, MI: OCSL Press.

Kahn, M. (2003). The seminar. From www.sonoma.edu.

Kolb, D. A. (1984). *Experiential learning: Experience as the source of learning and development.* New York: Prentice Hall.

Kolb, D. A. (1985). *Learning style inventory.* Boston: McBer & Co.

Kolb, D. A., & Fry, R. (1975). Toward an applied theory of experiential learning. In C. Cooper (Ed.), *Theories of group process.* New York: John Wiley & Sons, 33–57.

Lacoursiere, R. (1980). *The life cycle of groups: Group developmental stage theory.* New York: Human Sciences Press.

Lemann, N. (2004). Liberal education and the professions. *Liberal Education* (Spring), 12–17.

Royse, D., Dhooper, S. S., & Rompf, E. L. (2007). *Field instruction: A guide for social work students* (5th ed.). Boston: Allyn & Bacon.

Saltmarsh, J. (2005). The civic promise of service-learning. *Liberal Education, 91*(2), 50–55.

Schutz, W. (1973). *Elements of encounter: A body mind approach.* Big Sur, CA: Joy Press.

Sgroi, C. A., & Ryniker, M. (2002). Preparing for the real world: A prelude to a fieldwork experience. *Journal of Criminal Justice Education, 13*(1), 187–200.

Simon, E. (1989). Field practice survey results. In C. Tower (Ed.), *Field work in human services*: Council for Standards in Human Service Education (Monograph #6).

Sullivan, W. M. (2005). *Work and integrity: The crisis and promise of professionalism in America* (2nd ed.). San Francisco: Jossey-Bass.

Sweitzer, H. F., & King, M. A. (1994). Stages of an internship: An organizing framework. *Human Service Education, 14,* No. 1 (Fall), 25–38.

Sweitzer, H. F., & King, M. A. (1995). The internship seminar: A developmental approach. *National Society for Experiential Education Quarterly, 21*(1), 1: 22–25.

Tuckman, B. W., & Jensen, M. (1977). Stages of small group development revisited. *Group and Organizational Studies, 2,* 419–427.

Waddock, S., & Post, J. (2000). Transforming management education: The role of service learning. In P. C. Godfrey & E. T. Grasso (Eds.), *Working for the common good: Concepts and models for service-learning in management.* Sterling, VA: Stylus Publications, 43–54.

Williams, M. (1975). The practice seminar in social work education. In M. Williams (Ed.), *The dynamics of field instruction.* New York: Council for Social Work Education, 94–101.

Zull, J. E. (2002). *The art of changing the brain: Enriching the practice of teaching by exploring the biology of learning.* Sterling, VA: Stylus.

Essentials for the Journey

Throughout this book, we use the metaphor of a journey to describe your internship experience. In this chapter, we present our ideas about what you will need at the outset of your journey. These essentials run the gamut from an awareness of the assets and challenges you bring to the journey, to information you can easily get if you don't have it, to skills you need to start developing, to attitudes you need to cultivate, to resources you need in order to take an empowered stance. You may find that you already have some of what you need. You may also decide that your particular internship does not require some of the attitudes, skills, and knowledge that we describe. We encourage you to make those decisions in collaboration with your campus instructor or supervisor.

ESSENTIAL PERSONAL RESOURCES

The most valuable assets you have for this journey in learning are what you personally bring to it. By this we mean the wealth of your life experiences, your aspirations, your knowledge, your hopes, your expectations, your attributes, and your relationships. These qualities and resources are your personal reserve of vitality and endurance; they have the potential to give you good feelings and get you through difficult moments. We focus on three resources in particular because of their importance to the learning experience: learning style, life context, and support systems. These resources can be powerful assets for you if they are in place and you know how to use them; they can also be sources of challenge for you if they are not in place or you are not sure how to get the most from them.

Your Learning Style

Learning style means the way that you most effectively take in and process information, and knowledge of your learning style is critical for you, your campus instructor, and your site supervisor (Birkenmaier & Berg-Weger, 2007). There are many theories that examine learning style, and they are not necessarily mutually exclusive. You might be categorized as a concrete or abstract learner (D. A. Kolb, 1984), as random or sequential (Gregorc, 2004), or as connected vs. separate (Belenky, Clinchy, Goldberger, & Tarule, 1986). Your learning style is influenced by genetics, family norms and values, gender and other cultural factors, and your personal learning history. We will discuss some learning style theories in later chapters, but for now, just think about the ways that you learn best; these are your strengths as a learner. Perhaps you have already thought about this issue and you can answer the question in general terms. If not, think about learning experiences you have had that were successful and unsuccessful for you. What has been your experience with learning from reading? From group discussions? From hands-on, trial-and-error approaches?

Journal

Think About It

How Would You Describe Your Learning Strengths?
Here are some statements from our students about their learning strengths. Try developing one of your own:

- When learning something new, I understand it much better when I'm able to try myself. Instead of asking me questions and quizzing me, I'm much better at just showing you what I can do. I also do not learn when people stand right over me. The best way for me to learn is hands-on experience; please take the time to let me.

- I am an active learner. I learn best when observing or doing things hands on. While in the process of learning, I may write a few notes so that I can refresh my memory if need be. Being an active learner allows me to process information effectively while experiencing it. I also like to reiterate what I am hearing if I feel I do not fully understand a process or subject. This method allows me to figure out if I understand what was explained or what parts I misunderstood.

- I seem to be able to learn in a variety of environments and with a variety of styles. I tend to be an auditory learner. On most occasions, I sit and listen and take very few notes, but I can recall the highlights of the topic with minimal difficulty. Please keep this in mind, as I can attend numerous conferences and trainings that cater to many learning styles.

If you are involved in an internship that is not well matched to your style and strengths, you can still learn, but something needs to give. You may be able to take steps to change or augment the learning experience so that it is better suited to your strengths. On the other hand, you may need to try to stretch your learning repertoire and strengthen a style that does not come easily to you. In this way, a mismatch of styles can be an opportunity for growth.

Your Life Context

> *My biggest concern is stress, basically from outside the internship. The stress to get all the papers done on time, working many hours a week at my job to pay bills, and still put in thirty hours in the internship.*
>
> STUDENT REFLECTION

Your life context consists of all the other things going on in your life besides the internship. That context will vary according to your family situation, your social life, and the configuration of your academic program. You will have some responsibilities outside your placement, some expectations placed on you (or being continued), and possibly some stress. Here are some areas to consider:

ACADEMICS

In some programs, interns take very few classes during their internship semester besides their internship. In other programs, the internship can be the equivalent of only one course, leaving a full-time student with three or four other courses to carry. Some students even need to carry an overload because they must finish college in a certain time frame. The internship tends to be at least as demanding of your time as one course or even two. Our students also report that internships demand a good deal of psychological and emotional energy, much more than a traditional course.

The work can be exhilarating, and many interns find themselves thinking about their work when they are not at the placement, or putting in extra time to read and research issues relevant to their work. But the work can also be emotionally draining, and the situations some interns face with individuals and communities can be heartbreaking and frustrating. Many interns have told us that they find themselves thinking and talking about the internship a great deal, which sometimes makes it hard to focus on other classes.

EMPLOYMENT

Some internships are paid; many are not. If you have been holding down a job in addition to attending classes, you will want to think about whether you can and should give it up during the placement. Some interns cannot do this; their economic situation simply prohibits it. If you are going to be employed during your internship, think about the schedule you will have and how it will fit into the time demands of your placement and coursework.

ROOMMATES

If you live with other people, they are part of your life context. Schedules for housekeeping, sleep, quiet time, study, and entertainment are areas of negotiation with roommates, and you've probably already managed a way to live with others. However, you may need to do it again depending on the demands of your placement. You may be up earlier, home later, or not available for cooking or other chores at your expected time.

FAMILY

In this category, we are including your family of origin (your parents, siblings, and anyone else who lives with them) and your nuclear family (your spouse or partner and children). You may be living with any, all, or none of these people, but you undoubtedly have responsibilities to them. What are they? How flexible are they? Could they be changed for this semester if the demands of the internship warrant it?

INTIMATE PARTNERS

If you have a special partner in your life, whether or not you live with that person, this is another area of responsibility. It took time and energy to build that relationship, and it will take time and energy to maintain and nurture it during the internship. How much time have you been devoting to that relationship? What does that person expect of you, especially now that you have begun your placement?

FRIENDS

Although some friends are closer than others, every friendship is a responsibility too. Think about the amount of time you spend with friends, individually or in groups, and about how much time you hope to spend with them this semester. Think, too, about the demands those friendships put on your time and energy.

OTHER COMMITMENTS

Many interns make the mistake of just inserting an internship into an already busy life and expecting it all to work out somehow. Some of our busiest students fall prey to this misconception, especially those raising families, working full time, caring for a family member, or seriously involved with sports. We are going to encourage you to look a little more objectively at your life context. If you list all the people and activities that you will be committed to this semester and estimate the number of hours each week or month each of those commitments is going to take, you may be shocked to discover that you have committed yourself to more hours than exist in a week or a month. If so, just as when you are doing your financial budget, you will have to make some compromises. Often, this compromising involves talking with others to whom you now have commitments about reordering things just for the short run.

Many interns report feeling that they have less time than they ever had. When we ask them to conduct an inventory of their commitments, as just described, they find that they should have enough hours in the day and week, but they just do not feel as if they do. Sometimes that comes from poor estimating. For example, if a class lasts for an hour, you need to remember that you will spend some time getting to and from class. One strategy you can try is to keep careful track for a day or two of everywhere you go, how long it takes to get there, and how long you spend there. However, another reason for this feeling is that interns often underestimate the psychological commitment and emotional demands of the internship. Whether you realize it or not, your emotional energy is likely to be taxed by an internship, and that is going to affect how available you feel for other activities and responsibilities.

YOURSELF

Don't forget to consider the time and energy you currently devote to taking care of yourself. By taking care of yourself, we mean sleep, good meals, exercise, and hobbies. All of us have activities we engage in for pleasure and as a source of stress reduction. Hiking, playing the piano, painting, lifting weights, meditating, and reading are just a few examples. You will need to take the best care of yourself so that you can to meet the demands of the internship. You may be able to cut down some on hobbies if necessary, but you will need to be realistic about the time required to take proper physical and emotional care of yourself.

Your Support Systems

Balancing all the demands we were just discussing is your support system, which is made up of people who give you what you need to get through life's challenges. Charles Seashore (1982), who has written extensively about support systems, calls them "a resource pool drawn on selectively in order to support me in moving in a direction of my choice that leaves me stronger" (p. 49). Your support system will be an important part of helping you meet the demands of your internship and the other demands in your life. As one of our interns pointed out in a journal entry, "*Support teams are a way to relieve stress and frustration . . . they listen to you, give advice, and give support. It is very comforting to know that you have someone to fall on when needed.*"

Everyone needs some support. Some people need more than others, and you will need more at some times than others, but you do need it and you will continue to need it. Sometimes, people in the helping professions have trouble accepting that they need support. Hence, they do not seek it out or decline it when it is offered. They give to others unstintingly and enjoy being people that others can count on, but they are better at giving than receiving. Most of the time, sooner or later, they become exhausted and are of little use to anyone. It may be that they have an image of the perfect helper—someone who never needs help. "How can I help others," students have asked us, "if I have problems myself?" Others know they need support, but they seem unable to accept it. Accepting your need for support and developing a strong support system are critical skills that could be called into play during the internship.

Regardless of the nature of your internship, your support system is made up of many different people, and you will need different kinds of support at different times (Seashore, 1982). Here is a partial list of the kinds of support you might need. You may be able to add to it.

- **Listening** Sometimes you just want someone who will listen to you without criticizing or offering advice. The person listening should be someone to whom you can say almost anything and on whom you can count not to grow restless or frightened. Think of these people as your "sounding boards."
- **Advice** On the other hand, sometimes you need sound advice. You may not always follow it, but you need a source of advice that you can trust. Think of these people as your "personal consultants."

- **Praise** There are times when what you need most is for someone to tell you how great you are. If they can be specific, all the better. Think of these people as your "fans."

- **Diversion** Some people are friends you can count on to go out and play with. You don't have to talk about work, your problems, or anything else. You just have activities you enjoy together. Think of these people as your "playmates."

- **Comfort** When we were children and we became ill, there was nothing we wanted more than pure comfort. A comfortable place to rest, good food, music we enjoy, all without having to lift a finger! At times, ill or not, this is still just the kind of support we need. Think of the people in your life who comfort you as your "chicken soup people."

- **Challenge** There are times when challenge is the last thing you want. At other times, though, someone who will push you to do more, look at things in a different way, and confront problems or inconsistencies in your thinking is the best friend you have. Think of these people as your "personal coaches."

- **Companionship** It is good to have people in your life with whom you feel so comfortable that you can do anything, or nothing, with them. Sometimes you may not care what you are doing, only that you are not doing it alone. Think of these people as your "buddies."

- **Affirmation** Another kind of support comes from people who have some of the same struggles that you do. It is a lonely and depressing feeling to believe, or suspect, that you are the only one who is troubled by a particular issue or set of circumstances. Knowing that others feel the way you do, even if they can't change it, can be very helpful. Sometimes there is no substitute for someone who has been through what you are going through. Think of these people as your "comrades."

As we said earlier, you are not going to need all these kinds of support at the same time. You are also not going to be able to get all your support from the same person or group. As you read the list, you must be able to think of people who are helpful with one aspect of support but not with others. Both of us have friends who are fun to be with and provide wonderful diversion, but they do not listen well at all! You will know soon enough if you have called on the wrong person for support. For example, you may need good advice, but call on the listener. Even if there were one person in your life who could provide all these kinds of support, that person would soon become exhausted if you asked him or her to meet all your support needs. Similarly, some people who can provide a particular kind of support may not be available at any given time. Try not to become frustrated. You need to learn about your needs and about how different people can best help you meet them.

Think about these categories and add some of your own. Now think about how well each of those needs is currently met in your life. You may find that there are some gaps that you need to address. Your support system will be stronger if you have more than one person in each category. Remember that the internship may, by itself or in combination with other things occurring in your life, create a greater need for support than you now have. The internship may also strain your existing support system. If most of your friends are not involved in internships, for example, they are not going to know how you feel, and they may not understand what you do or how difficult it is.

"That's work?" they may say, or "You get credit for that?" They may not understand your need to get to sleep early, work on weekends, or cut back on your social life. Or they may have negative attitudes about the population you are working with, like the homeless, the addicted, or people with HIV/AIDS.

If you have discovered gaps in your support system, we urge you to take steps to fill them. Cultivate new friends; discuss your needs with friends and family. Like investing in self-understanding, it will pay off for you now and later.

ESSENTIAL ATTITUDES

As we mentioned in the previous chapter, many professions have lists of essential attitudes and values. What we present to you here are attitudes we believe are essential to success in any internship, regardless of the profession.

Being Open-Minded

You approach an internship with at least some preconceived ideas about the profession, the work you will be doing, your likes and dislikes, and so on. In some cases, the internship may confirm every one of those preconceptions. In most cases, though, some will turn out not to be true. It helps a great deal if you keep your mind open to the way things really are, as opposed to the way you wish they were or thought they might be. If you keep an open mind, you can also learn that there are many ways to do a good job, that many styles work well depending on the situation, and that norms and customs that may seem strange, useless, or even destructive to you in fact serve an important purpose.

Being Flexible

We hope that you and your campus support people have planned your internship carefully; if so, the internship site is clear with you about what will happen, when, and with whom. But in most organizations, situations change—and sometimes suddenly. Some environments are very fluid, and things change all the time; emergency shelters are a good example. Other settings, such as corporate settings, are more stable. No organization is immune, though, and when situations change for the organization, sometimes they change for you. The more flexible you can be, the better you will be able to respond. If you can keep focused on the essential learning opportunities that are there for you and that may arise while you are there, you will be less attached to the specific experiences by which you learn.

Being Receptive

It is imperative that you be receptive to constructive feedback on your work. If you are not, you will surely make a bad impression, but more importantly, you will miss out on great learning opportunities. Supervisors, co-workers, other interns, instructors, and even clients (if you have them) can be valuable sources of information about your strengths and weaknesses. We don't mean to say that you should accept all feedback at face value,

---

Think About It

Do Others See You as "Just an Intern"?
Is That How You See Yourself?
Many interns find relief in knowing they are not being treated as "just an intern." As a matter of fact, feeling like "just an intern" is one of the most prevalent fears and sources of anxiety that interns report having when they begin their internships. You might want to give some thought to these questions now so you can head off an attitudinal shift before it happens at the placement site, whether it's your attitude or those of site supervisors.

- What exactly does the phrase "just an intern" mean to you?
- How would you know if you were being treated as "just an intern"?

Have you been feeling this way? If so, what is it about your internship that cultivates this attitude in you? In your supervisors or co-workers? If not, what is it about your internship that leaves you feeling like more than "just an intern?"

---

only that you should consider it from a nondefensive position. If you think of feedback as criticism, you are apt to become defensive. If you react defensively, you will be less likely to get feedback in the future and even less likely to benefit from it.

Being Open to Diversity

It is very likely that you will meet people whose cultural backgrounds are different from your own, by virtue of their race, class, gender, ethnicity, sexual orientation, and in other ways as well. There is a natural human tendency to react with some suspicion to difference, and for some people, their first response is to equate difference with being somehow wrong. Others are tolerant of difference but assume that their way of doing things is really the best way and attempt to get others to change. Yet others take a "live-and-let-live" approach, appearing to accept diversity but not showing any real interest in it. All those stances result, again, in lost opportunities for learning and growth. If you are willing to learn about and from members of other cultural groups, if you look for the strengths and assets in their norms and values, then diversity will enrich your experience, making you a better professional and a better citizen.

ESSENTIAL SKILLS

Your attitudes form a solid base for you to move forward, while your skills are the hands-on tools you will need. Again, many more skills will be presented later, both in this book and at your placement. But here are some that we consider crucial for you to have right away.

The Skills and Habits of Reflection

Reflection to me is the connection that you make between an experience and all of your feelings that surround it. I believe more and more that reflection can only make someone better. Just looking at myself in regard to my internship, school, and personal life, if I never stopped to reflect on the emotions that I was feeling, the thoughts that I was having, and the knowledge that I was being taught, I would never learn and, essentially, never grow.

STUDENT REFLECTION

Reflection is a fundamental concept in any kind of experiential education, and the internship is no exception. It is also the hallmark of professional practice (Schon, 1987, 1995; Sullivan, 2005). In order to turn your experience into a learning experience, you need to stop, recall events, and analyze and process them. Although this may sound daunting, you actually do it all the time. If you are walking back from class and find yourself mulling over the remarks of a professor or wondering how a classmate came up with a particularly interesting comment, you are reflecting. If you are in the car and start to think over an argument you have had with your child, your partner, or a friend, trying to figure out what happened and how you could have handled it differently, you are reflecting.

These examples, though, are instances of spontaneous reflection; we want you to make reflection a deliberate and regular habit. Reflection means to look back, and there will come a time when we ask you to look back on your experience as a whole, but for your internship, reflection should start at the beginning and be integral to the process (NSEE, 1998). Developing the habit of productive reflection takes patience, practice, and discipline. It means setting aside quiet time to think because as one of our students put it, *"the best answers come from the silence within."* And it means resisting the temptation to just keep going from one activity to the other, in your internship or in your busy life; another intern said, *"The internship proceeded at such a fast pace that I often felt it was one step ahead of me."* Paradoxically, we have found that one of the best ways to stop the internship from getting ahead of you is to take time to stop and think.

There are lots of techniques for reflection; we will discuss some specific ones below, but it is not a comprehensive list, and you will need to find the one or ones that work best for you, your instructor, and your situation. Remember, too, that there is a difference between reflecting and recording, although the two can overlap. Your campus instructor, your placement site, or both may have specific ways that they want you to keep a record of what you have done. These records may be used for documentation and kept in official files. They may be used in supervision as well. The primary purpose of reflection, though, is to promote your growth and learning, and the primary audience for a reflective technique is yourself.

Eyler and Giles (1999) have offered important guidelines for selecting and assessing potential reflective techniques. These principles are called the "Five Cs": *connection, continuity, context, challenge,* and *coaching.* In keeping with these principles, you need to make reflection a habit; structure and connect that reflection to learning goals; and make sure you are challenged to reflect more deeply and through a wider range of lenses. You also need to work with your campus supervisor or instructor to ensure that your choice of techniques makes sense for your particular

Focus on THEORY

Eyler and Giles's Five Cs

The principle of *connection* refers to the importance of connections . . . between the classroom and the field, the campus and the community, your experience and your analysis of it, your feelings and your thoughts, the present and the future, you and your peers, the community, and your campus and site supervisors (Eyler & Giles, 1999).

The second principle is that of *continuity*, which refers to learning as a lifelong process for you, and the importance of developing a habit of reflecting on your field-based learning experiences, be they service-learning, practicum, internship, co-op education, or study abroad. The principle of *context* is the third principle and refers to the thinking and learning you will be doing in the field with the tools you are given, the concepts you are learning, and the facts of a given situation. This principle also refers to the importance of the style you use and the place where your reflecting takes place.

The fourth principle is that of *challenge*, and it refers to the need for you to be challenged in order to grow. It is the challenge of a new situation and new information, such as that of your internship, that creates an ambiguous situation, at odds with your perspectives, and requires that you resolve these differences by developing "more complex and adequate ways of viewing the world. . . ." (pp. 184–185).

The fifth principle is that of *coaching*, which refers to the emotional as well as intellectual support you will need during your internship; the sources of these forms of coaching could come from both your campus and site supervisors, your peers, and your personal support system. By emotional support is meant a safe space where you know that you can express your feelings and share your insights without criticism and with respect. By intellectual support is meant having the opportunities to ask questions, think in new and different ways about situations, and question in retrospect how you think about issues and situations that challenge you (p. 185).

context and that you get the coaching and guidance you need to use these techniques to your best advantage.

KEEPING A JOURNAL

I believe that class work and journals are critical to internships because they allow support from peers, feedback from teachers, and reflection on your own work and feelings.

STUDENT REFLECTION

One of the most powerful tools for reflection that we know of is keeping a journal. Your instructor may require you to keep a journal of some kind, but even if it is not required in your setting, we strongly recommend that you keep one. We also suggest that you write an entry following every day that you go to your internship. Although it may occasionally seem like a chore, if you put time into it, journal keeping will give you a way to see yourself growing and changing. It also forces you to take time on a regular basis to reflect on what you are doing. Many of the quotes you have seen and will continue to see throughout this book are drawn from student journals. A well-kept journal is a gold mine to be drawn on for years to come. It becomes a portfolio of the experience as well as a record of the journey.

Again, perhaps the most important thing you can do for your journal is to allot sufficient time to do it. Doing it over lunch on the due date is not a good approach! For many of you, it is going to take practice and focus to learn to write in your journals in the most effective and productive way. As you plan your days and weeks, leave at least thirty minutes after each day at your internship to write.

If you have learning challenges that make it impossible or difficult for you to write, or if writing does not come easily to you, your journal could be tape recorded instead. Your instructor can listen to the tape each week and respond to you on tape or in writing, whichever the two of you prefer. Of course, you will need to negotiate these arrangements with your instructor, but a little time and thought should yield a method that allows you to reflect comfortably on your experience and maintain a dialogue with your instructor.

For those of you who are doing your internship at a great distance from campus, or as part of a distance-learning program, the journal is even more important. In addition to the benefits already mentioned, the journal and responses to it are a way for you and your campus instructor to have a continuing conversation about your work and your reactions to it. Advances in web-based technology such as Blackboard and WebCT make it easy to send journal entries back and forth. If you do not have access to these technologies, e-mail can work just fine.

If you do decide to keep a journal, make sure you are very clear with your instructor, supervisor, and clients about the intent of the journal and issues of confidentiality. If your journal is for your personal use only, then there is no issue. You have full responsibility for its contents and for ensuring that what you write is for your eyes only. However, if you want, or are required, to show it to other interns, your instructor, your supervisor, or anyone else, you must be careful not to disclose information about clients, the placement, or even yourself that is supposed to be kept private (see Chapter 13 for a discussion of the privacy issues in the pervasive new Health Insurance Portability and Accountability Act (HIPPA) regulations). Discuss this issue with your instructor and your site supervisor before going too far with your journal. You may also be concerned that you cannot be completely candid in your journal if some of the people you are writing about are going to read it. Some interns keep their journals in loose-leaf format and merely remove any pages they wish to keep private before showing the journal to anyone else.

There are many different approaches to journal writing, and many different reflective techniques. Your instructor may have forms and techniques that you are required to use, but we would like to discuss just a few of the more common forms here.

Unstructured Journals   The simplest form of journal writing is just to take time after each day to think back on what stood out for you that day. Although there is no "right" length for these entries, they should record what you did and saw that day, new ideas and concepts you were exposed to and how you can use them, and your personal thoughts and feelings about what is happening to you. It may be helpful to divide what you learn at an internship into five categories: (a) knowledge, (b) skills, (c) personal

Focus on SKILLS

When You Don't Know What to Write

Many interns tell us they are afraid that there are going to be days when there is just nothing to say. Well, our experience is that you won't have that happen very often, but there may be some days when writing is difficult. For those days, here are some questions to consider, generated by a community service program a number of years ago that apply as well today as they did then (National Crime Prevention Council, 1988):

- What was the best thing that happened today at your site? How did it make you feel?
- What thing(s) did you like least today about your site?
- What compliments did you receive today and how did they make you feel?
- What criticisms, if any, did you receive and how did you react to them?
- How have you changed or grown since you began your work at this site? What have you learned about yourself and the people you work with?
- How does working at this site make you feel? Happy? Proud? Bored? Why do you feel this way?
- Has this experience made you think about possible careers in this field?
- What kind of new skills have you learned since beginning to work at this site and how might they help you?
- What are some of the advantages or disadvantages of working at this occupation?
- If you were in charge of the site, what changes would you make?
- How has your work changed since you first started? Have you been given more responsibility? Has your daily routine changed at all?
- What do you think is your main contribution to the site?
- How do the people you work with treat you? How does it make you feel?
- What have you done this week that makes you proud?
- Has this experience been a rewarding one for you? Why or why not?

From "Reaching Out: School-Based Community Service Programs," by the National Crime Prevention Council, 1988, p.101. Reprinted with permission.

growth, (d) career development, and (e) civic development. Try to include all these categories in your journal.

Other Kinds of Journals (Baird, 2007; Compact, 2003; Inkster & Ross, 1998) Some other forms of journals that have been used with interns include:

- **Key Phrase** journals are those in which you are asked to identify certain key terms or phrases as you see them in your daily experience. A more expanded version of this concept will be discussed later.
- **Double Entry** journals are divided into two columns. In one column, you record what is happening and your reactions to it. In the other, you record any ideas

and concepts from classes or readings that pertain to what you have seen and experienced.

- In **Critical Incident** journals, you identify one incident that stands out over the course of a day or a week and write about it in some depth.

PROCESSING TECHNIQUES

There are a number of processing techniques that you can use and that you may want to include in your journal. Some techniques are specific to a discipline, such as "verbatims" in pastoral counseling or "process recording" in social work. If you are not in one of these fields, you can use different ones at different times or you can use one consistently, depending on your preference or that of your instructor. There are merits to both choices. Switching techniques from time to time may let you see things you have been missing. On the other hand, using a consistent technique allows you to look back over several entries and look for patterns. Common techniques include process recording (Sheafor & Horejsi, 2003), SOAP notes, and DART notes (Baird, 2007). In two Focus on Skills entries in this chapter, you can read about two other techniques that are less common but that we think are very useful. Three-Column Processing is particularly useful for those just learning the skill of reflection (Weinstein, 1981; Weinstein, Hardin, & Weinstein, 1975). It provides a structured way of writing about incidents (both positive and negative) that you recall so that you can look for tendencies and patterns. The Integrated Processing Model (Kiser, 1998) is a particularly useful approach for developing the art and skill of Practical Reasoning, as it guides you back and forth between theory and practice.

Focus on SKILLS

Weinstein's Three-Column Processing

Gerald Weinstein (1981; Weinstein, Hardin, & Weinstein, 1975) developed a method of reflecting on events that may be helpful with your reflective journal. Take a moment at the end of the week to recall any events that stand out in your mind. Select one or two (they can be positive or negative). Divide a piece of paper into three columns. In the left-hand column, record each action taken by you or others during the event.

Record only those things that you saw or heard, such as "She frowned," "He said thank you," or "They stomped out of the room." List them one at a time. Now review the list and try to recall what you were thinking when the different actions occurred. When you recall something, enter it in the middle column, directly across from the event. For example, you may have been thinking "What did I do now?" when the people left the room. Finally, read the list again and try to recall what you were feeling at the time each action and thought occurred. Record what you recall in the right-hand column. For example, you may have felt embarrassed, confused, or angry when they walked out.

Focus on SKILLS

Kiser's Integrative Processing Model

The Integrative Processing Model (1998, 2008) consists of six steps:

- **Gathering Objective Data from Concrete Experience**   In this step, you select an experience that you have seen or been part of. You can use a written, videotaped, or audiotaped account of the experience.
- **Reflecting**   In this step, you record and assess your own reactions to the experience. You may respond to particular questions or you may use a less structured format.
- **Identifying Relevant Theory and Knowledge**   Here, you seek out or recall ideas that can help you make sense of the experience in a variety of ways.
- **Examining Dissonance**   Now you review all the ways you have looked at the experience to see whether there are any points of conflict. These conflicts may be between or among competing theories; between what the theory says should happen and what actually did; between what you believe and what the agency seems to value; or between any two or more aspects of the experience. Sometimes this dissonance is resolvable, and sometimes it is not.
- **Articulating Learning**   Here, you look back over your writing and thinking and write down the major things you have learned from thinking about this experience.
- **Developing a Plan**   This comes next, and here is where you consider the next steps in your learning and your work. You may identify areas you need to know more about and places to pursue that knowledge. In addition, you may identify new goals or approaches you plan to use in your work. Taking these next steps can be another new experience, and the cycle can begin again.

PORTFOLIOS

Another use for your journal is as part of a portfolio of your experience. You can use a portfolio as a personal record, but many academic programs require them. A portfolio is often used to document and reflect on your entire academic journey in your major, and your internship experience can make a valuable contribution. There are many types of portfolios, and a discussion of them is beyond the scope of this book; however, there are resources listed at the end of the chapter if you want to explore portfolios further. Portfolios can also be valuable assets in an interview for a job, for graduate school, or even for another internship. If you are interested in using portions of your journal for this purpose, start planning for and discussing that project with your instructor. There may be items you will want to remove, such as highly sensitive comments about a person or organization. There may also be items you want to include, such as samples of your work, photographs, brochures, and so on. Your instructor can help you think about this now so that you know what sorts of items to seek out and save and so that you avoid violating confidentiality.

Communication Skills

You may have already studied the skills of effective communication in your courses. There is a wide range of such skills and we cannot cover them all here. Once again, we have chosen two that we believe are essential for an internship.

ACTIVE LISTENING

We all listen to others every day, but really listening well is an art, a skill, and a gift to the other person. Active listening means giving the other person your full attention. It means making sure that you understand what the person is trying to tell you. It means avoiding the impulse to rush in with advice or to tell a story about something similar that happened to you (Johnson, 2006). There are, of course, many other ways to respond to someone, and they all have their strengths and weaknesses. The Focus on Theory box in this chapter presents one way to categorize and think about responses.

The most important component of support is listening, so start there. It sounds simple, but think about how rare it is that someone really wants to listen to you, especially when you are struggling. Attention wanders, small talk intrudes, unwanted advice is given. It is a wonderful experience just to have someone listen quietly, attentively, and empathically to whatever you want to say. It is also a wonderful gift to give another person. As one student put it: *"Knowing that others are feeling similar feelings doesn't make those feelings go away, but it does make me feel better about having them. I feel more comfortable now opening up to my classmates and feel better equipped in encouraging them."* Listen carefully, and listen actively. Use skills such as paraphrasing to be sure that you understand what the person is trying to say. You will often find that these techniques also help the person to elaborate even more.

Once you have listened, and listened carefully, we suggest that you ask the person what she or he needs at the moment. Advice, analysis, problem solving, or reassurance can be wonderful if they are wanted. Or it may be enough just to be listened to.

GIVING FEEDBACK

You may be in situations at the site as well as in an internship seminar where you are asked to give feedback to a peer, a co-worker, or even a supervisor. You have almost certainly studied the principles of effective feedback, but they bear repeating. Effective feedback is specific and concrete, as opposed to vague and general; it should refer to very specific aspects of the situation being discussed. It is descriptive rather than interpretive. You can tell the person how you feel about what they did or did not do, but it is not your job to tell them why they did it. It is usually best delivered using an "I" statement rather than a "you" statement. "I had a hard time understanding what was happening when you were describing the situation" is better than "Your story was really confusing." If you have feelings about what was said, state them directly and attach them to a specific statement or portion of the story. And finally, feedback should always be checked with the receiver to see whether you have been understood and whether you have understood the other person.

Focus on THEORY

Modes of Responding

Theories about and guidelines for effective communication abound in the helping professions. You may very well have read some of them. We offer here, as one way to think about responding, categories of responses suggested by David Johnson in his extraordinary book *Reaching Out* (2006). Imagine that you or a peer has just told a story about something at the internship. A patient has relapsed, a co-worker or supervisor has been very harsh in their criticism, a community meeting turned into a shouting match, or whatever other event you want to imagine.

Advising and Evaluating   If you have taken a helping-skills class, you probably know that giving advice is often not the best approach, but when a friend or classmate is struggling, it can be awfully tempting to offer your heartfelt suggestions. And it can be a great relief as well. But it is not always helpful, and it certainly does not empower your peer to confront and resolve the situation—not to mention the next one! We have found that it is usually best to hold your advice until and unless it is asked for.

Analyzing and Interpreting   Here, the listener uses theory to interpret what has happened. It may be an opinion about the underlying psychological dynamics of the people involved, a sociological analysis, or any number of other interpretations. These thoughts can be contagious as other students move into the intellectual realm they know so well. But they can also be very distracting, especially if that is not the direction in which the person telling the story wants to go.

Questioning and Probing   There is a difference here between asking clarifying questions so you are sure you understand and asking questions that take the conversation in a different direction. So, if a student tells a story about a customer and someone asks, "Is this the same customer you spoke about last week?" that is clarifying, but "Who referred this customer to your agency?" is not.

ESSENTIAL KNOWLEDGE

In addition to essential personal resources, attitudes, and skills, there is also knowledge that is essential to have as you begin this journey in learning. In this section, we will introduce the sorts of knowledge you may need as well as some issues you could encounter in your role as an intern that could have professional, ethical, and/or legal implications.

Information About the Site

You will learn a lot about your internship site as the internship progresses, but there are some things you should know before you begin. Depending on how the placement process at your school works, you may already know most or even all this information. If not, it's important that you find out.

You need to know where the site is located and your options for getting back and forth. If you are planning on using public transportation, be sure to find out how far you will have to walk to and from the bus or subway and what sort of neighborhood(s) you will be walking through. You also need to know the norms of dress and behavior at your placement site. If you arrive either over- or underdressed, that will be a momentary embarrassment that is easily corrected, but with a little research, you can avoid it. If you start out addressing the others who work with you, including your supervisor, in an informal way (using their first name, using slang when you talk) and the norm of the workplace is to be more formal, it can be more than a brief embarrassment. Finally, you need to understand the rules and conventions about confidentiality. In medical and social service settings, this is a huge issue, but it can be an issue in other settings as well. You may want to come home on the first day and tell your friends and family all about something that occurred or you may want to take a file home with you. And that may all be fine, but it may not be, and crossing those lines can be considered a serious ethical transgression.

Knowing What is Expected

It is important that there is a clear, preferably written, and mutual understanding among all parties—yourself, the site, and the academic program—about what your start and end dates will be, your hours, those days (if any) that you will not be there (for example, during semester breaks) or that the site will be closed, and your basic responsibilities. There should also be a clear understanding of any other conditions or requirements of you, the internship site, or the academic program. In many programs, this document takes the form of an agreement or contract, which is signed by all parties and kept on file. If your program does not use such a document, you should ask how you can get some sort of written understanding of these issues.

Think About It

The Rights of Interns
As an intern, you may feel like you don't have any special rights, but you do. Are these rights being respected in your placement?

- The right to a field instructor who knows how to supervise, i.e., has been adequately trained and skilled in the art of supervision
- The right to a supervisor who supervises consistently at regularly designated times
- The right to clear criteria when being evaluated
- The right to growth-oriented, technical, and theoretical learning that is consistent in its expectations

---

Munson, cited in Royse, Dhooper, & Rompf, 2007

On-Site Resources

Most interns have assigned supervisors on site, and we believe that supervision is a critical component of a successful internship. We will be discussing supervision in detail in a later chapter, but as you begin, you should know who will be supervising your work, providing you with feedback, and answering your questions. It may be a single person, a team, or a rotating group of individuals. You also need to know who you will be meeting with your supervisor(s), how often, and what the general format of those meetings will be. In some cases, in addition to a supervisor, you will be assigned a co-worker to support and assist you with direction.

Campus Resources

If you are attending an internship seminar, the instructor is one of your campus resources. However, that person may or may not have arranged the placement and may or may not be able to troubleshoot with you if problems arise. You may also have a university staff or faculty member who visits you at your site, reads your journal, papers, and assignments, and provides you with feedback. Other campus resources might include an internship or co-op office, service-learning office, or career services office, and, of course, your peers.

Liability Insurance

We should note that there is considerable risk and liability involved in working in a direct service internship, be it with patients, employees, students, customers, or clients. Many campuses require liability insurance coverage for all students who are involved in off-site learning, such as internships; service-learning, course-based practica; and co-op education. If you don't know whether your campus has a blanket liability insurance policy that covers your work in the field, it is important to ask your campus instructor so you are fully informed about your coverage before you begin your work in the field.

Professional, Ethical, and Legal Issues

There some aspects of your internship that you may encounter early on and could necessitate discussion, usually with your supervisors or your co-workers. For your needs at this point, we find it's important to become aware of the breadth of the issues as well as how they are named. If you are particularly interested in knowing more about these issues than we cover here, as well as information about issues with professional, ethical, and legal dimensions, you will want to peruse Chapter 13, where we discuss these issues in substantial detail. The following list lays the groundwork for knowledge about your role as an intern and the three related issues of academic integrity, competence, and supervision.

- *Academic integrity* issues include a quality field site, responsible acceptance and learning contracts, and a seminar class that ensures a "safe place" for reflective discussions (Rothman, 2000).

- *Competence* issues include knowing your limitations and finding a balance between challenging work and a realization that you have exceeded your level of competency. It is important that you know the limits of your skills and seek help as needed (Gordon, McBride, & Hage, 2005; Taylor, 1999, p. 99).

- *Supervision* issues include the assignment of an appropriate supervisor who knows how to supervise interns in particular and can appropriately deal with such complex issues as client abandonment, the dynamic of attraction in the supervisory relationship, and quality evaluations of the intern.

Table 2.1 contains an even more detailed list of issues. As you look over the list of issues common to interns, there is no need to feel overwhelmed by it. You do not need to remember the issues identified nor are you expected to know what each one means. Knowing what they mean in practice will develop naturally over time

TABLE 2.1 Intern Role Issues Across Disciplines	
• Right to quality supervision	• Deportment
• Responsibility to confront situations in which educational instruction is of poor scholarship and nonobjective	• Negligence
	• Malpractice
• Disclosure of risk factors to all potentially affected parties (to site supervisor about intern; to intern about site supervisor)	• Implication of federal funds and related statutory and regulatory requirements
	• Use of college work-study funds for interns
• Behaving consistently with community standards and expectations	• Grievance processes
• Awareness of risk status of agency	• Informed consent in accepting the internship
• Active involvement in the placement process and consideration of more than one placement site	• Respecting the prerogative and obligations of the institution
• A clearly articulated learning contract that identifies mutual rights, responsibilities, and expectations	• Responsibility to confront unethical/illegal behaviors
	• Public representation of self and work
• A service contract with the agency that defines the limitations of the intern's role	• Disclosure of status as intern
• Liability insurance	• Boundary awareness
• The prior knowledge clause	• Boundary management
• Assurance of work and field site safety	• Personal disclosures
• Assumption of risk as limited to ordinary risk	• Criminal activities
• Employer-employee–independent contractor relationship	• Political influences/corruption
	• Subversion of service system
• Compensation: stipend, scholarship, taxable/tax-free	• Office politics

as you experience your internship. We will revisit this list in Chapter 13, where we have a more in-depth discussion about the issues and cite examples of dilemmas that interns have faced. For the time being, you are only responsible for reading over the list and taking time to consider those issues that have meaning for you.

Information About Personal Safety

Depending on where you are interning, you may be concerned about your safety. This is not a pleasant topic to discuss at a time when we are trying to buoy your enthusiasm, but it is important. In our experience, there are interns who worry when there is no realistic cause for concern, and there are interns who perceive no risk—and hence take no precautions—when the risk, while manageable, is very real. Whether you have these concerns or not, it is important that you make sure that you or someone at your program helps you assess your level of risk and develop a plan to minimize it.

ASSESSING YOUR LEVELS OF RISK

You should be made aware of any and all potential risks to your safety before you start your internship, which can come from the organization, the community, and the client population. If you are not made aware of these risks, then it is important to ask about them.

Client Risk Levels   When it comes to clients, three factors should be considered: the client's developmental stage, motivations, and immediate situational factors and conditions (Baird, 2007, p. 172). Details on these and other client issues related to personal safety can be found in Chapter 9.

Site and Community Risk Levels   Location and hours of operation are another set of variables. Some sites are located in risky neighborhoods (i.e., at risk for violence or crime), especially for someone who does not live in that community. Some placement sites have conducted as complete an assessment of the risk factors involved in their work as they can; others have been less attentive. Some have thorough procedures for staying safe and for handling risky situations, and others don't or might not be nearly as thorough. Agencies have the responsibility to be thorough in their assessment of risk factors and levels and to respond accordingly with policies and procedures for staying safe. Your responsibility is to determine how well your site meets these criteria for safety.

Birkenmaier & Berg-Weger (2007) offer excellent suggestions to assist you in your fact-finding efforts, including finding out the number and nature of violent or abusive incidents that have occurred at your placement in the past, arranging a tour of the surrounding neighborhood, and gathering information about other neighborhoods you may be visiting or traveling through.

MINIMIZING RISK LEVELS

If you in any way feel unsafe or otherwise not comfortable traveling to and from your site due to location, hours of operation, or security measures in or near the building, consult with the site supervisor and co-workers immediately to determine how they ensure their personal safety working at the site.

If you suspect the possibility of violence by clients or their family/friends, it is critical that you consult with others immediately, starting with the campus placement coordinator who arranged your placement, your campus supervisor or instructor, your site supervisor, and co-workers as to the history of violence at the field site. If there is a history of violence, it is time to meet with your supervisor and instructors and develop a safety plan.

If you have not already been informed about the relevant policies and procedures for safety, crisis management protocol, and reporting policies and procedures, then it is time to ask your supervisor for them. And when you get them, ask to copy them so that you can spend some time studying and understanding them. In that way, you know ahead of time just what to expect of your co-workers and supervisor should a safety emergency develop. You'll also know what is expected of you in that situation.

If there is ANY possibility of client violence against you, you want to be trained in physical restraint that is safe, nonviolent, and respectful of the client. The staff at the site already has the training and your supervisor can refer you to the next class. If you find out from your peers that there are better choices for the training, then this is not the time to be shy. Ask about it and ask if and how you can attend. You want the best training possible for yourself.

Again, if there is ANY possibility of client violence against you, be sure to request training in Universal Precautions for contact with bodily fluids such as blood or vomit. In addition, you want to ask if there is special training you can or should receive, such as cardiopulmonary resuscitation (CPR), basic first aid, and hygiene management for communicable diseases (Baird, 2007).

The Developmental Stages of an Internship

As we noted earlier, how you learn in the field differs considerably from how you learn in a traditional classroom on campus. The five developmental stages of an internship—*Anticipation, Disillusionment, Confrontation, Competence,* and *Culmination,*—reflect your phenomenological experience while you are learning. These stages are described in the next chapter and discussed in detail in subsequent chapters of the book.

ESSENTIALS FOR EMPOWERMENT

We have identified five sources of personal power, combining attitudes, skills, and knowledge, that will ensure the vitality and endurance you need to make your way through the challenges of learning experientially in your internship. They are the powers of *positive expectations, perspective, discussion, humor,* and *mindfulness*—which, if used effectively, will empower you in the ways you go about your work.

The Power of Positive Expectations

The power of expectations in your internship cannot be underestimated. Do you recall the classic play *Pygmalion* by George Bernard Shaw? The heroine in the story, Eliza Doolittle, who is "transformed" by Professor Higgins from an uneducated and poor street girl selling flowers into a woman of culture, reminds the professor that the difference between the two Elizas is not how Eliza behaves but, rather, how Eliza is treated. Eliza, like all of us, tended to live up to what was expected of her. Years after this Irish playwright wrote the play, which became the basis for the film *My Fair Lady*, Rosenthal at Harvard and Jacobson, an educator in San Francisco (Rosenthal & Jacobson, 1968), tested this premise among schoolchildren and published their findings in their classic text *Pygmalion in the Classroom*. These researchers found that children whose teachers believed they were intelligent in fact did "improve in their school work to a significant degree," whereas the other children who were not believed to be as smart achieved significantly less. Their research on the accuracy of interpersonal predictions found that one's expectations of another's behavior can become an accurate prophecy simply for its having been made (1968, page vii). This self-fulfilling prophecy, as it's come to be known, occurs when people live up to the labels and expectations of others. What we expect of others is in fact is what we get because of how we treat others, just as Eliza Doolittle wisely observed to Professor Higgins. In your internship, what you expect of your supervisor, co-workers, community leaders, clients, and others will affect how you interact with them and, in turn, how they work with you. So, too, will their expectations of you.

The Power of Perspective

Have you noticed that when you look at a glass that is half-filled with liquid and half-empty of liquid, you respond to this classic exercise in perspective based on whether you are more optimistic or more pessimistic in your thinking that day? That is because your perspective is affected by your experiences, your needs and values, and your feelings. Do you recall the story of *The Wise Men and the Elephant*? In that story, six blind men each argued about what an elephant looked like. They based their perceptions of the elephant on their "blindness," hence the concept of blind spots affecting one's perceptions and thoughts.

We all are prone to blind spots because we bring our self in its entirety to bear on our perspectives. In our seminar classes, we use exercises that demonstrate how not everyone perceives illustrations similarly—case in point, the glass that is half-filled or half-empty. In your internship, using the power of your perspectives to question your possible blind spots and to question how those blind spots can contribute to biases, prejudices, or oppressive actions will make the difference between an empowering experience and a disempowering one. What should you be striving for in terms of your perspective taking? Bradley (1995) identified the capacity to view situations from multiple perspectives and place them in context as a high level of perspective taking. So, too, is having the capacity to perceive conflicting goals within and among individuals while recognizing that the differences can be evaluated.

The Power of Discussion

One of the most challenging aspects of the internship for many students is the necessity to talk at times with supervisors and other colleagues about frustrations with their work or supervision. Would you be one of those interns? If so, you are hardly alone.

However, how you think about having that conversation makes the difference between moving beyond your frustrations or being stuck in them. Many interns tend to believe that having such a conversation must be confrontational in nature, so the conversation rarely happens. However, the issues that frustrate you are real issues, such as lack of supervision or quality supervision, sexual approaches by a co-worker, ethical issues that suddenly surface, perhaps even legal infractions, and so on. Think about it for a moment. You probably have experienced some frustrations in your employment or in volunteer work. How did you deal with them at that time? Would that way be the best way to handle them in this instance? Because of the negative connotations associated with the term *confrontation*, we ask our interns to think in terms of a discussion instead. Discussions allow perspectives of situations to be brought into the open and talked about. Confrontations can set up turf and battle lines that are difficult to rise above and even more difficult to move beyond. Discussions, on the other hand, allow you to work *with* your supervisor to address the issues and develop realistic resolutions. If you think these sorts of sensitive discussions will be difficult for you, ask your campus instructor or supervisor to create scenarios for you and your peers to role-play so that you can move beyond the confrontation trap into the power of discussion.

The Power of Humor

Some people just love to laugh. Others don't. But if you are one of those that enjoys a good laugh and has a good sense of humor, then you have an invaluable tool at your disposal to help you through tough times during your internship. Humor used appropriately has been shown to lift us above feelings of fear, despair, and discouragement. It helps us cope, especially with anxiety, and gives us strength to get through adversity; it keeps us "in balance" when all seems to be falling apart. Its physiological benefits have been well documented: It's good medicine for the heart and for the mind. It heightens and brightens mood, releases tensions, and leaves us feeling uplifted, encouraged, and empowered. Most importantly, it allows us to transcend predicaments, be flexible, and see alternative ways of looking at situations. It allows us to deal with high stress that we can't escape from by making fun of it, removing us from our pain and providing us the strength we need to get through difficult times. Its therapeutic values have been recognized for some time. If we can laugh at our setbacks, we can't feel sorry for ourselves. That is a very empowering way to move beyond crises (Fanger, 1999).

The Power of Mindfulness

Mindfulness is a centuries-old practice central to contemplative traditions, poets, and religions such as Buddhism (Schatz, 2004). Thich Nhat Hahn, a contemporary Buddhist who travels the world introducing communities to the practices of mindfulness, extols the power of mindfulness in serving those in need (Hahn, 1976). Mindfulness meditation is described as the capacity to be fully present in the moment, that is, to pay attention to what is going on in the actual moment in time without allowing our thoughts to drift to the past, to the future, or to other aspects of the present. Mindfulness is being used more and more in mainstream health care in this country because of the growing evidence that it can increase our enjoyment in life, expand our capacity to cope, and possibly improve our health, emotionally as well as physically (Schatz, 2004).

By now you are well familiar with the media's attention to the issue of multitasking and the research findings that our brains cannot do two independent things at the same time that both require conscious thought. We know that when we are mindful, our attention is focused on doing one thing well at a time; being mindful fits well with mono-tasking. Multitasking, on the other hand, means doing *more than one thing* at a time. Can one be mindful and multitask at the same time? According to Rebecca Shafir (Personal correspondence, 2007), author of *The Zen of Listening: Mindful Communication in the Age of Distraction* (2006), the answer is both *no* and *yes.*

Shafir suggests redefining multitasking within a context of mindfulness. It is possible with mindfulness practice to become quite adept at *shifting* from task to task, seamlessly and with greater efficiency, so that by the end of the day, you can return home from the internship knowing that the day's outcomes were as productive as possible, having carried out each task in mindfulness. By reframing what it means to multitask within the context of mindfulness is like adding more hours to your day. If you are mindful, time slows down. You get more done, enjoy things more, and feel less stress (R. Shafir, personal correspondence, 2007).

Practicing mindfulness doesn't mean that you won't be able to take on a number of tasks at the same time, or push through a project with intensity, or allow yourself to experience many different emotions. Practicing mindfulness does mean that you can make the multitasking expected of you a conscious choice and do it deliberatively, that you'll be aware that you are pushing through the project while paying attention to salient details, and that you will have insight into your feelings and be more aware of your choices in managing as well responding to them. Mindfulness meditation can be learned on your own or by taking a brief course on your campus, through your local hospital, or at the local community center (Schatz, 2004, p.1).

SUMMARY

Just as you probably feel like there is so much to learn, we feel like there is so much we want to tell you and so much we want to call to your attention. As the book and your internship progress, you will use a wider array of tools and look at your experience through a wider and more sophisticated set of lenses. For now, though, these are the basics.

For Contemplation

PERSONAL REFLECTION:
SELECT THOSE INQUIRIES THAT ARE MOST MEANINGFUL TO YOU.

1. A concise statement about your strengths as a learner will serve you well in a variety of contexts. Practice developing one, consulting the examples in this chapter for guidance.

2. Review your life context as described in this chapter. Which components of your life context seem to be assets for you as you begin your internship and which may be liabilities?

3. Review the essential attitudes discussed in this chapter. What are your strengths and weaknesses with regard to each one? Can you think of specific examples of times when you have or have not displayed these attitudes?

4. As you begin your journal, you need to decide whether it will be handwritten, tape recorded, typed, or computer generated. Then you will need to make sure you have the materials you need. It is also time to clarify the larger purposes of your journal, who will have access to it, and how you want to set it up to meet those needs.

5. Review the kinds of journals and reflective techniques discussed in this chapter. Which ones seem interesting and useful to you? Are any of them required for you?

6. Consider using your journal as a portfolio of your internship. What sorts of things might you include? If this is not appropriate in your particular case, you can always keep a portfolio that is separate from your journal.

7. What are some of the personal safety risk factors involved in your internship? Have you gathered the information you need? If not, where can you get it?

8. How do you feel about the level of risk you may be facing? If you are concerned, what can you do to feel more prepared?

SEMINAR SPRINGBOARDS

1. Choose a real or hypothetical (written-out) experience at an internship and practice the categories of responding. Have one intern tell the story and then have others take turns responding, each in a different category. One gives advice, one analyzes, and so on. Then ask the person who told the story how each response felt. Talk about which modes of responding were easier and which ones you may need to work on.

2. Devote a class, or a portion of the class, to practicing effective feedback. Again, you may use a real or scripted situation. Have each person practice giving feedback, and have others use a checklist or rating scale to point out strengths and areas for development. This exercise could also be done in triads or small groups.

3. As a class, make a list of the populations that you are working with. Brainstorm a list of the more common stereotypes that society has about each group. Remember that you are not being asked whether you subscribe to these stereotypes, just what they are. Where do you think these stereotypes come from? You may want to revisit this list later in the internship to see how these stereotypes have held up to the light of experience.

4. As a group, put aside some time to talk about the safety issues that each of you faces. Frame your discussion by focusing on risk factors, risk levels, and safeguards that need to be in place. Remember that all risk levels that warrant safeguards need to be brought to the attention of your campus supervisor. An important piece of that discussion is acknowledging the feelings that the risk levels evoke and how you will manage them so you thrive instead of survive each day in placement.

For Further Exploration

Baird, B. N. (2007). *The internship, practicum and field placement handbook: A guide for the helping professions* (5th ed.). Upper Saddle River, NJ: Prentice Hall.

Excellent coverage of safety issues as well as a variety of reflective techniques.

Birkenmaier, J., & Berg-Weger, M. (2007). *The practical companion for social work: Integrating class and field work* (2nd ed.). Boston: Allyn & Bacon.

Particularly thoughtful discussion of safety issues and how to advocate for yourself. Coverage of intrapersonal and interpersonal issues, including communication skills.

Collison, G., Elbaum, B., Haavind, S., & Tinker, R. (1999). *Facilitating online learning: Effective strategies for moderators.* Madison, WI: Atwood.

Informative, useful guide for netcourse instructors for moderating a web-based course such as the internship seminar.

Corey, G., & Corey, M. (2005). *I never knew I had a choice* (8th ed.). Belmont, CA: Brooks/Cole.

Eyler, J., & Giles, D. (1999). *Where's the learning in service-learning?* San Francisco: Jossey-Bass.

Thorough coverage of the theory and practice of reflection.

Johnson, D. W. (2006). *Reaching out: Interpersonal effectiveness and self-actualization* (9th ed.). Boston: Allyn & Bacon.

Excellent discussion of feedback and modes of responding, with useful exercises.

Kabat-Zinn, J. (2005). *Full catastrophe living: Using the wisdom of your body and mind to face stress, pain, and illness.* NY: Random House.

Anniversary edition of this classic work on mindfulness, meditation, and healing; tapes available to accompany book, which can be used for self-teaching.

Kiser, P. M. (1998). The integrative processing model: A framework for learning in the field experience. *Human Service Education, 18*(1), 3–13.

Kiser, P. M. (2008). *Getting the most from your human service internship: Learning from experience* (2nd ed.). Belmont, CA: Brooks/Cole.

A more complete explanation of the Integrated Processing Model.

Shafir, R. (2006). *The zen of listening: Mindful communication in the age of distraction.* Wheaton, IL: Quest Books.

Inspirational work on the practice of mindfulness for listening more fully.

Zubizarreta, J. (2004). *The learning portfolio: Reflective practice for improving student learning.* San Francisco: Jossey-Bass.

A useful collection of articles on portfolios.

References

Baird, B. N. (2007). *The internship, practicum and field placement handbook: A guide for the helping professions* (5th ed.). Upper Saddle River, NJ: Prentice Hall.

Belenky, M. F., Clinchy, M., Goldberger, N. R., & Tarule, J. M. (1986). *Women's ways of knowing: The development of self, voice and mind.* New York: Basic Books.

Birkenmaier, J., & Berg-Weger, M. (2007). *The practical companion for social work: Integrating class and field work* (2nd ed.). Boston: Allyn & Bacon.

Bradley, J. (1995). A model for evaluating student learning in academically based service. In M. Troppe (Ed.), *Connecting cognition and action: Evaluation of student performance in service learning courses.* Denver: Education Commission of the States/Campus Compact.

Compact, Campus. (2003). *Introduction to service-learning toolkit: Readings and resources for faculty* (2nd ed.). Providence, RI: Campus Compact.

Fanger, M. T. (1999). *Humor and social work: Are you serious?* Boston: Chapter, National Association of Social Workers.

Gregorc, A. F. (2004). *The Gregorc style delineator.* Columbia, CT: Gregorc Associates, Inc.

Hahn, T. N. (1976). *The miracle of mindfulness*: Boston: Beacon Press.

Inkster, R., & Ross, R. (1998). Monitoring and supervising the internship. *NSEE Quarterly (Summer),* 10–11: 23–26.

Johnson, D. W. (2006). *Reaching out: Interpersonal effectiveness and self-actualization* (9th ed.). Boston: Allyn & Bacon.

Kiser, P. M. (1998). The integrative processing model: A framework for learning in the field experience. *Human Service Education,* 18(1), 3–13.

Kolb, D. A. (1984). *Experiential learning: Experience as the source of learning and development.* New York: Prentice Hall.

National Crime Prevention Council. (1988). *Reaching out: School-based community service programs.* Washington, DC.

NSEE (1998). Foundations of experiential education, December 1997. *NSEE Quarterly,* 23(3), 1: 18–22.

Rosenthal, R., & Jacobson, L. (1968). *Pygmalion in the classroom.* New York: Holt, Rinehart and Winston.

Schatz, C. (2004). The benefits of mindfulness. *Harvard Women's Health Watch,* 11(6), 1–2.

Schon, D. A. (1987). *Educating the reflective practitioner.* San Francisco: Jossey-Bass.

Schon, D. A. (1995). *The reflective practitioner: How professionals think in action* (2nd ed.). Aldershott, UK: Ashgate Publishing.

Seashore, C. (1982). Developing and using a personal support system. In L. Porter & B. Mohr (Eds.), *Reading book for human relations training.* Arlington, VA: National Training Laboratories, 49–51.

Shafir, R. (2006). *The zen of listening: Mindful communication in the age of distraction.* Wheaton, IL: Quest Books.

Sheafor, B. W., & Horejsi, C. R. (2003). *Techniques and guidelines for social work practice.* Boston: Allyn & Bacon.

Sullivan, W. M. (2005). *Work and integrity: The crisis and promise of professionalism in America* (2nd ed.). San Francisco: Jossey-Bass.

Weinstein, G. (1981). Self science education. In J. Fried (Ed.), *New directions for student services: Education for student development.* San Francisco: Jossey-Bass.

Weinstein, G., Hardin, J., & Weinstein, M. (1975). *Education of the self: A trainer's mdnual.* Amherst, MA: Mandella.

CHAPTER *3*

Framing the Experience: The Developmental Stages of an Internship

Internship is like a diamond, in that it is multifaceted; it is also like a roller coaster with its highs and lows.

Allowing the stages to happen allows the intern to learn and have positive learning experiences.

STUDENT REFLECTIONS

Each intern's experience is unique, and yours will be too. You may have a different experience from other interns at the same placement or from any previous field experiences you have had. Placement sites differ too, and you may be in a seminar with peers who are doing very different work with very different groups of people. We continue to be amazed and enriched by the diversity of experiences that interns have as well as the diversity of their personal, professional, and civic development; it is one of the factors that makes working with interns gratifying, even after many years. Over time, we have noticed some similarities that cut across various experiences. Some of the concerns and challenges that interns face seem to occur in a predictable order. Our experience, our review of years of student reflections, plus our study of other stage theories have yielded a developmental theory of internship stages that helps to guide the thinking behind this book (Lacoursiere, 1980; Sweitzer & King, 1994, 1995).

THE DEVELOPMENTAL STAGE MODEL

As I read through the stages, I was comforted in knowing that I was not the only one experiencing all the various emotions. Knowing that others have felt the same way calmed my fears a bit to know that (what) I was experiencing was "normal."

The focus (of the week) has been for me to normalize my feelings and allow the process to happen.

STUDENT REFLECTIONS

We have identified five developmental stages that students tend to experience in an internship: Anticipation, Disillusionment, Confrontation, Competence, and Culmination. The learning in each stage is driven by concerns that the intern experiences; the concerns reflect what is most meaningful to the intern at that time in the internship. Once the concerns are no longer the focus of the intern's energy and attention, the intern is able to move forward in the journey of learning. The stages are not completely separate from each other; rather, concerns from earlier and subsequent stages can often be seen, in a less prominent way, during the current stage. Certain concerns and issues are apt to be particularly prominent during a designated stage, along with associated feelings and/or affect.

Each of the stages has its own obstacles for you to deal with and its own opportunities for you to grow through. There are concerns you will have during each stage, and to some extent, those concerns must be resolved for you to move forward and continue learning and growing. The process of resolving the concerns is also a learning experience in and of itself. In each stage, there are important tasks that will help you address the concerns. We cannot predict how quickly you will move through the stages; we can only predict the order in which the stages will occur.

An important distinction to keep in mind as you read and think about each of the stages in this model is what is meant by the terms *morale* and *task accomplishment* (Lacoursiere, 1980), both of which describe important aspects of your internship experience in each of the stages. The term *morale* refers to the interpersonal and intrapersonal tone of your experience at the agency. High morale is characterized by positive feelings about yourself, your work, and the agency. The tone is one of hope, optimism, and enthusiasm, and there is movement toward goals, even in the face of obstacles. As much as you would probably like to have high morale at all times, that is not usually what happens. The good news is that morale can often be recovered when it drops, and there is great learning in the process that only occurs if you fully experience both the drop and the recovery.

The term *task accomplishment* refers not so much to the specific tasks assigned by the placement site but to the attitudes, skills, and knowledge that you hope to acquire. Of course, there may be considerable overlap between the two. Here again, you might hope that the growth of this dimension would follow a steady, linear, upward path, but our experience suggests that this is not the case. There will be periods when you are learning and growing at an incredible pace. There will also be periods where you feel stuck, and you may be tempted to think you aren't ever going to get where you want to go or that you aren't learning anything. You are always learning, though, or at least the opportunity is always there, and paying attention to what you are learning, rather than dwelling on what you are not, will help you get back on track. Keep in mind that your rate of progress through the stages is affected by many factors, including the number of hours spent at the agency; previous internships or field experiences; your personality; the personal issues and levels of support you bring into the experience; the style of supervision; and the nature of the work.

Stage 1: Anticipation

As you look forward to and begin your internship, there is usually a lot to be excited about. Interns often look forward to the internship for several semesters, and it is your best chance to actually get out there, do what you want to do, and make a contribution to others. For most interns, however, along with the eagerness and hope, there is inevitably some anxiety. It may not be very visible, even to you, but there are enough unknowns in the experience to cause some concern and anxiety in anyone.

For interns, this anxiety generates the first set of concerns, which generally center on the self, the supervisor, co-workers, the community, and in some cases the clientele, such as customers, patients, contractors, citizen groups, or clients. We often refer to this as the "What if . . . ?" stage because interns wonder about things like: What if I can't handle it? What if they won't listen to me? What if they don't like me? or What if my supervisor thinks I know more than I really do? You will probably be concerned about what you will get from the experience and what it is really like to work at this particular site. Many interns wonder whether they "really can do this" and what will be expected of them.

Some interns report fears that they are not competent, that they have gotten this far only by great good luck and that in their internship, they will surely be found out. You may also wonder about your role; you are not in a student role while in the placement, but you are not a full-fledged employee either. Depending on your personal situation, you may also be concerned about your family and the effect that such a demanding experience will have on them.

In Their Own Words   **Voices of Anticipation**

I had a lot of anticipation going into my internship. Even at training, I asked if it was normal to feel as nervous as I did. The trainer reassured me that if I didn't feel nervous, something would be wrong.

This stage reminded me of my first year in high school and college. I just wanted to be accepted and didn't know how to do that. Although I am starting to gain a sense of what is expected from me in the internship, I am still wondering what staff members and clients think of me. Do they think I am stupid, lazy, ignorant, etc.?

You are going to interact with a number of people during your internship, and it is natural to wonder what to expect from them and whether they will accept you in your new role. You may, for example, be unsure of the role and responsibilities of the site supervisor, and you will be unusual if you don't wonder what your supervisor will think of you and whether he or she will care about you. Those working directly with clientele inevitably wonder about how they will be perceived and accepted by them and just what kinds of demands, behaviors, or problems they are going to exhibit. You may also wonder how you are going to manage the other responsibilities in your life and who is going to be there to support you. And finally, most interns are concerned about how they are received and treated by staff in the role of an intern.

The level of task accomplishment at this time is often relatively low, meaning that you may not be learning the specific things you went there to learn, and that can be frustrating. What is most important at this stage, however, is that you learn to define your goals clearly and specifically and begin considering what skills you will need to reach them. You must also develop a realistic and mutually agreed on set of expectations for the experience. Since you have not yet actually experienced the internship but have probably thought and maybe heard a lot about the agency where you will be working, it is inevitable that you will make assumptions, correctly or incorrectly, about many aspects of the internship (Nesbitt, 1993). Some of these assumptions come from stereotypical portrayals in the media of certain client groups (such as the mentally ill) or organizations (such as detention centers, corporations, law offices, green businesses etc.); others may come from your own experience with certain issues or problems. As much as possible, these assumptions and expectations need to be made explicit and then examined and critiqued. Key to moving beyond this stage is feeling accepted by and developing good relationships with supervisors, co-workers, the community, and/or the clientele.

You might be thinking to yourself that this doesn't really apply to you and that you have had some uncomfortable feelings but nothing you would label as anxiety—at least not yet. And that might well be the case. We have found that students who are in required internships for their academic majors and whose internships are graded do tend to experience this stage with much more intensity than do students in internships that are not graded or are pass/fail and are not related to their academic majors. The pressure of a graded internship that is essential for graduation can be much more stressful than an optional internship selected because of personal interest.

Stage 2: Disillusionment

Sooner or later, you are probably going to reach a time when you are not as certain or as positive about your internship as you would like to be. You may find that you are having some trouble getting up and going to the internship or that you are mumbling under your breath or complaining to friends about it. This experience is by no means universal, but many interns experience some kind of disappointment with their internship.

When the shift in tone happens, as it so often does, one reason for it is that there is almost always a difference between what you anticipated about your internship and what you really experience.[1] The size of this gap, and hence the dip in your morale, will depend on how successfully you accomplished the tasks of the Anticipation stage. If the concerns of the Anticipation stage have been adequately addressed, you will be less likely to encounter a wildly different reality from what you expected. As this intern noted: "*I have never, so far, complained about it, even though it is still so early in the semester. I truly hope that I skip this stage.*" Most likely, there will be some discrepancies, and some of them will be troubling. Furthermore, issues may arise that you simply never considered. For some of you, the change will be subtle and barely noticeable; for others, it *will be* profound and overwhelming. It is important to keep in mind that *having problems* is a necessary and essential element of an internship; otherwise, growth and development will not occur. This is quite different from believing that your *internship* is the problem; this belief will tend to define your entire internship experience and affect all aspects of it as well as affect your sense of self as an intern along with your feelings.

If the Anticipation stage was the "What if . . . ?" stage, then the Disillusionment stage is the "What's wrong with my internship?" stage. Concerns in this stage center on many familiar aspects of the placement: supervisors, the organization, the "system," the community, the clientele, such as customers, contractors, patients, citizen groups, or clients, or yourself. However, feelings associated with these concerns often include frustration, anger, sadness, disappointment, and discouragement. You may find yourself directing any or all these *feelings* toward any of these people, the community at large, or the organization itself.

This stage is the onset of what we refer to as *a crisis of growth*. It is possible to become stuck in this stage, and that can have unfortunate consequences. At best,

---

In Their Own Words   Voices of Disillusionment

At a certain point, the internship was not what I expected it to be.

I was still unsure of what I would be doing next, and getting up early was very difficult for me. I found myself wanting to go back to bed rather than getting up and going to my internship.

The most critical learning incident of the week was by far the decision to review the stages . . . most particularly stage 2 . . . dissatisfaction . . . it was right on. It was important because it reassured me that my feelings of confusion, frustration and the emotions were normal . . . essential in keeping the intern in touch with reality during (the) internship. It allows me to relax and allow the experience to be true and legitimate. Previously I had thought of "where am I going with this . . . how will this work?"

---

[1] In some cases, the intern's primary focus is not the work of the internship but the alluring location of the internship site. For example, some students use an internship in part as a way to be in a major city, like New York, Chicago, or Washington, D.C., to be part of a cultural experience, such as Hollywood, Wall Street, or Walt Disney Pictures, or to work or study internationally. In the case of such geospecific internships, the Disillusionment stage, if it occurs, is likely to focus on some of the unexpected realities of life in this new and intriguing place. Of course, some interns are focused equally on the work and the location. For them, there may be two possible sources of feelings of disillusionment in the internship experience.

learning and growth will be limited; at worst, the placement may have to be renegotiated or even terminated. On the other hand, letting yourself feel the impact of these issues and working through them present tremendous opportunities for personal and professional growth and development.

Stage 3: Confrontation

As the saying goes, "The only way around is through." The way to get past the Disillusionment stage is to face and study what is happening to you. Some interns resist acknowledging any problems, even when their level of task accomplishment is dropping. You may fear that any problems must somehow be your fault or that you will be blamed for them. You may think that "really good" interns would never have these problems. Paradoxically, though, it is the failure to acknowledge and discuss problems that can diminish your learning experience, your performance, and your evaluation by supervisors on site and on campus (Blake & Peterman, 1985).

Moving through this stage often involves taking another look at your expectations, goals, skills, and role. Although you may have set goals that were reasonable at the time, experience may have shown you that some of them are not realistic or the opportunities may have changed. Most importantly, having a clear sense of your role in the organization and what is expected of you in that role is critical to feeling grounded

TABLE 3.1 Developmental Stages of an Internship		
Stage	**Associated Concerns**	**Response Strategies**

Anticipation	Positive expectations
	Acceptance
	Anxieties
	  Self
	  Supervisor
	  Co-workers
	  Field site
	  Clientele
	  Community
	  Life context
		Examine and critique assumptions
		Develop key relationships
		Acknowledge concerns
		Clarify role and purpose
		Make an informed commitment
		Develop a learning contract

Disillusionment	Unexpected emotions
	  Frustration
	  Anger
	  Confusion
	  Panic
	Adequacy of skills
	Breadth of demands
	Relationship with clientele
	Operating values of
	  organization
	Disappointment with
	  supervisor/co-workers
		Acknowledge and clarify feelings
		Acknowledge gap between
		  expectations and reality
		Normalize feelings and behaviors
		Acknowledge and clarify specific
		  issues

Stage	Associated Concerns	Response Strategies
TABLE 3.1 *(continued)*		
Confrontation	Achieve independence Gain confidence Experience effectiveness Changes in opportunities Interpersonal issues Intrapersonal blocks	*Reassess goals and expectations Reassess support systems Reassess role and purpose Develop specific strategies*
Competence	High accomplishment Investment in work Quality supervision Ethical issues Worthwhile tasks Home/self/career issues	*Develop coping strategies Share concerns openly*
Culmination	Redefine key relationships Termination with patients or clients Transfer of case management Collegial gatherings End of studies Post-internship plans	*Set final supervisor meeting Identify feelings Engage in introspective/reflective writing*

and having a sense of purpose in your work. This is also a time to reexamine and perhaps take the necessary steps to bolster your support system—your lifeline when the going gets tough as well as when the going is great.

There may be interpersonal issues between you and your supervisor or co-workers that are getting in the way or you may be at odds with the way the community or the clientele is approaching a problem or project. What may be needed is some help clarifying these issues and developing a strategy for resolving them. You will need to consider intrapersonal factors, such as mounting personal problems or unexpected crises in your outside life. There may also be aspects of your personal makeup that are contributing to the problems. For example, it may be that your reactions to some typical features of an internship (such as negative

In Their Own Words Voices of Confrontation

I recognized, sensed, and felt the great struggle exploring change can initiate within others as well (as me).

I chose to confront this situation and figure out just what was happening to me that made me feel unsatisfied.

Once I confronted these anxieties, though, everything worked out. I am beginning to confront this situation by making something good of this.

feedback, authority, or speaking in or before a group) reflect patterns evident throughout your life that are being exacerbated by the internship. There are many strategies for dealing with these intrapersonal issues, and we will explore them in Section Three of this book.

As the issues raised in the Disillusionment stage are resolved, morale begins to rise, as does task accomplishment. Your task at this stage is to keep working at the issues raised; ways to do just that are discussed later in the book when we explore in greater depth how to manage the necessary confrontation. This is a time in your internship when you may be tempted to "freeze the moment" and resist raising any more issues for fear of spoiling the progress you have made. The temptation is normal, but if you give in to it for long, you may find yourself stagnating or even regressing. However, with each new round of confrontation, you will feel more independent, more effective, and more empowered as a learner. You will have a sense of confidence that comes not just from what you have accomplished, and not from denying problems, but from the knowledge that you can grapple with problems effectively.

Stage 4: Competence

As your confidence grows, you will forge ahead into a period of excitement and accomplishment. This is the stage that every intern looks forward to—the reason for the internship. Morale is usually high, as is your sense of investment in your work. Your trust in yourself, your site supervisor, and your co-workers often increases as well. You may find yourself thinking of yourself less as an apprentice and more as a professional. You may even wonder why you are not being paid!

As an emerging professional, you have a solid platform from which to expect, or even demand, more from yourself and your placement. You may find that you want more than you are getting from your assignments, your instructor, or your supervisor. Many

---

In Their Own Words   **Voices of Competence**

The Competence stage consists of being confident in myself. I am looking forward to that aspect of my internship. It's not that I have never been competent in a task before. It is just that the sense of professionalism will be much greater.

(The site supervisor) gave me some good guidance and lots of space to create what I wanted. This is usually a good thing for me. To be left alone to do what I wish. I felt intimidated by my audience, so I want to be perfect in what I present and not look stupid. So, I have had some difficulty getting it all together.

I feel that all my years of schooling have been for this exact purpose. The minute I walked in the door, I felt at home . . . and in my element. . . . It's like nothing I have ever felt before.

While I was aware that I had experienced a difficult day (at work), my focus at the internship at night was to be more attentive. It meant putting into perspective all that occurred during the day in order to be effective at night. Balancing the work during the day continues to be the hard work of negotiating the intellect and the affect.

interns also report that during this time, they are better able to appreciate the ethical issues that arise in their placements and are more willing to confront them. These are all positive developments. If taken too far, though, they can lead to perfectionism. You may begin to apply unreasonable standards to those around you, to yourself, or both. Excellence, not perfection, is your goal in this stage.

Another issue that can arise during this time is the stress of juggling your life outside the internship with your increasing commitment to the work. You may feel pulls on your time and loyalties throughout your internship; however, they may feel stronger now. That is because your earlier anxieties and roadblocks may have demanded too much of your attention to think about these conflicts at the time. Now that you have moved beyond the earlier crises, conflicts between home, campus, internship, and friends can surface more easily. This can be overwhelming, especially if you strive for perfection rather than excellence in all these aspects of your life.

Stage 5: Culmination

This stage occurs as your internship approaches its ending date. The end of the internship, coupled with the end of the semester and in some cases with the end of the college or university experience, can raise some significant issues for you. You may experience a variety of feelings as this time approaches. Typically, there is both pride in your achievements and some sadness over the ending of the experience. If you are in direct service, you may also feel guilty about not having done enough for the clientele or concern that no one will be as effective with certain customers or clients as you have been (and you may be right!).

For those of you who are ending your college career, you may be concerned with continuing your education, employment, or economic survival. Relationships with friends, family members, lovers, and spouses that have been organized around your role as a student have to be reorganized. In any case, there are many goodbyes to be said. Goodbyes are never easy, and for some people, they are very difficult.

Often, interns find ways to avoid facing and expressing these feelings, particularly the negative ones. Avoidance behaviors may include joking, lateness, or absence. Some interns may devalue the experience—they begin saying it hasn't been all that great or find increasing fault with the placement site, the community, the supervisors or the clientele. Many interns find themselves having a variety of feelings and reactions—some of them conflicting and changing by the hour. This can be very confusing and upsetting.

To address the concerns of this stage, you need to focus on your feelings (whatever they may be), have an appropriate place to express them, and find satisfying

---

In Their Own Words   Voices of Culmination

I looked forward to entering the Competence stage, but I am not looking forward to stage 5—when placement ends.

The end of this experience will be sad . . . everything is ending . . . but the reflections will be great.

ways to say goodbye to staff, supervisors, community groups, customers, patients, clients, and, in some cases, other interns—both at the site and in an internship seminar on campus.

Of course, if you do not pay attention to the concerns of the Culmination stage, the internship will end just the same. However, you could be left with an empty and unfinished or unfulfilled feeling. In some cases, interns struggling with the Culmination stage actually sabotage their internship by allowing their discomfort about ending to color their perceptions of the entire experience and affect others' perceptions of their work.

SUMMARY

Reading about (the) stages, I was amazed to see how much in common I had with the student reflections.

Reflecting on a recent confrontational situation at my site . . . I began to read (about the stage), but after a few pages, I stopped. I had two immediate reactions: One, I wish I had this material a few weeks ago. It would have helped me greatly. . . And two, I would prefer to read (this) text from start to finish . . . (even now that my internship has ended) . . . to follow the internship process "postmortem."

I am aware that my stages and concerns are not unique to me . . . One strategy was to respond to the concerns of each of the stages. The vocabulary (of the stages) gave me a way to express myself . . . to work to normalize my feelings and behaviors . . . This week, I acknowledged feelings, reassessed goals, expectations, and support systems, began to develop strategies, and I shared concerns and started to identify unfinished business. These experiences spurred me to make decisions—(to) choose behaviors I would not have in the past.

STUDENT REFLECTIONS

Now you have a sense of what is ahead in your internship and in this book. And you have an awareness of the resources, qualities, and essential skills that you will need to navigate the course ahead. Remember that even though these stages may hold true for many or even most interns, especially when viewed from the outside, both the pace with which you move through the stages and the phenomenological experience of being in each of them will vary a great deal from individual to individual. As you move through the experience of your internship, you will find that the chapters that follow explore each stage in more depth and also encourage you to remain focused on aspects of yourself, which you will explore in the next chapter.

For Contemplation

PERSONAL REFLECTION

As you read the description of the stages of an internship, did anything seem remotely familiar? Did the stages remind of you of any other experiences you have had? As you read the stages, did any one in particular stay in your mind or attract your attention? Why?

SEMINAR SPRINGBOARDS

Think about the issues raised for you by your understanding of the stages and discuss your thoughts with your peers. Are there issues that the stages do not obviously address? What are your thoughts about knowing about these stages at this point in your placement?

For Further Exploration

Birkenmaier, J., & Berg-Weger, M. (2007). *The practicum companion for social work: Integrating class and field work* (2nd ed.). Boston: Allyn & Bacon

Thorough, contributive fieldwork text for the social work student.

Baird, B. N. (2007). *The internship, practicum and field placement handbook* (5th ed.). Upper Saddle River, NJ: Prentice Hall.

Comprehensive and especially useful to graduate students in the helping professions.

Boylan, J. C., Malley, P. B., & Reilly, E. P. (2001). *Practicum & internship: Textbook and resource guide for counseling and psychotherapy* (3rd ed.). Philadelphia: Bruner-Routledge.

Takes a comprehensive approach to many aspects of graduate counseling internships.

Chiaferi, R., & Griffin, M. (1997). *Developing fieldwork skills: A guide for human services, counseling and social work students.* Belmont, CA: Brooks/Cole.

Offers a developmental framework of stages for the intern-supervisor relationship.

Chisholm, L. A. (2000). *Charting a hero's journey.* New York: The International Partnership for Service-Learning.

Based on Joseph Campbell's tale of the hero's journey, this is a guide for reflective journal writing in a cross-cultural context, for students involved with study abroad and service.

Cochrane, S. F., & Hanley, M. M. (1999). *Learning through field: A developmental approach.* Boston: Allyn & Bacon.

Takes a developmental approach to the fieldwork experience in social work education.

Faiver, C., Eisengart, S., & Colona, R. (2000). *The counselor intern's handbook* (2nd ed.). Belmont, CA: Wadsworth.

Takes a focused, pragmatic approach to counseling field experiences.

Gordon, G. R., McBride, R. B., & Hage, H. H. (2005). *Criminal justice internships: Theory into practice* (5th ed.). Cincinnati, OH: Anderson Publishing.

Offers the undergraduate criminal justice intern a comprehensive guide to issues specific to a criminal justice internship.

Grobman, L. M., Ed. (2002). *The field placement survival guide: What you need to know to get the most from your social work practicum.* Harrisburg, PA: White Hat Communications.

A collection of previously published articles from *The New Social Worker* that identify and address important issues across the duration of the fieldwork experience.

Kiser, P. M. (2008). *Getting the most from your human service internship: Learning from experience* (2nd ed.). Belmont, CA: Brooks/Cole.

Offers a framework for processing and integrating previous learning while engaging in field experiences.

Lacoursiere, R. (1980). *The life cycle of groups: Group developmental stage theory.* New York: Human Sciences Press.

Explains Lacoursiere's theory in detail and discusses its application to many kinds of groups.

Milnes, J. A. (2003). *Field work savvy: A handbook for students in internship, co-operative education, service-learning and other forms of experiential education.* Enumclaw, WA: Pleasant Word/Division of Winepress Publishing.

A nuts-and-bolts guide filled with advice that leads the student through the many tasks of a structured experience in field-based learning.

Schutz, W. (1967). *Joy.* New York: Grove Press.

A group development theory that has had an effect on our view of internships. Talks a great deal about acceptance, inclusion, and control issues.

Sweitzer, H. F., & King, M. A. (1994). Stages of an internship: An organizing framework. *Human Service Education, 14*(1), 25–38.

Gives more details of how Lacoursiere's and Schutz's works informed our thinking about our model developmental stages of an internship model.

Sweitzer, H. F., & King, M. A. (1995). The internship seminar: A developmental approach. *National Society for Experiential Education Quarterly, 21*(1), 1, 22–25.

Discusses our general approach to working with interns in a seminar class from a developmental perspective.

MODELS THAT FRAME FIELD EXPERIENCES

Chiaferi, R., & Griffin, M., (1997). *Developing field work skills.* Belmont, CA: Brooks/Cole.

Chisholm, L. A. (2000). *Charting a hero's journey.* New York: The International Partnership for Service-Learning.

Cochrane, S. F., & Hanley, M. M. (1999). *Learning through field: A developmental approach.* Boston: Allyn & Bacon.

Gordon, G. R., McBride, R. B., & Hage, H. H. (2005). *Criminal justice internships: Theory into practice* (5th ed.). Cincinnati, OH: Anderson Publishing.

Grant, R., & MacCarthy, B. (1990). Emotional stages in the music therapy internship. *Journal of Music Therapy, 27*(3), 102–118.

Grossman, B., Levine-Jordan, N., & Shearer, P. (1991). Working with students' emotional reaction in the field: An educational framework. *The Clinical Supervisor,* (8), 23–39.

Inkster, R., & Ross, R. (1998/Summer). Monitoring and supervising the internship. *National Society for Experiential Education Quarterly,* (23), 4, 10–11, 23–26.

Inkster, R., & Ross, R. (1998). *The internship as partnership: A handbook for businesses, nonprofits, and government agencies.* National Society for Experiential Education: www.nsee.org.

Kiser, P. M. (1998). The integrative processing model: A framework for learning in the field experience. *Human Service Education, 18*(1), 3–13.

Lamb, D., Barker, J., Jennings, M., & Yarris, E. (1982). Passages of an internship in professional psychology. *Professional Psychology,* (13), 661–669.

Kerson, T. (1994). Field instruction in social work settings: A framework for teaching. In T. Kerson (Ed.), *Field instruction in social work settings.* New York: Haworth Press (pp. 1–32).

Michelsen, R. (1994). Social work practice with the elderly: A multifaceted placement experience. In T. Kerson (Ed.), *Field instruction in social work settings.* New York: Haworth Press (pp. 191–198).

Rushton, S.P. (2001). Cultural assimilation: A narrative case study of student-teaching in an inner-city school. *Teaching and Teacher Education,* (7), 147–160.

Siporin, M. (1982). The process of field instruction. In B. Sheafor & L. Jenkins (Eds.), *Quality field instruction in social work.* New York: Longman (pp. 175–198).

Skovholt, T. M., & Ronnestad, M. H. (1995). *The evolving professional self: Stages and themes in therapist and counselor development.* New York: John Wiley & Sons.

Wentz, E. A., & Trapido-Lurie, B. (2001). Structured college internships in geographic education. *Journal of Geography, 100,* 140–144.

References

Lacoursiere, R. (1980). *The life cycle of groups: Group developmental stage theory.* New York: Human Sciences Press.

Nesbitt, S. (1993). The field experience: Identifying false assumptions. *The LINK (Newsletter of the National Organization for Human Service Education), 14*(3), 1–2.

Sweitzer, H. F., & King, M. A. (1994). Stages of an internship: An organizing framework. *Human Service Education, 14* (1), 25–38.

Sweitzer, H. F., & King, M. A. (1995). The internship seminar: A developmental approach. *National Society for Experiential Education Quarterly, 21*(1), 1, 22–25.

CHAPTER 4

Understanding Yourself

I have been in my placement for several weeks and have challenged my own philosophies many times. It frightens me to think that the very foundation on which I have based my life is being challenged by clients who I believed were going to be textbook cases. Not that I assumed that I was entering a vacuum, but I didn't think that my own beliefs could be shaken in such a short period of time. Maybe I am making no sense at all. Maybe I am trying to make too much sense.

STUDENT REFLECTION

INTRODUCTION

As you become more involved in your placement, what would you like to know? What knowledge, if you had it, would make you feel more prepared? When we have asked students these questions over the years, we have gotten all kinds of answers, including those focused on clients, co-workers, intervention techniques, community dynamics, and the organization's rules. Those are all good responses and important aspects of the internship to think about. What we hear less of, though, and what is just as important, is that you need to know about yourself, especially if you have chosen to work in the helping or social professions. In Chapter 2, we talked about the personal resources and qualities that prepare you for your journey and can leave you "feeling good" throughout your internship. Understanding yourself—that is, being aware of *who* you are as a person and as an intern and why you are the way you are—will allow you to realize the value of your strengths and the resources you have at your disposal so you can use them effectively when the time comes. It will also allow you to anticipate and prepare for some challenges that could arise as a result of your interactions.

Meaning Making

All of us are constantly engaged in a process of trying to understand what is happening to us; we struggle to make sense of the world in which we live, the people in it, and the experiences we have. Robert Kegan, a developmental psychologist, believes that the most fundamental human activity is "meaning making—that is, making meaning or making sense of our lives (Kegan, 1982, 1994). David Kolb, whose work we have already discussed, also refers to learning as making meaning (A. Y. Kolb & Kolb, 2005; D. A. Kolb, 1984). Perhaps nowhere is this activity more important than in the helping professions. If you are preparing for a career as a nurse, counselor, or advocate, you have been encouraged to consider the complex factors that lead to human problems and their solutions. Now that you have begun your internship, you are aware that it stops being an abstract exercise; you aren't thinking about human problems in general anymore but about the ones that confront you daily. Even if you are not entering the helping professions, problem solving will be an important part of your job; it is an important part of *every* job. You must consider an additional factor in thinking about problems, and that is *you*. You must pay attention not only to the sense you are making of what you encounter but of *how* you are making sense of it. Doing so will make you a more effective intern, a more effective practitioner in the future, and a more sensitive individual as you journey through the stages of an internship.

Dealing with Difference

You will deal with many people whose experiences and values are very different from your own. In those cases, self-understanding will help you avoid two major pitfalls: professional myopia and the tendency to regard difference as deviance (Sweitzer & Jones, 1990). At the heart of combating these tendencies is the ability to see your own reactions

---

Think About It

Making Sense of Daily Encounters

Suppose you are confronted with the following situations at your internship and must make sense of them:

- Your supervisor, in correcting your work, makes some insensitive statements.
- One of your co-workers is too aggressive in advancing some ideas.
- One of your clients is backsliding.

What factors do you consider? What is the problem and who or what needs to change? The very words we just used to describe the situations imply that the problems, and hence the solutions, lie in other people. However, you are part of the problem too and not just in what you may have done to evoke another person's response. For example, you think your co-worker is being too aggressive. Too aggressive for whom? Some people would not find that behavior too aggressive or even describe it as aggressive at all. Why did you? What needs to happen now? Does your co-worker need to tone it down? Do you need to learn to be more tolerant? Do you need to see things from other perspectives?

and views as only one among many possibilities. For this reason, understanding some of the sources of your personal characteristics and responses is critical. Understanding how your family, for example, influenced your beliefs and personality, coupled with knowledge of the diversity in family patterns and dynamics, will let you be more objective in your assessment and opinion of others.

Because everyone has their own system of meaning making, not only will two people see the same thing differently, but they will often see two different things altogether. If you are not aware that your point of view is only one of many, you may fail to consider other possibilities before making decisions about causes and solutions; we call this *professional myopia*. For example, suppose you are an intern in a predominantly White high school and you hear a Black student complain of feeling out of place and uncomfortable, perhaps even unable to work to capacity. Part of the problem may be that certain features of the school itself, such as staffing patterns, menus, and recreational opportunities, are suited to middle-class White students. If you are not sensitive to these dynamics, you may describe the problem as one of adjustment or even cultural deprivation. Your efforts, then, will be aimed at helping the Black student feel better about a situation that may in fact be unfair or discriminatory.

When people are aware of differences between themselves and others, they sometimes confuse *difference* with *deviance*. They do not describe it as something different; they describe it as something wrong. You are not going to like every client nor everyone you work with, but making the distinction between difference and deviance can help you accept and empathize with a wider variety of people. The tendency to assess difference as deviance can also lead to poor choices in using power and influence. If you do not see your perceptions as one of many possibilities, you may react to different perspectives by trying to change them. A trait assessed as a flaw in a client or co-worker's personality, for example, may in fact be a cultural difference (Axelson, 1999). A family may choose to sacrifice some degree of their children's comfort and education to allow elderly grandparents to live in relative comfort. The family then becomes the target of a complaint by the school system. Their decision may reflect a cultural value about the relative responsibility families have to children and elders. If you do not see the decision in this light, you might condemn the family for their treatment of the children, thereby damaging your relationship with them. Furthermore, you might then choose to concentrate on changing the family, never considering the option of urging the community to accommodate a wider variety of family customs and structures.

Recognizing the Stages

Finally, the more you understand about yourself, the better you will be able to recognize the stages of the internship as they happen. It will not be enough to know, or suspect, that you are in stage 1 or stage 2. What does that mean for you? What strengths can you draw on and what personal traps should you avoid? To move through the stage you are in, are there aspects of yourself, as well as aspects of the internship, that you must confront and try to change? If you can discuss these issues in your journal and talk about them with peers, instructors, and supervisors, you will recognize the challenges sooner and move through them more smoothly.

It is important to realize that self-knowledge is not like other knowledge. Once you have learned a fact or a skill, you have it (unless, of course, you forget it or it gets rusty). But you are changing all the time, and your system of making meaning is changing with you. In addition, just as there is always something new to see and appreciate in a work of art, there is always more to know about yourself. So, self-understanding is a process to which you must become accustomed and committed. It takes work, but it will pay you big dividends.

The Special Relevance of Self-Understanding for the Helping Professions

Self-understanding also plays an important role in helping human service professionals fulfill their responsibilities effectively and responsibly. As an intern in a helping profession, regardless of your specific responsibilities, you are usually trying to form relationships with clients. In the context of those helping relationships, you attempt to respond, or help clients respond, to a variety of human problems.

Although you may never have thought of it this way, in fulfilling these functions, you occupy a position of power and influence over clients (Brammer, 1985). Not only do you form opinions and make judgments, but you also make those opinions known to others in individual and group sessions, reports, and staff meetings. Furthermore, you may be asked to recommend for or against various kinds of intervention. Sometimes the internship site controls vital resources, in some cases as basic as food and shelter. Increased understanding of the feelings, beliefs, tendencies, and styles that make up your system of meaning making will help you be more effective in all your functions. It will also help you be sure you are using the power inherent in your role for the welfare of your clients.

A number of authors have commented on the role of self-understanding in forming effective relationships (Brill & Levine, 2004; M. S. Corey & Corey, 2006; Schram & Mandell, 2006). If you can clarify and discuss your feelings and personal patterns, you can serve as examples for clients who are struggling to do the same. Furthermore, if you have confronted and modified aspects of yourself that you didn't like, you are more likely to communicate a belief in the capacity for change (M. S. Corey & Corey, 2006). Finally, awareness of ways in which your experiences are similar to those of clients can help establish empathy and trust.

Self-understanding will also help you avoid projection. Projection, as we are using the term, refers to the unconscious tendency to believe that you see in others feelings and beliefs that are actually your own, but you are unaware of them because of their unconscious nature. This tendency can affect your ability to understand, accept, and empathize with another person. If a client brings up an issue with which you are uncomfortable in your own life, you may avoid the subject or decide that the client is being difficult (G. Corey, Corey, & Callanan, 2006). If you are angry at a client and unaware of that anger, you may decide that the client is angry at the world; if you have trouble with assertiveness, you may react negatively to an assertive client, supervisor, or co-worker. Projection can also lead to poor decisions about client needs. You may fall into the trap of inadvertently using clients to meet your own needs (G. Corey et al., 2006). For example, if you have trouble with assertiveness, you may push clients to be assertive in part because it fulfills your need to be assertive vicariously. Increased self-awareness will help you make more conscious choices in all these situations.

COMPONENTS OF SELF-UNDERSTANDING

Of course, self-understanding is an enormous area to consider and one that requires a continuous commitment on a personal level. We have chosen several topics within the realm of self-understanding to concentrate on in this chapter. Our goal is to introduce you to these areas, to give you some things to think about, and to provide you with resources to further explore these areas. We are trying to summarize some complex theoretical information, some of which may be new to you. You and your instructor will need to decide which of these areas merit your time and energy. That decision may be made for your class as a whole or different class members may choose to pursue different areas. In any case, we hope this chapter is a resource for you throughout your internship.

Knowing Your Values

A value is an idea or way of being that you believe in strongly—something you hold dear and that is visible in your actions. In the previous section, we argued that self-examination is important; that is a value for us. You may believe strongly in taking care of your family, in serving your country in some way, or in the existence of a deity. You are probably very aware of some of your values, especially ones that have been challenged, debated, or highlighted in the media (e.g., values about abortion, euthanasia, or freedom of speech).

Others, though, are so much a part of you and are shared by so many people in your life that they don't seem like values—they seem like truths. For example, both of us always placed a high value on punctuality (especially professionally) and even assumed at one time that people who were habitually late for professional appointments were irresponsible. We have come to understand, however, that not all cultures share the same view of time. It is important that you take the time to clarify your values and give some thought to how you will respond when faced with co-workers, clients, or members of the public who do not share those values.

You will almost certainly encounter someone at your internship who does not share some of your values, and you should think about how you might respond should issues arise. It might be a client, a co-worker, or even a supervisor whose value system is at odds with yours. Discussions about values can be lively and interesting if both parties are open to the discussion. But discussion is not always advisable or appropriate, especially in an internship. If there are some values that you cannot or will not accept in a client, a co-worker, or a supervisor, that is something you want to know as soon as possible and perhaps to discuss with your supervisor or campus instructor.

Think About It

A Values Check

Values permeate your life; we could never list all the important areas of value. However, here are some that may be especially important in your internship. To help you clarify your values in these areas, we ask you to think about select questions and answers to these questions, especially those that describe how you believe things "should" be, as these are important clues to your values. Think also in terms of how these values affect the work of your internship.

- **Sexuality**   How do you feel about homosexuality? Bisexuality? Heterosexuality? Teenage sex? Premarital sex? Monogamy? Extramarital sex? Various sexual practices?

- **Family**   How should a family be structured? Are single-parent families okay? Should one parent stay home with the children? Should there be a "boss"? If so, who should it be? How should decisions be reached? Should grandparents or other relatives live with the family?

- **Religion**   How important is religion to you? How do you feel about other religions? Would you ever do something that a leader of your religion said was prohibited?

- **Abortion**   How do you feel about abortion for yourself? For others? Are there circumstances under which you believe it is morally wrong? Morally justified? Should teens be able to choose on their own? Should the father of the fetus have a say?

- **Euthanasia**   Should a person be able to choose to end his or her own life? If so, under what circumstances?

- **Self-disclosure**   What kind of things is it okay to tell someone you hardly know? For example, would you tell that person your financial problems? Your family difficulties? Your income? Which of these things would you tell a close friend? Are there things you would never tell anyone?

- **Honesty**   How do you define this word? Are there different kinds of lies? Is it ever okay to lie? If so, when and to whom?

- **Autonomy**   Do you think people should normally make their own choices and accept responsibility for their lives? Or do you tend to see people as more responsible for one another? How large a role do you think fate plays in a person's life?

- **Work**   How hard do you think a person should work? What do you think about people who work only enough to get by and no more? What about people who seem to have no desire to find a better job or make more money? Those who always push themselves to work harder, no matter what? How about people who don't want to work at all?

continued

Think About It *(continued)*

- **Acceptance** How important is it to you that people from different cultural traditions be able to do things in a way that is consistent with that culture, even though it may be unusual by mainstream standards?
- **Hygiene** How often do you think a person should bathe? Should a person use deodorant? What would you think of a person who showed up at your office with dirty hands? A dirty face? Dirty feet? Dirty clothes?
- **Freedom of Speech** Do you think people's speech should be limited or restricted if it is hurtful to members of a cultural group, such as women or Muslims? What about statements disagreeing with U.S. military policy? With democracy as a basic form of government?
- **Time** How important is it for you to be on time? That others are punctual?
- **Alcohol** Is moderate use of alcohol okay? Is it ever okay to get drunk? If so, how often? What about binge drinking—is it a dangerous practice? Harmless fun? Is it up to the individual? Do you think alcoholism is a disease?
- **Drugs** Are there some illegal drugs that you think are okay to use in moderation? If so, is it also okay to sell them? What are your thoughts about the use or abuse of over-the-counter drugs or prescription medications?

It is important to ask yourself how strongly you feel about these values in your own life and about them in general and whether you are open to changing them. You might also want to think about how you came to have these values. Did you choose them consciously after careful thought? Did they come to you from your family, your friends, or from a cultural group to which you belong? Did you accept them more or less uncritically? Perhaps you are not really sure where they came from and why you hold them. That is fine, but you may want to think critically about them before they are challenged.

Recognizing Reaction Patterns

Reaction patterns are ways that you respond—your thoughts, feelings, and actions—to particular kinds of situations. Some patterns are helpful and work well. Others are distinctly unhelpful; they do not appear to get you what you want, and yet you repeat them in spite of yourself. One of the most frustrating, stressful experiences people can have is one in which they find themselves doing something they don't want to do or reacting in a way they don't want to react. Here are some examples:

- A friend has seen your in-class presentation and offers some constructive feedback. As the conversation goes on, you find that you are getting angrier and angrier and are having a hard time listening. You keep coming up, mentally or verbally, with defenses for every perceived criticism, and you imagine telling your friend off.

- You are at a party and people are talking about some hot topic. You have something to say but can't seem to say it. Since others are very vocal, it is easy, although frustrating, for you to just sit there. Later, someone says exactly what you were going to say and everyone seems impressed.

- You are struggling with an intimate relationship. A good friend asks you how it's going with that person, and to your surprise, you hear yourself saying that things are fine.

- A friend calls late at night and asks that you meet him right away. The matter does not seem like an emergency, but you leave your homework undone and go off to meet your friend.

These are not situations in which you later find, after much reflection, that you made a mistake. They are situations in which you know immediately afterwards or even during the situation that you are not responding the way you want to. In fact, almost as soon as it is over, you can think of several ways to handle the situation that would have been better.

Think about situations like this that have happened to you. Jot a few of them down. You may find that these are isolated incidents or that they occur with just one person. You may also find, however, that these responses are part of a pattern for you. You may find that, in general, you are defensive about criticism, unable to speak in groups, unable to say "no" even when you want to or to ask for help. Gerald Weinstein (1981) describes these tendencies as dysfunctional patterns. Please note that, in this case, dysfunctional does not mean useless nor does having a dysfunctional pattern make you a dysfunctional person. On the contrary, these patterns are clues to some important aspects of yourself. We both have them and so does everyone else. The reason we are asking you to think about these patterns is that you may "bump into" them during your internship. We will talk more about what patterns you might encounter and what to do about them in Chapters 10 and 11.

Appreciating Your Learning Style

In Chapter 2, we discussed the importance of knowing how you learn. As you progress in your internship, that knowledge will continue to be important, so we discuss it in a bit more depth here. As we mentioned earlier, knowledge about your learning styles can help you be an advocate for yourself in designing and modifying learning experiences at the internship. If you are working with people as clients, this knowledge can help you be more effective as you help them learn to do new things, such as change their behavior, locate resources, or improve their parenting skills. If you can learn to adapt interventions to a variety of learning needs, you will be more successful in this attempt. If you understand diversity in learning styles, you are less likely to view those different from your own as deviant or deficient. In this section, we are going to focus on one learning style theory that does not get as much attention as others and that we think has important implications for you as an intern.

Think About It

Everyone Has Dysfunctional Patterns—Even *You*!

See if you can identify any such dysfunctional patterns in your life by using the format suggested by Weinstein:

Whenever I'm in a situation where _____, I usually experience feelings of _____. The things I tell myself are _____, and what I typically do is _____. Afterward, I feel _____. What I wish I could do instead is _____.

Here is an example to help you:

- Whenever I am in a situation where I feel angry at a friend, I usually experience feelings of anxiety and self-doubt. The things I tell myself are: "Take it easy. It's not that bad. There's probably a good explanation, and besides, you don't want to upset him." What I typically do is smile, joke, or protest very weakly. Afterward, I feel as if I let us both down. What I'd like to do is find a clear, respectful way to tell my friend what is upsetting me.

And here are two from student journals:

- I am trying to build my self-esteem and confidence. It is hard for me to hold back my emotions when something upsets me, especially given my fear of doing something wrong or failing. I appear to be happy and cheery, but it takes so much out of me that I become negative and sometimes grumpy. This gets in the way of me liking myself.
- I feel guilty that I need to rest and refuel my batteries, but this is my life. I dislike that I want to do everything perfectly.

SEPARATE AND CONNECTED KNOWING

During the 1980s and since, a body of work in psychology has focused on the elaboration of two basic approaches to the world: separation and connection (Gilligan, 1982; Gilligan, Ward, & Taylor, 1988; Lyons, 1983). Separation is an approach to the world that emphasizes autonomy and abstract principles. Connection, on the other hand, emphasizes relationships and the importance of context. Much of this work was originally undertaken to fill a gap in developmental psychology caused by the use of predominantly male samples in psychological research and observation. By focusing on the experience of women, scholars were able to uncover important aspects of human development. Because of these origins, this work is often referred to as work on *women's development*. However, it appears that both orientations are found in men and women, although it also appears that the connection orientation is more often found in women.

The ideas of separation and connection were applied to learning in the book *Women's Ways of Knowing* (Belenky, Clinchy, Goldberger, & Tarule, 1986). *Separate* and *connected knowing* are terms used to describe opposite ends of a spectrum. Most people show some evidence of both styles, and most also lean distinctly in one direction or the other. Each way of knowing has strengths and limitations. In describing these styles to you, we will emphasize those aspects that seem most pertinent to your work as an intern. As you read the descriptions of these two styles, see whether you can recognize yourself.[1]

Separate knowers try to understand through analysis. They apply the principles learned in various disciplines. The scientific method, which you have probably studied, is an example of such a set of principles that is used to try to uncover scientific truths. You may also have learned methods for analyzing a poem, a play, or a painting.

Principles of analysis let us compare and evaluate. For example, two works of art may use different media, come from different times in history, and have been created in very different emotional contexts, but they can be compared and judged as works of art. Separate knowers, then, tend to separate themselves from too many specific details and look instead to organize experience according to abstract principles.

UNDERSTANDING PEOPLE

Turning now to the understanding of people, disciplines such as psychology and sociology offer separate knowers different lenses through which to make sense of human behavior. Each theory within these disciplines offers a different lens as well. Theories of human development (such as those by Piaget, Gilligan, and Erikson) as well as psychological theories (such as cognitive behaviorism, Gestalt, and transactional analysis) offer distinct ways of examining a person's behavior and reactions. A separate knower tries to understand a person by applying such theoretical frameworks.

Connected knowers, on the other hand, believe that truth is personal, specific, and located in experience. They attempt to understand not by analyzing but by empathizing. Rather than apply abstract principles, they immerse themselves in specifics. Connected knowers approach an idea or person with the belief that there is something of merit there, and they try to find it. In attempting to understand people, connected knowers try to share the experience of the person they are talking with or thinking about and understand what created that experience. They are less drawn to theories as a way of making sense of a person's experience.

The field of the helping professions in particular is an arena where connected knowing is valued, as opposed to being a liability, as it may be in many learning contexts (Belenky et al., 1986). Using this model, it is readily apparent that you will need to use both separate and connected approaches in your internship. What is important is to know *where* your natural strengths lie, what to use when, and where you will need to stretch and grow. Table 4.1 (see following page) summarizes the key differences between these two ways of knowing and approaching the world.

---

[1] *Women's Ways of Knowing* is a complex and interesting book. It identifies five epistemological positions taken by women (and presumably by some men as well). The distinction between separate and connected knowing is made within one of these positions, called *procedural knowledge*, which is a position commonly found among college students. We urge you to explore this work further.

TABLE 4.1

TABLE 4.1 **Two Ways of Knowing**		
	Separate	Connected
Mode of Understanding	Analysis Abstract Principles	Empathizing Immersion
Understanding People	Psychological & Sociological Theory	Experiential Logic of Each Individual
Relationship to Ideas	Skeptical Critical	Looking for Merits

APPLICATIONS TO YOUR INTERNSHIP

Many of your courses up to now probably emphasized the learning of principles and theories. In your internship, though, you are going to be immersed in experience and in specific situations. The theories you know will be only as good as your ability to use them in service to clients and the organization. We have known many students who struggle some with theory but are outstanding in the field and vice versa.

Theoretical knowledge is important in an internship. When you engage a client, conduct an interview, plan a change strategy, or organize a project, theories help you know how to proceed. Connected knowers' tendency to want to "get into the head" of the theoretician may help them achieve a richer and fuller understanding of the theory. However, when it comes to using that theory to analyze a specific person or event, they may struggle.

In working with others, a theory may help you understand, but it will not necessarily help you to connect nor to develop trust and empathy. After all, you are not interacting with a developmental stage or a psychiatric syndrome; you are interacting with a person. Most people wish to be seen as unique, not reduced to a stage or a diagnosis. The more you discover the experiential logic of that particular person (Belenky et al., 1986), the less strange that person's behavior and reactions will seem and the better you will be able to communicate empathy and acceptance. On the other hand, if you use only the client's frame of reference, you lose the opportunity that theories provide to get a fresh look and to step away from repetitive, destructive patterns and rationalizations.

APPLICATIONS TO SEMINAR CLASS

In your seminar class or support teams, there is a time and place for both separate and connected approaches. If you or someone else is struggling with a problem at your placement, a theoretical analysis can help you look at it in a new way. At the same time, a connected approach helps ensure that each person's unique experience is attended to and honored. Understanding the difference between separate and connected knowing may also help you understand your supervisor and co-workers better. Their expectations of and reactions to you may be based in part on their approaches as separate or connected knowers. We will return to these themes later in the book.

Recognizing Family Patterns

For many people, their family of origin is the group with which they spend the most time until they form their own family or leave home for some other reason. Your experience in your family of origin is a powerful influence on who you are. Each family has its own way of doing things. Often, we are so accustomed to the way our family is that we assume that everyone's family is that way. For example, while both of us come from families where eating dinner together was very important, they differed considerably in how that time was spent. For one of us, conversation at the table was quiet and polite, and interruptions were frowned upon. For the other, the dinner table was lively and often loud, with many conversations taking place at the same time. Imagine the shock for one of us visiting the other's family for dinner or for either of us visiting a family whose members came and went from the table or ate in different parts of the house.

We are going to encourage you to think about two common features of family life: rules and roles. Although very few families have a list of rules posted on the wall, they all have unwritten rules that tell everyone what they can and cannot do. As a child, you usually learn the rules by breaking them; someone reprimands or disciplines you, and you gradually figure out what is and is not acceptable. Sometimes families have rules that they are not even aware of. For example, one of us once listened to a family over the course of a weekend as they talked about all their relatives, living and deceased. There was one who was never mentioned though. The family was quite surprised when this observation was shared but upon reflection agreed that they almost never talk about that person. In observing this informal tradition, they were obeying a rule, even though they would never have called it that. Some family rules are a combination of individual family patterns and cultural norms (which we will explore more a little later on).

Here are some examples of family rules. Use this list to stimulate your thinking about the rules in your family:

- Don't talk about sex.
- Keep family business in the family.
- Never question a decision or disagree with an opinion from someone older than you.
- If your brother or sister picks on you, handle it yourself. Don't go crying to your parents.
- Guests are always welcome for dinner—you don't have to ask.
- Grandparents must always be consulted on important decisions.
- Mom and Dad need fifteen minutes to relax after work before you ask them anything.
- No one is entitled to privacy, except in the bathroom.

Family roles tell you who in the family performs which functions. Some of them are pretty formal. One parent, for example, may pay the bills, do the cooking, or handle the discipline. There are other kinds of roles though. There might be a family jokester who is counted on to make people laugh or the mediator who tries to help settle conflict between family members. Some children get the role of the "good" child, and their mistakes are often overlooked or treated lightly, whereas others get the role of "bad" child and are treated more strictly. In some families, these roles are shared or certain people have them only in certain situations. Here are some questions to help you think about roles in your family:

- Who has the final say in an argument or dispute?
- Who is in charge when the parent or parents are not home?
- Which child is the smart one? The talented one? The athletic one?
- Who can you count on to help you out of a jam?
- Which child gets the most leeway from Mom? From Dad?

In your internship, you will undoubtedly meet people whose family patterns and norms are different from your own. That can be pretty confusing as you try to make sense of their behavior and reactions. Understanding the sources of some of your behavior and feelings will, again, help you look more thoroughly and empathically at others.

Also, if you are not careful, you may carry a rule or a role from your family into your internship that is not helpful in that context. For example, your role as the jokester in your family, while it may serve to ease tension at home, may be annoying in a staff meeting, especially one where there is tension. Similarly, your role as the comforter may make it difficult for you to let a client struggle through his or her own problems.

Giving Thought to Your Cultural Identity

One very important part of you, your feelings, thoughts, and behavior is your culture. The term *culture* is defined in many different ways, but for our purposes, we use the definition offered by Donna Gollnick and Philip Chinn (2005) that culture is a shared and commonly accepted set of beliefs, practices, and behaviors. They go on to point out that there are both microcultures and macrocultures. The macroculture consists of those beliefs and practices shared by the majority of citizens. So, for example, in the United States, a belief in democracy is shared by most (although not all) citizens and would be considered part of the macroculture. Microcultures, on the other hand, are those beliefs and practices that come from membership in a smaller group or subgroup. There are, for example, attitudes and beliefs that are more accepted by women than by men and vice versa. Your cultural identity consists of these subgroups, the degree of identification you have with them, and your level of privilege as a member of dominant or subordinate groups.

YOUR SUBGROUP MEMBERSHIP

You are a member of many subgroups. Some of them are temporary; you can move in and out of them as you choose. For example, you are now a college student, but you could stop being one at any time, and eventually, you will no longer be a student. Other subgroups are relatively permanent; your membership in them is determined primarily by accident of birth. Your race, gender, and ethnicity are examples. Your age will change constantly, of course, but you will always be part of an age group. You are also part of an age cohort, sometimes called a generation. There are pieces of music and historical events that have enormous power for people of your particular age. Some have argued that sexual orientation is determined at birth; others believe it can change. However, your status as a heterosexual, homosexual, bisexual, or transgendered is a relatively permanent feature of your life. Social class is another important subgroup, and you need to consider both the social class in which you were raised and the one to which you now belong (Payne, 2005).

DEGREE OF IDENTIFICATION

Knowing the subgroups you belong to is only part of the picture. Another important part is how strongly you identify with that group. There are Jews, for instance, who identify very strongly with their Jewish heritage. They observe the traditions and holidays, obey dietary laws, and travel to Israel. There are also Jews who do none of these things. Individuals have varying degrees of identification with subgroups to which they belong. Although both of us are very aware of issues that confront us as a man or woman, we vary considerably in our identification with other subgroups. Fred, for example, knows almost nothing about his ethnic heritage, which is a mixture of Irish, English, and German, and is aware of very little influence from that subgroup. Mary, on the other hand, identifies strongly with both her Irish and Lithuanian heritage and has a good sense of how those traditions affect her as a person; however, she still struggles at times to fully appreciate the effects of her race (Caucasian) in her personal and professional life.

ATTITUDES TOWARD OTHER GROUPS

Do you see yourself as a prejudiced person? Do you hold any stereotypes about people of a different gender, race, ethnicity, sexual orientation, or religion? If you are like most of our students, you will answer these questions "no" or "not anymore." If you have qualified for an internship, you have probably been exposed to the core values of tolerance, pluralism, and respect for individuals that are at the heart of many professions. And naturally, you want to see yourself as a person who adheres to those values. If a part of you suspects that you do have some lingering prejudices, you may be keeping them hidden for fear that your peers or your professors will think badly of you or block your progress in your program.

Here is what we believe and invite you to consider. Everyone carries some stereotypes and prejudices, including us. If you work at it, you can learn to see your prejudices and make progress in overcoming them. However, at the same time, you will surely discover other, more subtle ones. The first step, though, toward being a nonprejudiced person is to confront and accept the prejudices you have. If you hide or deny them because you are ashamed or think that a good person would never have any, they will never change. Listen to what this intern had to say: *"Throughout my college years, I held opinions of people, groups, and cultures. I never once stopped to think about how my preconceived notions affected the way I acted toward people. I truly thought I was an open-minded person until I caught myself acting in this way."*

| Think About It |

You've Got Stereotypes!

Try this exercise. Pick a subgroup that has been the target of some discrimination and of which you are not a member. Depending on your own subgroup membership, some examples are Blacks, Muslims, lesbians, blue-collar workers, Native Americans, and many others. Now think about all the stereotypes you know about that group. Remember, we are not asking which ones you believe. Just write down as many as you can think of. You might want to get together with a classmate or two and expand your list. Now look at your list. Where did the stereotypes come from? If you don't believe them, how did you come to "know" them?

Most students have little trouble coming up with a substantial list. The reason stereotypes are so easy to recall is that they are literally all around us. Think about the group you picked. How many members of this group do you actually know? In some cases, the answer will be few or none. So where do your impressions of them come from? Some, of course, come from family and friends. Another part of the answer, though, is the way members of these groups are depicted in the media. Usually, members of various subgroups are portrayed in very limited, stereotyped roles. Think, too, about the books you read in school. How many members of these groups were in those books? How were they portrayed? The point here is that we have all "learned" harmful stereotypes about other groups (and even about our own).

Naomi Brill and Joanne Levine, in their excellent book *Working With People* (2004), caution professionals to be sensitive to their use of "the paranoid 'they.'" If you find yourself thinking that "they" are too aggressive, too concerned with money, or too standoffish, that is a key to a stereotype or prejudice that you hold. We encourage you to explore and face your prejudices, and we recommend some resources for doing so at the end of this chapter.

Once you have identified some prejudices, you can work on changing them. For example, through an experiential exercise like the one we just encouraged you to do, Fred discovered that he held some stereotypes about Hispanic men:

> *I had to admit that I felt uncomfortable in groups of Hispanic men and that I thought of them as being in general more violent and temperamental than I. Some reading about Hispanic cultural and family life helped me to get a more balanced picture of this group, with whom I had very little contact.*

Mary recalls that there was a time when she was somewhat apprehensive about working with same-sex couples in therapy. She attributed the discomfort to being generally uninformed about the homosexual culture:

> *After reading a good deal and attending workshops, it became apparent that my discomfort stemmed from such stereotypical beliefs as gay individuals*

are more maladjusted, unhappy, and pathological than their heterosexual counterparts. It was only after an exploration of these beliefs that I could better understand the culture as well as the basis for my own homophobic thinking.

As you get to know yourself, you need to consider your membership in various groups and subgroups in society and the positive and negative associations you have with your own and other subgroups.

THE NOTION OF PRIVILEGE

Privilege is an important and challenging concept when considering your cultural identity and the cultural identities of others (Doane, 2003; McIntosh, 1989). It may be that one or more of the groups you listed earlier is often referred to as a *minority group*. This term has been used to describe not only a group's relative numbers but also its position in society. Because numbers do not always equal position (e.g., there are more women than men in our society), we prefer to use the terms *dominant* and *subordinate* to describe various subgroups (Hardiman & Jackson, 1997). Dominant groups are those who hold the majority of resources and power. They are more often in control of major institutions in society and are more often in positions of power in government, education, and business. So, for example, while women outnumber men in the general U.S. population, the vast majority of CEOs, members of Congress, college presidents, state governors, and so on are men. Thus, with regard to gender, men are the dominant group and women the subordinate because men hold the majority of the positions of power. Using the same logic, in the United States, Whites are dominant with respect to race, heterosexuals with respect to sexual orientation, upper and upper middle class with respect to social class, Christians with respect to religion, and so on.

If you think about it, you will see that it is a mistake to refer to a person as dominant or subordinate. At any given time, most of us are members of some subgroups that happen to be dominant and others that happen to be subordinate. If you are a White, heterosexual, Jewish female from a working-class background, you are a member of two dominant groups and three subordinate groups. Go back to your list of subgroups and think about your place in each one of them in terms of being in a dominant or subordinate position.

If you are hearing these terms in this context for the first time, you may have some trouble accepting them; many people do. After all, if you are a woman, you may not feel subordinate to men, and if you are White, you probably have no desire to dominate people of color. The terms do not necessarily describe *your* behavior. You are, however, a *member* of groups that, as groups, are relatively more or less powerful in our society. More importantly for our purposes, in your internship, you will be working with co-workers, supervisors, and groups of individuals who are members of different dominant and subordinate groups than you are. And in general, the experience of members of the dominant group tends to be pretty different from that of subordinate groups.

Dominant groups are afforded certain privileges by society. You may find that hard to believe, given all the publicity that affirmative action and other programs receive, but many of the privileges given to members of the dominant group are subtle.

For example, both of us are White, heterosexual, and Christian. We can go into a department store and wander around as long as we want; friends of ours who are Black or Hispanic have told us that they are frequently followed by store detectives or questioned by clerks after only a few minutes. Wherever we have gone to school or have worked, the major holiday in our religion is at least two different days off for us and with school breaks accommodating the holy days. When we are out in public with our partners, we can hold hands, hug, or even kiss without worrying about who may be watching or disapproving. These two instances are not the case for our Jewish friends or our gay or lesbian friends, respectively.

Why have we devoted so much time to the issue of cultural identity? Many of the values, attitudes, and reaction patterns you examined earlier are strongly influenced by your membership in subgroups and the attitudes you learned about your own and other groups. Understanding the sources of these aspects of your personality will help you see them as only one of many possible ways to be in the world. Furthermore, in your internship and your life, you will surely interact with people whose cultural identity is different from your own. You need to understand your reactions to them and theirs to you. Understanding cultural identity will help you accept others for who they are and avoid displacing feelings or prejudices you have about subgroups onto your clients and co-workers. It is an important part of your personal, professional, and civic development.

For Helping and Service Professionals: Grasping Your Psychosocial Identity Issues

Particularly if you are preparing for a career in the helping professions, you have probably had a course in developmental psychology in which you had some exposure to the ideas of Erik Erikson (1980). In fact, you may have had a lot of exposure to Erikson and are now thinking to yourself, "Erikson? Again?" Well, yes, but just a little bit and with a different twist. Although some of Erikson's ideas have been questioned and criticized since he first advanced them, some of the basic tenets of his work are valuable tools for self-understanding (see Sweitzer, 1993, for a full explanation). There are issues in each stage that are not always obvious from the familiar names of the stages (trust vs. mistrust and so on). Your early experience with these issues often reverberates into your later life, and we find that some of these issues are especially relevant in internships. Here, we will review those issues and encourage you to think about them and their relevance to your work. You may find that this theory does not seem relevant or helpful to you, and there are lots of other developmental theories you can explore. If your work in the helping or service professions does not involve close interaction with people, you may want to skim or skip this section of the chapter.

TRUST VS. MISTRUST: GETTING YOUR NEEDS MET

Regardless of your level or preparation and skill, there will be times during your internship when you need help and support. Sometimes you will get it, and sometimes you won't. Your reaction to these events may be rooted in the Eriksonian stage of Trust vs. Mistrust.

Children who lean toward a sense of basic trust believe that even though their needs are sometimes not met, fundamentally they can count on the world to satisfy them. Note that they do not believe that their every need will be met; children with that belief are in for disappointment. Children who lean toward a sense of basic mistrust, on the other hand, believe that even though their needs are sometimes met, fundamentally they cannot trust themselves or others to meet their needs.

For children with a sense of trust, a negative experience (one in which their needs are not met) does not disconfirm their fundamental view of themselves and the world. For children with a sense of mistrust, however, those events actually confirm the view that they cannot, and perhaps should not, have their needs met; they cannot trust their world.

You may be interested in exploring this part of yourself, trying to find out about your experiences as a child. What is more important, though, is exploring this part of you as an adult. What is it like for you now when you really need something and don't get it? No one likes that experience, but some people find it devastating. You may also be a person who doesn't ask for much from others because you are afraid you won't get it.

In your internship, you may find yourself needing various things from your co-workers or supervisor. If they disappoint you, it is natural to be slightly upset, but a person with a shaky sense of trust may find these experiences unduly distressing. You may also find that, fearing your needs will not be met, you hide them, trying to appear confident and refusing to ask for help.

AUTONOMY VS. SHAME AND DOUBT: HANDLING YOUR IMPULSES

Impulses and impulse control may seem like childhood issues, and indeed, they are, although you will meet clients who struggle significantly in this area. But an internship is often powerful, emotional work in which you are not sure what will happen next and yet have to react quickly. Sometimes you will handle these situations with ease and clarity. Other times you will find that your first impulse is to do something that in your calmer moments you know is not appropriate. We all have these moments; how you react to them may be rooted in your experience with this stage.

In Eriksonian terms, children who end up toward the basic autonomy end of the continuum feel that even though they have to be controlled or their impulses curbed sometimes, fundamentally their impulses are good, and they are capable of independent action. Instances where they are controlled (by others or by themselves) do not disconfirm their fundamental picture of themselves. Children who lean toward the other end of the continuum—basic shame and doubt—feel that even though they can sometimes do what they want—complete an impulse—their impulses are suspect and shameful, and they must not surrender to them.

Adults have impulses too, and your experience in this Eriksonian stage may influence your feelings about impulses today. The issue is how you feel when you have an impulse, regardless of whether you act on it. Consider the example of an intern who is insulted in a particularly hurtful way by a client. Her first impulse is to insult him right back and put him in his place. She stops herself, though, and sets a limit with the client in a firm and calm manner. But her drive home, her journal, her conversation with peers, and her supervision time are dominated by how tempted she was. Rather than being proud that she controlled the impulse, she feels guilty for having it. The impulse, not the control, has confirmed her fundamental view of herself.

Even if this example does not sound like you, if this is a vulnerable area, you may be struggling with perfectionistic tendencies. Perfectionism can take a number of forms that you would not expect. For example, with one particular type of perfectionism, if you know how you should handle a situation and handle it that way but are tempted to behave otherwise, you become upset with yourself, not for what you did but for what you almost did. If you are one of the many interns who are prey to the perfection trap, we suggest you try to let go of perfection and settle for excellence, even if just for the length of the internship. It will make your life and the lives of those around you a lot easier!

INITIATIVE VS. GUILT: FINISHING WHAT YOU START

Picture this: You are in a meeting at your internship and you offer an opinion on an approach to use with a group of clients. Your supervisor thinks it's a great idea and asks you, in front of the rest of the staff, if you would like to develop your idea further, design some activities, and cofacilitate them with another staff member. You say yes, but inside, you are terrified and can think of hundreds of reasons why this is a very bad idea.

Or imagine you have an idea for a project in the community and mention it to your supervisor. Your supervisor thinks the idea has some merit but does not think that the timing is right or that you are quite ready to handle it. Later you find yourself angry and unable to let go of the anger. You feel like your supervisor is never going to "turn you loose" (even though this is really the first time you have asked). Both of these issues may be touching the Eriksonian issue of initiative.

Children with a basic sense of initiative believe that even though they sometimes may not be able or allowed to complete what they start, fundamentally they are capable of doing so and enjoy the inquisitive, adventuresome part of themselves. At the negative end of the continuum is a sense of guilt, which is felt by children who believe that even though they sometimes do complete what they start, their desire to explore and experiment is not a positive thing and fundamentally they or the world will find a way to block these initiatives. The thwarting of an initiative, no matter how infrequent, confirms the fundamental belief these children have about themselves and reinforces feelings of guilt.

Ask yourself how you feel now when you have an idea, make a plan, or need to take charge of something. Does it feel like an adventure or more like a minefield? Think about your current reaction to being told "no" or having your desires frustrated in some way. Might some of these issues come up in your internship?

INDUSTRY VS. INFERIORITY: FEELING COMPETENT

Your internship is your chance to shine—to see if you are really suited for work in this field or with this population. You may decide after the experience is over that it is not for you, but you want to feel like you *can* do it. A sense of professional competence is important to most helping professionals we know and to most interns. Odd as it may seem, though, the trick is to be able to separate feeling competent from your successes and failures.

The internship offers countless opportunities for success and failure, and you will experience both. An interaction with a client, an attempt to defuse an argument, a group activity, a phone conversation, and more could all go awry. No one likes when such occurrences happen, but individuals with a sense of inferiority (in Eriksonian terms) may find them especially troublesome and difficult to overcome. You may also see clues to your position on this continuum in your reaction to criticism or evaluation.

Children who are praised only when they succeed, regardless of how much or little effort was required, may come to believe that success equals competence. Since no one can succeed all the time, their sense of competence may be tenuous, regardless of how much success they experience; they have a sense of basic inferiority. On the other hand, children who are praised for their efforts rather than the results, and who are encouraged to view failure as a learning experience and a natural occurrence, may develop a much more unshakable sense of competence. These children emerge with a sense of basic industry; they feel competent even in the face of failure. Children with a sense of basic inferiority, however, feel incompetent even in the face of success.

Remembering What Motivates You

At some point in your education, someone has probably asked you to think or write about the reasons you are considering a particular profession as a career. Perhaps, too, you have been asked by a placement coordinator or a prospective placement why you are interested in a particular kind of internship. Understanding and reminding yourself of these motivating factors can be a source of strength. Yet every one of them can be a liability as well, especially in social service settings or other settings in which you are working closely with people in need. Corey and Corey (Corey & Corey, 2006) discuss several possible reasons for entering the helping professions or needs that workers bring with them, including the need to have an impact, the need to care for others, the need to help others avoid or overcome problems they themselves have struggled with, the need to provide answers, and the need to be needed. They also point out that each one of these motivations can cause problems.

It is an important to make a distinction here between wanting and needing. All the reasons just listed are good ones for entering the helping professions, provided you substitute "want" for "need." For example, wanting to care for others is fine, but as Corey and Corey (2006) point out, it is not fine if you always place caring for others above caring for yourself. Sometimes what you need conflicts with what the placement wants or needs. For example, you have worked a very full week and are very tired. Your supervisor explains that someone is out sick and asks if you would mind working over the weekend. If you choose to say "yes" to these requests from time to time, that is fine. However, if you find that you cannot say "no," ever, then you may be someone who needs to put others' concerns first in order to feel good about yourself. That can happen for many reasons, but none of them offset the price you will pay if you don't learn to attend to your own needs as well as those of your clients.

Providing answers is great too; coming up with a solution for a problem is a wonderful feeling. However, there are going to be problems you cannot solve. Furthermore, there will be times when it is better to let someone struggle to find an answer than to jump in with a lot of advice, even if your advice is on target and would solve the problem faster. If you need to be the one with the answers, these situations are likely to be difficult for you. One intern realized in a journal entry, *"There is some part of me that has a need to rescue these women."*

If you have struggled with a particular issue in your life, such as depression, substance abuse, or an eating disorder, and feel like you have emerged from that struggle, pay special attention to your motivation as it relates to this issue. Often, past struggles are the reason people choose the helping professions (Collins, Fischer, & Cimmino, 1994). They want to help others deal with or avoid problems that they experienced and feel that their experience will be an asset. In many cases, it is. It can be a powerful motivator for you, and it can be a source of hope for both you and your clients. However, it can also be a trap. You may have learned a lot about your own struggle and your own path, but there is always more to learn about how to help others deal with the same issue. The techniques that worked for you may not work for others, and newer, more effective approaches may have been developed. You need to remain open to that possibility. You also need to remember that your path is not necessarily the best path for others. Just as people have different learning styles, they have different healing styles, and your job is to fit their needs. If you have clients, they may need to struggle, just as you did, and it may be that you cannot and should not "save them" from that struggle.

Finally, some interns try to use the placement to resolve personal issues that they themselves have not yet resolved. If you encourage or insist that a client do what you cannot (e.g., be assertive or show love) or rush to protect a client from a critical parent (in a way that you never were), you are falling into this trap.

Take some time to think about your own motivations for entering the sort of placement you have chosen. If those desires turn into needs, what kinds of problems could that create for you?

Considering Unresolved Issues

Each of us has struggled with different personal issues in our lives, and this self-examination may have uncovered or reminded you of some. There are other kinds of issues that can linger as well. You may have had a prolonged struggle with one of your parents during adolescence. You may have been a victim of abuse or assault. These struggles can leave us all with issues that are unresolved or partially resolved. For example, you may have overcome your adolescent struggle, but the memory of those days may still be very painful. These issues could be thought of as your "unfinished business."

This unfinished business does not have to come from some traumatic event or a particular struggle such as those just mentioned. Family patterns are another source.

For example, if you were always the "mediator" in your family—the one who stepped in and calmed people down and helped them resolve their differences—you may have some very strong feelings about conflicts. It may be hard for you to see someone in a conflict and not step in to help, even though it is sometimes best to let people work it out for themselves.

Your unfinished business can also come from your membership in certain societal subgroups. Your vulnerable areas are not just the result of your personality or of your childhood and family experiences. They are also the result of your experience in society as a member of racial, ethnic, gender, and other subgroups. If you are a member of a group that has been discriminated against, you may have a difficult time with clients who express prejudice, especially toward that group. If you are a member of a dominant group, such as males, Whites, or heterosexuals, and have thought about issues of discrimination, you may feel guilty, or very hesitant, around members of corresponding subordinate groups.

Even if your unfinished business is not part of the reason you selected human services as a career, human service work can often stimulate those issues. Suppose you are working with a substance abuse population. As you work with a particular client, he begins to discuss his struggle with an overcritical father. Even though you have not had the experience of being a substance abuser, you have had a similar struggle with your father or some other parent figure. You are likely to be touched by this client in ways that seem mysterious at first. You may find yourself thinking about him when you are at home or when you wake up in the morning. This kind of preoccupation can be very draining.

In addition, some clients have an uncanny sense for your sore spots and will use them to manipulate you. If you have a strong need to be liked, for example, the client may withhold approval as a way to get you to relax a rule or go along with a rationalization. Your unfinished business can show up with colleagues and supervisors as well. If your father or mother was hypercritical of you, you may decide that in the internship, you are not going to accept criticism without standing up for yourself.

The point here is not that you should be free of unfinished business; all of us have some. However, you need to be as aware as you can of what that business is and how those areas of vulnerability may be touched in your work.

SUMMARY

This chapter has encouraged you to begin, continue, or develop the habit of self-examination, and we have suggested specific areas that will be important to your internship. Making an investment in self-understanding will help you see your particular issues as you move through the stages of an internship. It will also help you identify and overcome obstacles and deal more effectively with clients, supervisors, and co-workers. Perhaps you will also see the benefits in your life outside the placement. There are some other issues about yourself that you need to explore. These issues, which pertain primarily to your functioning as an intern, are the subject of the next chapter. For now, we leave you with a quote from John Schmidt (2002):

> By seeking assistance for oneself and exploring our own development, we illustrate our belief in the helping process and demonstrate the same courage we expect of those who ask for our assistance" (2002, p.12).

For Contemplation

PERSONAL REFLECTION:
SELECT THOSE INQUIRIES THAT ARE MOST MEANINGFUL TO YOU.

1. Review the areas of values discussed in this chapter. How many of them are areas about which you have strong feelings? Are there other of your core values that are not listed? Have your feelings about any of these issues changed over the years? Why do you think that happened? You might want to discuss your answers with one or two other interns.

2. Which of these values do you think will be important in your internship? How will it feel to encounter others who do not share those values?

3. Are you more attracted to and comfortable with using theories to understand people, as a separate knower does, or are you more likely to approach each person as an individual?

4. Can you give examples of both separate and connected knowing in your life? How will your style be a strength and a liability in your internship?

5. Think about recent times when you have found yourself responding to a situation in a way you later regretted. Were any of these incidents typical of you? Do you think they are indicative of a response pattern? Are you aware of any other patterns that you struggle with? Try to write them out using the format on p. 70. Which of these patterns might prove a challenge at your internship?

6. What are some of the important rules and roles in your family (both formal and informal)? How do they affect you? In what ways might they affect you in your internship?

7. Do getting your needs met, handling your impulses, finishing what you start, or feeling competent seem like particularly important areas for you? If so, do you think this may have something to do with the Eriksonian stages described in this chapter? How might these issues arise at your internship, and what will you have to be careful of?

8. Make a list of all the subgroups you belong to (race, gender, class, etc.). Next to each one, note whether it is a dominant or subordinate group and rate it 1–5 (low to high), depending on how strongly you identify with that group. Explain each of your ratings.

9. Are there subgroups in society about which you have some stereotypes? Will you be encountering any of these groups in your internship?

10. What are the major reasons you chose the academic and/or professional field that you have? This internship in particular? Can you see any way that those motivations could be troublesome for you in your internship?

11. Have you struggled with a particular issue or problem in your life? How might that struggle be an asset to you in your current placement? What traps does it represent for you?

12. What unfinished business do you think you have at this time? How might that business arise during your internship?

For Further Exploration

ON MEANING MAKING

Kegan, R. (1982). *The evolving self: Problem and process in human development.* Cambridge, MA: Harvard University Press.

Kegan, R. (1994). *In over our heads: The mental demands of modern life.* Cambridge, MA: Harvard University Press.

> A full elaboration of Kegan's work on the centrality of meaning making, and on the five qualitative stages that meaning making takes as people progress through the lifespan.

ON THE IMPORTANCE OF SELF-UNDERSTANDING

Brill, N., & Levine, J. (2004). *Working with people: The helping process* (8th ed.). New York: Longman.

> Excellent book on self-understanding.

Corey, M. S., & Corey, G. (2006). *Becoming a helper* (5th ed.). Belmont, CA: Brooks/Cole.

> A very readable and thought-provoking book, devoted entirely to self-understanding and effectiveness in interpersonal relationships.

Shafir, R. Z. (2006). *The zen of listening: Mindful communication in the age of distraction.* Wheaton, IL: Quest Books.

ON VALUES

Corey, G., Corey, M. S., & Callanan, P. (2007). *Issues and ethics in the helping professions* (7th ed.). Belmont, CA: Brooks/Cole.

> The issue of values is woven throughout this book, but Chapter 3 in particular discusses the issue of imposing vs. exposing values.

ON REACTION PATTERNS

Weinstein, G. (1981). Self science education. In J. Fried (Ed.), *New directions for student services: Education for student development.* San Francisco: Jossey-Bass.

> Describes an approach to uncovering and interrupting these patterns (pp. 73–78).

ON LEARNING STYLE

Belenky, M. F., Clinchy, M., Goldberger, N. R., & Tarule, J. M. (1986). *Women's ways of knowing: The development of self, voice and mind.* New York: Basic Books.

> An excellent and thorough treatment of this topic, with summaries of other research as well.

ON FAMILY PATTERNS

Goldenberg, I., & Goldenberg, G. (2007). *Family therapy: An overview* (6th ed.). Belmont, CA: Brooks/Cole.

Extremely readable text on families and family therapy.

Goldenberg, I., & Goldenberg, H. (2003). *Family exploration: Personal viewpoints from multiple perspectives* (6th ed.). Belmont, CA: Brooks/Cole.

The accompanying workbook serves as a guide for you to explore your own family dynamics.

ON USING ERIKSON

Corey, G., & Corey, M. (2005). *I never knew I had a choice* (8th ed.). Belmont, CA: Brooks/Cole.

An excellent book on self-understanding based on Erikson and other theorists.

Sweitzer, H. F. (1993). Using psychosocial and cognitive behavioral theories to promote self-understanding: A beginning framework. *Journal of Counseling and Human Service Professions, 7*(1), 8–18.

Combines Erikson and reaction patterns and discusses implications for human service work.

ON CULTURAL IDENTITY

Adams, M., Bell, L. A., & Griffin, P. (Eds.) (1997). *Teaching for diversity and social justice*. New York: Routledge.

A collection of readings that describes workshops on many forms of oppression. Useful as a resource for trainers, with exercises to work through.

Adams, M., Blumenfeld, W. J., Castenada, R., Hackman, H. W., Peters, M. L., & Zuniga, X. (Eds.) (2000). *Readings for diversity and social justice: An anthology on racism, anti-semitism, heterosexism, ableism, and classism*. New York: Routledge.

Diller, J. V. (2004). *Cultural diversity: A primer for human services* (2nd ed.). Belmont, CA: Thomson.

Doane, A. (Ed.) (2003). *White out: The continuing significance of racism*. New York: Routledge.

Green, J. W. (1998). *Cultural awareness in the human services: A multi-ethnic approach* (3rd ed.). Upper Saddle River, NJ: Prentice Hall.

Covers both theoretical and practical aspects of working with diverse populations.

Lum, D. (2007). *Culturally competent practice: A framework for understanding diverse groups & justice issues* (3rd ed.). Belmont, CA: Thomson.

Payne, R. K. (2005). *A framework for understanding poverty* (4th ed.). Highlands, TX: aha! Process, Inc.

Rothenberg, P. S. (2000). *Invisible privilege: A memoir about race, class and gender*. Lawrence, KS: University Press of Kansas.

Rothenberg, P. S. (Ed.) (2004). *White privilege: Essential readings on the other side of racism* (2nd ed.): New York: Worth Publishers.

References

Axelson, J. A. (1999). *Counseling and development in a multicultural society* (3rd ed.). Belmont, CA: Brooks/Cole.

Belenky, M. F., Clinchy, M., Goldberger, N. R., & Tarule, J. M. (1986). *Women's ways of knowing: The development of self, voice and mind.* New York: Basic Books.

Brammer, L. M. (1985). *The helping relationship: Process and skills* (3rd ed.). Englewood Cliffs, NJ: Prentice Hall.

Brill, N., & Levine, J. (2004). *Working with people: The helping process* (8th ed.). New York: Longman.

Collins, R., Fischer, J., & Cimmino, P. (1994). Human service student patterns: A study of the influence of selected psychodynamic factors on career choice. *Human Service Education, 14*(1), 15–24.

Corey, G., Corey, M. S., & Callanan, P. (2006). *Issues and ethics in the helping professions* (7th ed.). Belmont, CA: Brooks/Cole.

Corey, M. S., & Corey, G. (2006). *Becoming a helper* (5th ed.). Belmont, CA: Brooks/Cole.

Doane, A. (Ed.) (2003). *White out: The continuing significance of racism.* New York: Routledge.

Erikson, E. H. (1980). *Identity and the life cycle.* New York: W. W. Norton.

Gilligan, C. (1982). *In a different voice.* Cambridge, MA: Harvard University Press.

Gilligan, C., Ward, J. V., & Taylor, J. (1988). *Mapping the moral domain.* Cambridge, MA: Harvard University Press.

Gollnick, D. M., & Chinn, P. C. (2005). *Multicultural education in a pluralistic society* (7th ed.). Upper Saddle River, NJ: Prentice Hall.

Hardiman, R., & Jackson, B. W. (1997). Conceptual foundations for social justice courses. In M. Adams, L. A. Bell, & P. Griffin (Eds.), *Teaching for diversity and social justice.* New York: Routledge (pp. 16–29).

Kegan, R. (1982). *The evolving self: Problem and process in human development.* Cambridge, MA: Harvard University Press.

Kegan, R. (1994). *In over our heads: The mental demands of modern life.* Cambridge, MA: Harvard University Press.

Kolb, A. Y., & Kolb, D. A. (2005). Learning styles and learning spaces: Enhancing experiential learning in higher education. *Academy of Management Journal of Learning and Education, 4*(2), 193–212.

Kolb, D. A. (1984). *Experiential learning: Experience as the source of learning and development.* New York: Prentice Hall.

Lyons, N. (1983). Two perspectives on self, relationships and morality. *Harvard Educational Review, 53,* 125–145.

McIntosh, P. (1989). White privilege: Unpacking the invisible knapsack. *Peace and Freedom, July/August,* 10–12.

Payne, R. K. (2005). *A framework for understanding poverty* (4th ed.). Highlands, TX: aha! Process, Inc.

Schmidt, J. J. (2002). *Intentional helping: A philosophy for proficient caring relationships.* Upper Saddle River, NJ: Prentice Hall.

Schram, B., & Mandell, B. R. (2006). *An introduction to human services: Policy and practice* (6th ed.). Boston: Allyn & Bacon.

Sweitzer, H. F. (1993). Using psychosocial and cognitive behavioral theories to promote self understanding: A beginning framework. *Journal of Counseling and Human Service Professions, 7*(1), 8–18.

Sweitzer, H. F., & Jones, J. S. (1990). Self-understanding in human service education: Goals and methods. *Human Service Education, 10*(1), 39–52.

Weinstein, G. (1981). Self science education. In J. Fried (Ed.), *New directions for student services: Education for student development*. San Francisco: Jossey-Bass.

DISCOVERING
THE FIELD

The previous section was the most theoretical and, for some students, the most challenging of the book. We have given you a lot of theory to understand and think about. Now we get to the business of applying that theory to your internship, moving through the stages of the internship, and helping you meet the challenges of each one. The beginning of an internship can be overwhelming emotionally. Experiencing the internship is much different than reading about it or imaging it or even doing role plays and simulations.

As you will see, it is also a very exciting time with great intellectual as well as emotional challenges. There is a lot to think about and a lot to prepare for. This section of the book will help you have as smooth a beginning as possible and set the foundation for further progress.

The rest of this section of the book will help you focus on the tasks of the Anticipation stage. In Chapter 5, we focus on the major concerns that beginning interns often have about themselves, including acceptance, role clarification, and competence. We also guide you in developing clear goals and objectives for yourself as you learn to write your learning contract. In Chapter 6, we help you to look at your assumptions about and relationships with your supervisors and co-workers. Chapter 7 helps you become aware of and address concerns about the organization as a whole from both a systems perspective and an organizational development perspective. In Chapter 8, you spend time thinking about the dynamics of the community in which your fieldwork is being conducted and the relationship between that community, the field site, the work you are doing, and you as an emerging civic professional. And finally, in Chapter 9, those of you who are working in direct service roles will find important issues and suggestions for beginning those relationships.

CHAPTER 5

Experiencing the "What Ifs":
The Anticipation Stage

> *Many of us seem to have a lot of self-doubt about our capacities and what is manageable for us. It would seem that after all the required academics, we would not have these feelings but apparently not.*
> STUDENT REFLECTION

Your internship is probably something you have been looking forward to for a while. Some of you have been waiting just a semester or two, others have waited their entire college careers, and some have been waiting a lot longer than that! You have been hearing about internships for some time; many of you have worked hard and sacrificed to get here, and now here it is. But as your friends and family ask whether you are excited, and you answer "yes," you may also feel some anxiety creeping in. It may be a little or a lot; it may be visible to your friends, or you may keep it hidden. You may even hide your anxiety from yourself; some interns become aware of it only after it starts to dissipate (Wilson, 1981). In any case, you are entering an unknown experience, and that's always at least a little scary.

Stop a moment and think about other new experiences in your life. Do you remember your first day of school? Of high school? Of college? How about summer camp or a trip to see relatives you didn't know? An excursion into an unfamiliar neighborhood? You were probably excited about these things too—and nervous. You have probably heard lots of stories from students who have already completed their internships; that is both a blessing and a curse. It builds your excitement, but you may also hear a little voice inside saying, "Am I going to be as good at this as they are?" You may have also heard some "horror stories" about internships that went sour in one way or another; even though these are usually few and far between, the stories seem to have great staying power.

We call this first stage of an internship the "What if . . . ?" stage. Many interns express concerns during this stage and wonder whether they will be able to handle it when their worst fears come true.

Think About It

The What Ifs?

Listed below are some of the "what ifs" we have heard. Although some apply to those in the helping professions, most are common across a variety of settings.

What if . . .

- I can't pick things up fast enough?
- They all think I know more than I really do?
- I make a mistake — and something really bad happens as a result?
- I don't grow up mentally and emotionally in time for graduation?
- I'm asked a question and don't know the answer?
- My supervisor doesn't like me?
- I don't like my supervisor?
- They do things differently from the way I was taught?
- The other employees resent or ignore me?
- I get sued?
- I hate it?
- I fall apart?
- I don't feel my service is valid?

- I can't handle all my responsibilities at home and at the internship?
- I don't enough time for my loved ones?
- I don't have time for those who need me?
- A client insults me?
- A client confronts me?
- A client gets physical?
- A client attempts to sexually assault me?
- A crisis happens and I don't know what to do?
- A client lies to me?
- A client falls apart?
- My clients don't get any better?
- Clients ask me personal questions?
- I say something that offends a client?
- I have a client I can't stand?
- I have to discipline a client who won't listen?

Some of these concerns may seem silly, and no one (so far) has worried about all of them. Your particular anxieties will be shaped by your personality, your knowledge of your placement site, and your past experiences. Every one of the entries on the list, however, is something we have actually heard. You can probably add to them. We encourage our students to share their anxieties with one another, and they are often surprised and relieved to find that they are not alone in their fears and concerns. Every one of those concerns is perfectly normal, and every one can be addressed or at least the attendant anxiety can be lessened.

THE TASKS AT HAND

I have come to realize that we are all in the same boat, although we are not all physically in the same site. And we are all going through the same feelings, worries, and anxieties. Knowing this and seeing it in black and white has helped me tremendously.

STUDENT REFLECTION

Even though you may be a little nervous, you are probably eager to get going. After all, you didn't sign up for this experience to sit in a class, read a book, or write in a journal. You want to start doing the work you signed up for, acquiring new knowledge and skills and making a difference. However, at the same time as you start these processes, we are going to encourage you to attend to some other tasks:

- Examine and critique your assumptions.
- Develop key relationships.
- Acknowledge and explore your concerns.
- Clarify your role and purpose.
- Develop a learning contract.

These tasks, if done well, will help address the concerns discussed earlier and will form a solid foundation from which to learn and grow. Without that foundation, you may falter sooner and harder. We discuss these tasks in detail in the paragraphs that follow, with the exception of the learning contract. Because of the centrality of this contract in the success of your internship, as well as its complexity, we devote a separate section of the chapter to it.

Examine and Critique

Interns begin their placements with certain expectations, which are products of assumptions made, correctly or incorrectly, about many aspects of the upcoming or beginning internship (Nesbitt, 1993). As we mentioned in Chapter 4, these assumptions may come from stereotypical portrayals in the media of certain groups or agencies or from your own experience. Your previous experience, however, does not always predict what the future holds, although it is natural to generalize. You probably have at least some assumptions about supervisors, the field site, co-workers, and the populations with whom you'll have contact. Making these assumptions explicit and subjecting them to critical examination will help you develop the most realistic picture possible of the internship you are beginning.

Develop Relationships

We noted in Chapter 1 that relationships are the context or the medium for learning during the internship. You will be involved in many relationships with supervisors, instructors, co-workers, other interns, and clients during this time, and many of these relationships will be new. You are in the beginning stage of those relationships, and a normal concern at this stage is acceptance. No book can tell you how to achieve acceptance with all these people or even with any one of them. Each of these people is unique; furthermore, it is an interactive process, and half of the interaction is you. All the facets of yourself that you explored in Chapter 4, and probably more, are part of the acceptance process. A relationship is a process created by you and the other person and, as such, is unique. However, there are some things we can help you consider and be aware of as you explore these relationships.

Acknowledge Concerns

In Chapter 3, we discussed at some length the kinds of concerns interns typically have at this stage. The table on the next page summarizes those concerns, and we will discuss just a few of the more common ones here.

One of the greatest concerns for beginning interns is competence (Brust, 1986). Naturally, you want to do a good job and be recognized for it. Furthermore, if you think about all the examples of professionals at work that you have observed, read about, or seen in videos, you have probably seen very few examples of mediocrity; the idea was to show you how effectiveness looks in action (Yager & Beck, 1985). However, a distortion effect can occur. When we go to the movies, we see the scenes that were kept, not the ones that were scrapped or had to be reshot (again and again). Even the most brilliant baseball players fail to hit safely three out of five times!

Yet, we never think about the failed attempts; we concentrate on the successes. Similarly, watching those videos and reading those cases make it easy to forget that everyone undergoes a learning process and that everyone makes mistakes. We imagine that we could never do quite as well and recall our own fumbling efforts. We often see only one of the paths to success and imagine we may never find it while fearing the many paths to failure. So the question is: Can I really do this?

You may have been quite successful in your classes, even in those that focused on skill building. Your previous field experiences may have gone quite well. But interns often tell us that this is the big one, and some feelings are new to them. Some interns, for example, struggle with what has been called the *imposter syndrome* (Clance, 1985). This phenomenon is a cognitive distortion preventing a person from internalizing any sense of accomplishment (Gravois, 2007, p. A1). They are vulnerable to believing that whatever success they have achieved is due to incredible good fortune rather than competence. At each step, they fear being exposed for the pretenders they imagine themselves to be. In the case of the internship, they imagine that they did well enough to qualify only through remarkably good luck and that they will surely be found out at their placement site (Clance, 1985).

The good news is that there is a cure. The bad news is that, in large measure, the cure is time. You will almost certainly feel more confident and competent as time goes on. In the meantime, if you check with your peers, you will see that you are not alone and that realization can have remarkable healing powers.

TABLE 5.1	
Concerns of the Anticipation Stage	
Associated Concerns	**Response Strategies**
Positive expectations	
Anxieties	
Self	
Role	
Being in authority role	
Abilities	
Appropriate disclosure	
Feeling "just an intern"	
Supervisor	
Acceptance	
Expectations	
Expectations of disclosure	
Perceptions and acceptance	
Supervisory style	
"Just an intern"	*Examine and critique assumptions*
Co-Workers	*Develop key relationships*
Organizational structure	*Acknowledge concerns*
Standards of behavior	*Clarify role and purpose*
Acceptance	*Make an informed commitment*
"Just an intern"	*Develop a learning contract*
Field Site	
Philosophy, norms, values	
Workload	
Hiring potential	
Clientele	
Acceptance and perception	
Needs and presenting problems	
"Just an intern"	
Community	
Acceptance by the community	
Understanding norms of the community	
Life Context	
Responsibilities	
Support system	

Clarify Role and Purpose

Another concern for many interns is the nature of their role in the field experience. As we have explored this issue further with students, it seems that it has two parts. First of all, interns are concerned about how they will be seen by others at the site and how they will come to see themselves. Are you a student? A volunteer? An observer? A researcher? A staff member? In fact, you will probably perform all these functions, but none of these roles by itself define who you are at the internship. The role of an intern is unique. Initially, you may spend a good deal of time observing. Later, you will be a more full-fledged participant, but it is important to retain some of your observer status (Gordon, McBride, & Hage, 2005). Because you are there to learn, not just to work, you must reserve some time to engage in and complete the experiential learning cycle we discussed in Chapter 1. You must take time to observe and reflect as well as take action and have experiences. And, importantly, in your role, you must have a clear sense of purpose that contributes to the work of the organization.

There may be volunteers at your internship site, and that can cause some confusion for you as well as for the staff. Volunteers and interns do some of the same tasks, but their roles and responsibilities are very different. Volunteers are essentially there to help out; you are there to learn. You will have specific learning objectives the volunteers do not have. As a learner, you should be doing, but you also have the responsibility to reflect, analyze, and critique, which a volunteer does not have. Furthermore, although you may perform some of the same tasks as a volunteer, as an intern, you have a responsibility to learn why a particular task is done, why it is done in a particular way, and how it relates to the bigger picture in the organization (Royse, Dhooper, & Rompf, 2007).

Next, interns wonder what it is they will be asked to do. Will they have to dive right in? Will they be given boring, mechanical tasks like copying, filing, answering the phone, or driving? Your learning contract, negotiated with your campus and site supervisors, should help allay some of these fears. However, you should expect to be given some of what is often called *grunt work*. In many agencies, almost everyone pitches in and does some of the drudgery (Royse et al., 2007). You will probably be asked to do your share. You may even be asked to do a little more than your share so that more skilled staff can focus on other kinds of work. It should be clear, though, to everyone concerned exactly who can assign tasks to you. If you feel you are being asked to do too much of this sort of work or being asked by too many different people, you should speak with your campus and/or site supervisors.

However, as Bruce "Woody" Caine (1994) points out, there is actually quite a bit you can learn from performing these tasks. Often, the learning is not in the task itself, but in the organizational meaning of the task. You get a chance to learn what it really means to be an entry-level worker in that agency. You become familiar with these tasks, which are critical if not always interesting, and how long it takes to complete them (which is easy for people who don't have to do them to forget). You can also learn how much it costs the organization, in material and human resources, to get this work done.

A word to the wise. . .   Regardless of how conscientious you are in examining your assumptions, developing key relationships, becoming aware of your concerns, or clarifying your role and purpose in it, the concerns that drive your learning and the anticipations you experience will not resolve until you have made an *informed* commitment to the internship. An informed commitment means that you are emotionally owning this internship as *yours* and are willing to commit to its work. It takes a conscious choice on your part to examine the various aspects of the internship and say to yourself, "Yes, this is what I want to do." It is a conscious choice as well to be actively involved in the energy and quality of the internship. It means taking on responsibilities and being assertive in defining your field experience. As you move beyond the Anticipation stage, it is important that you are doing what *you* want to do in this internship.

THE LEARNING CONTRACT[1]

*I never thought that I would feel this way, but I am very happy I took the time
to work on the contract. . . . it helped enormously to define my role.*

STUDENT REFLECTION

At some point in the internship or the placement process, you, your site supervisor,
and your campus supervisor or instructor will negotiate the goals and activities of
learning for the field experience. This written agreement is a document that articulates in detail what you will be learning, how you will go about learning it, how
you will determine your progress, and how you will be evaluated on achieving your
goals. This document may be referred to as a learning plan or agreement, depending on your academic discipline or campus policies. We refer to it as the *learning
contract.*

The Importance of a Learning Contract

The learning contract is perhaps best described as a master plan for your field placement. It is your map to success in the field. Developing the right kind of contract is
essential to maximizing your learning and ensuring that your learning needs will be
met. For one thing, the learning contract minimizes the chance of misunderstandings.
It also will keep you focused on your internship commitments and provide you with a
sense of achievement. It is an opportunity to exercise imagination and creativity and to
consider new possibilities. Most importantly, the contract will ensure the integrity of
your academic work (Royse, et al., 2007).

Some of you may be feeling a bit lost in your internship right now, not really
knowing what you are going to be doing over the course of the field placement or perhaps concerned that the work you will be doing will be insignificant to you or the site.
If that is the case, it may help to know that the learning contract will clarify your role
and responsibilities and will give you the opportunity to make sure that your work is
meaningful to you and to the site. That translates into you having a sense of purpose
in your fieldwork. Without a sense of purpose, you will begin to question your value to
the site. If you question your value as an intern, your overall morale and investment in

---

[1] Many campuses use two documents for their internships: a *placement* or *internship agreement* and a
learning contract. The placement agreement details the responsibilities of the involved parties (campus, internship site, and intern) and describes in general terms what those parties will do to carry out
the terms of the agreement. The learning contract articulates what an intern is expected to learn in
the course of the internship and how that will happen (learning goals and objectives). This contract
typically involves the campus supervisor and intern and may involve the site supervisor as well. On
some campuses, one document is used that reflects the intent of both the placement agreement and
the learning contract. The question of whether these documents are legally binding is not simple to
answer. Each of the documents may have a different intent depending on the campus, with different
parties involved in agreeing to the specified conditions—on campus as well as at the internship site.
To make matters more complicated, the documents may or may not be written in ways that make
them legally binding. The best resource for answers to this question is the legal counsel for your
campus.

your internship will inevitably be affected—and not in positive ways. We cannot stress enough how important it is for you to be involved in developing this document and to work conscientiously on it, as it ensures that you have a most challenging and rewarding internship.

Get Involved in Your Contract!

There are three key individuals involved in creating this master plan: you, your site supervisor, and your campus supervisor or instructor. Most agree that an internship is most effective when all three parties work together to develop a field experience that meets everyone's needs and provides an opportunity for all to learn (Hodgson, 1999). What that means is that it is important that you be actively involved in writing your learning contract, taking a more active role in designing your learning than you have done in traditional courses. This involvement may seem a little odd at first; however, our experience tells us that once you become involved in the process, you will get beyond these feelings, share in the process, and be more invested in the outcome.

There are three factors in a contract that we consider in order to maximize the possibility of you having a fulfilling field experience (King, 1988). First, your work in the field must be *worthwhile* for you to do it. That translates into you being given opportunities to accomplish tasks, be acknowledged and respected for your work, create your own work, push your limits to touch your potential, and enjoy what you are doing. Second, your supervisor and co-workers must be capable of demonstrating *responsible relationships* in working with you. That translates into competent supervision and collegiality, which we deal with in the next chapter. And, last, your supervisor must be willing to cultivate your *active* and *conscious involvement* in the internship. That translates into encouraging you to take on responsibilities that may be quite challenging initially and be assertive in defining your field experience.

The Timing of the Learning Contract

When you, your instructor, and your supervisor negotiate the contract tends to be a matter of policy—either that of the academic program, the campus, or the field site. Sometimes, that becomes a matter of debate. There are programs we know of that develop their agreements before the field placement begins—sometime between when the site/campus agreement is negotiated and when the internship begins. In programs with year-long internships, students can have up to a month in which to develop their learning plans, while interns in programs of a semester or less may have from one to two weeks to develop the contract.

Many agree it is important that you have some understanding of your field site, the populations you are working with if you are in the helping or serving professions, and your learning tasks before negotiating the contract (Rothman, 2000). We generally develop our contracts within the first three weeks of the placement, regardless of the length of the field experience or hours in placement, as we believe you can take an active and informed role in the contract process by the third week.

Fundamentals of the Learning Contract

There are three general components to the learning contract, and they are pretty straightforward: learning goals, activities to reach the goals, and assessment measures. Some learning contracts also include objectives. How carefully these components are developed and how well matched they are to your expectations makes all the difference in whether you having an empowering, transforming field experience or a mediocre one.

In addition to these components, there are a number of details that also need to be included in the contract. For example, you will want to be sure to note the time spent on a weekly basis in the internship (days per week, hours per day); supervision specifics (methods/names—individual/peer/group; frequency; length of time); case specifics (numbers, types, time frame); documentation and recording procedures; special arrangements; and the date the contracted was drafted (Rogers, Collins, Barlow, & Grinnell, 2000; Royse et al., 2007).

Remember that it is best if the contract is written in such a way that it can change as needed and adapt to shifts and modifications in the internship. Just imagine what would happen if an opportunity for an exciting project develops after you negotiate the contract and your contract is nonnegotiable. You could miss out on what might be *the* defining aspect of your internship! It is important that you are able to renegotiate the contract to ensure the best fit with your overall learning goals. By the same token, if a project you agreed to take on does not materialize, it is equally important that you are able to replace it with a more viable learning opportunity.

Now we turn to the three interlocking pieces of the contract: goals, activities, and assessment. Although we will discuss them in that order, and we suggest you approach them in that order initially, it is not at all uncommon for goals to be rethought once activities are considered or for both goals and activities to be rethought once you are focused on assessment.

The Goals of the Learning Contract

There are many kinds of learning contracts, ranging from simple goal statements that apply to all internships to highly specific and individualized plans (Wilson, 1981). Your academic program may already have a format for developing learning contracts. If so, it probably takes the form of one of two general approaches, one of which is a *goals-focused contract* and the other of which is an *objectives-focused contract*. Both of these approaches begin the contract process by identifying the *goals of learning* and end the process by setting forth *assessment criteria* to determine how well the goals are met. These two general approaches also identify *activities* as ways of achieving the goals. Differences between the approaches arise in whether or not *learning objectives* are used to guide the learning contract (Garthwait, 2005; Rothman, 2000). It is generally accepted that the more succinct approach—the goals-focused contract—is easier to implement and provides sufficient direction to be a worthy document for students and agency personnel to use (Garthwait, 2005). Regardless of the approach used, the learning goals should reflect in part the answer to the question: *What is your purpose of being an intern in this organization?* In the sections that follow, we discuss some considerations in writing goals and give examples in each to guide your thinking. We have included a section on objectives (see "Using an Objective-Based Approach" on page 103) for those of you who want or need to write a contract from that perspective.

The contract itself should be demanding and ambitious, stretching you while enhancing your knowledge. The contract should also be grounded in realistic time frames as well as in realistic expectations of the work and your capabilities. Keep in mind when negotiating your contract that the goals you set for the end of the placement should seem lofty at the time, and feeling intimidated by them is not uncommon. Be reassured, though, that the feelings of intimidation will pass and you will attain your goals, but it will take time and work to make that happen.

CATEGORIES OF GOALS

Before you start drafting your goals, we encourage you to think broadly about the learning opportunities in front of you. In Chapter 1, we discussed at length the sorts of things you can learn from an internship; that abstract discussion can now be concrete, since you have experience to refer to. Remember that your goals can focus on the work that you do, on your understanding of what it means to be a professional, and on the human context and public relevance of your work, including the community and organizational contexts. We suggest two ways of categorizing goals for your contract. First, goals can focus on *knowledge, skills,* and *attitudes or values.*

Knowledge Goals focus on learning and understanding factual information, terminology, principles, concepts, theories, and ideas of the profession. Typically, they involve learning new information that builds on previous learning, such as:

- Become familiar with how the criminal justice system in this state operates.
- Learn the network of services in this city available to people who are physically challenged.

Skill Goals focus on things you want to learn to *do,* such as:

- Learn to write a case report.
- Develop oral presentation skills.

Attitude and Value Goals focus on attitudes and values you believe to be important and that you want to improve in yourself, such as:

- Being more patient.
- Being less defensive about criticism.
- Being more open to multiple points of view.

PROFESSIONAL, PERSONAL, AND CIVIC DEVELOPMENT

The second way to categorize goals will sound familiar to you; goals can focus on professional, personal, or civic development. In each of these categories, it is possible to have goals that focus on knowledge, skills, or attitudes/values.

Professional Development Goals focus on how you want to grow as a professional or on what you want to learn about a profession. For example:

- Learn specific skills of the profession, such as process recording, case management, or project management.
- Understand career options in the field.
- Become familiar with the major ethical issues in the profession.
- Strengthen your understanding of various theories.
- Become more adept at the art of Practical Reasoning (see Chapter 1).

Personal Development Goals are those that may be applicable to your profession but are also of value to you whatever profession you choose, and in your personal as well as professional life. For example:

- Become a more effective writer.
- Conduct an effective interview.
- Develop the skill of critical thinking.
- Discover more about your personal strengths and weaknesses.
- Become more able to operate in a constantly changing, uncertain environment.

Civic Development Goals focus on the knowledge, skills, and values of citizenship in a democracy, and on the understanding of the public relevance of the profession in which you are interning. Jeffrey Howard, in his very useful *Service-Learning Course Design Workbook* has developed a sophisticated chart showing examples of civic knowledge, skills, and attitudes (2001):

Purposeful Civic Learning Objectives

Goal Categories for Purposeful Civic Learning	Knowledge	Skills	Values
Academic Learning	Understand root causes of social problems	Developing active learning skills	There is important knowledge only found in the community
Democratic Citizenship Learning	Becoming familiar with different conceptualizations of citizenship	Developing competency in identifying community assets	Communities depend on an active citizenry
Diversity Learning	Understanding individual and institutional "isms"	Developing cross-cultural communication skills	Voices of minorities are needed to make sound community decisions

continued

Purposeful Civic Learning Objectives *(continued)*			
Political Learning	Learning about how citizen groups have effected change in their communities	Developing advocacy skills	Citizenship is about more than voting and paying taxes
Leadership Learning	Understanding the social change model of leadership	Developing skills that facilitate sharing leadership roles	Understanding that leadership is a process and not a characteristic associated with an individual or a role
Inter- and Intrapersonal Learning	Understanding one's own multiple social identities	Developing problem-solving skills	Learning an ethic of care
Social Responsibility Learning	How individuals in a particular profession act in socially responsible ways	Determining how to apply one's professional skills to the betterment of society	Responsibility to others applies to those pursuing all kinds of careers

From *Service-Learning Course Design Workbook* (www.umich.edu/~mjcsl). Copyright © 2001 by University of Michigan.

TOOLS FOR WRITING GOALS

Developing goals can be tricky business. One technique that we find useful is sentence-completion exercises like this.

For each of the categories (personal, professional, and civic development), complete the following question:

- If I could make it happen, what _____ goal would I like to achieve by the time I complete my internship? (deShazer, 1980, cited in Curtis, 2000).
- If I were given a magic wand, what would I like to see happen in my internship in the area of _____ goals?

Also essential to formulating a goals statement is knowing the language to use to frame the learning goals. The following verbs are frequently used in goals statements and may help you in writing your contract (Garthwait, 2005, p. 22).

- acquire
- analyze
- appreciate
- become
- become familiar with
- comprehend
- develop
- discover
- explore
- know
- learn
- perceive
- synthesize
- understand
- value

Focus on SKILLS

Using an Objective-Based Approach

Learning-objective statements are behavioral in nature and tend to be more detailed, concrete, and precisely worded than goal statements. Learning objectives describe specific actions and activities; e.g., to demonstrate the ability to match needs of clients to available resources. It has been our experience that developing learning objectives is both an art and a skill. The art is reflected in the vision of the objective; the skill is reflected in the ability to incorporate the goals, activities, and assessment measures in one statement and in realistic and heuristic ways. The following verbs are typically used in writing learning objectives (Garthwait, 2005, p. 23). If you are using the learning objectives contract, this list may be helpful.

answer	*decide*	*obtain*
arrange	*define*	*participate in*
circulate	*demonstrate*	*revise*
classify	*direct*	*schedule*
collect	*discuss*	*select*
compare	*explain*	*summarize*
compile	*give examples*	*supervise*
conduct	*list*	*verify*
count	*locate*	*write*

Rothman (2000) offers excellent examples of how to write and not to write learning objectives—advice that will serve you in good stead. For example, she frames the contract in such a way that you move smoothly from the general (overall goals), to the specific (objectives), to the most specific (approaches, tasks, and methods) (p. 95), and then she suggests that you reverse the process when it is time to review your progress (p. 95). Rothman also stresses the importance of using measurement criteria that are both relevant and measurable (p. 99), and she recommends that your objectives be grouped by time frame so that you can easily review what you have done in the designated time periods. She offers an excellent example of a statement using both relevant and measurable criteria: the "ability to use empathy appropriately as demonstrated in five instances recorded in process recordings of three client interviews in one month" (p. 99). This example will guide you well developing assessment objectives.

RESOURCES FOR GOALS

Before settling on goals, there are several resources that you, your supervisor, and your instructor may find useful to consult in considering the goals. Some of these resources are familiar to your supervisors, and some may be new to them.

Program-Specific Goals   Many academic programs have learning outcome statements for their students, and sometimes even for the internship in particular. This is a resource that you and your campus instructor are probably familiar with, but your site supervisor may not be. Many professional organizations have published statements about what professionals should know and be able to do. For example:

- **The National Society for Experiential Education** has published the document *Eight Principles of Good Practice for All Experiential Learning Activities*, which identifies the principles that underlie the pedagogy of experiential education and the responsibilities of the student and facilitators in the learning process (NSEE, 1998). In addition, the NSEE has published a series written by Inkster and Ross titled *The Internship as Partnership*, the first in the series being *A Handbook for Campus-based Coordinators* and the second being *A Handbook for Businesses, Nonprofits, and Government Agencies*, both of which have value for faculty and other campus supervisors in developing effectiveness in working with placement sites (1995, 1998).

- **Council for the Advancement of Standards in Higher Education** (CAS) has published the standards and self-assessment guides for Internship Programs and for Service-Learning. This resource is intended for the campus instructor and offers a book of standards (2006a), *Frameworks for Assessing Learning and Development Outcomes* (2006c), and *Self-Assessment Guides* (2006b) to ensure institutional effectiveness, student learning, outcomes assessment, and quality assurance (www.cas.edu).

- **The Council for Standards in Human Service Education** has a detailed list of knowledge, skills, and attitudes (www.cshse.org) expected of human service education programs and/or their students.

- **The Service-Learning Field** The literature in service-learning focuses, among other things, on how to use service activities to develop specific academic and civic outcomes. Of particular note is a series edited by Edward Zlotkowski (2006), which contains twenty-one individually edited volumes each focusing on a separate academic discipline; the *Michigan Journal of Community Service Learning* (www.umich.edu/~mjcsl); and materials from Campus Compact (2003).

Choosing Activities

Strategies, activities, methods, and *approaches* are all terms to describe ways of accomplishing your goals or objectives. Sometimes they are used interchangeably. Technically, objectives are *strategies* for achieving goals, but we like the word *activities*. They should be described as specifically as possible, specifying what you will do, where, when, and how. As you probably suspect, there is considerable room for personal preferences when it comes to selecting ways to meet your goals; just be sure the resources are available to carry out the activities. As with goal setting, choosing activities should ideally be a three-way process, involving active participation from you, your supervisor, and your instructor. In some settings, the site is actually quite prescriptive about what activities interns will engage in; there may even be a detailed position description. In those cases, the choosing of activities

Focus on THEORY

The Levels of Learning as a Guide to Activities

Part of the process of developing activities for a learning contract is paying attention to different levels of learning that need to take place along the way (King, Spencer, & Tower, 1996). Learning takes time, and your activities must be organized so as to build on your attitudes, skills, and knowledge as they develop and change over time. We find it useful to think in terms of three sequential levels of learning: (a) orientation, (b) apprenticeship, and (c) mastery.

The *orientation* level is a time when you become acquainted with the work of the internship. Days at the site are marked by lots of observing, reading, shadowing, and inquiring. These are far from aimless activities. You need to have goals during this time because you are building a foundation for a successful internship (Feldman & Weitz, 1990). The *apprenticeship* level is an intensive teaching and learning period, with much learning-by-doing under very close supervision. The third level is a period of *mastery* during which you will be engaged in your own work and actually structuring and organizing your schedule to meet timelines and agency goals. You are more self-directed at this level of learning, and your need to understand subtle aspects of the work and yourself in relation to the work is heightened. Typically, you will need less direction and may or may not need much support from your supervisor.

will be less collaborative; however, it is no less important that they be matched to explicit learning goals.

Each of your activities should connect clearly and specifically to a goal. You should make sure you have at least one activity listed for every goal. Sometimes, an activity may be appropriate for the pursuit of more than one goal. For example, if you are going to be writing project or case summaries, that may connect to a number of knowledge and skill goals. We also encourage you to think hard about opportunities for civic development. If you are assigned case management duties in a social service agency, for example, you can use that opportunity to find out more about the community context of your clients. You might have to suggest that as a goal, though, or an extension of the activity.

It is possible that you will have some activities assigned to you that are not connected with a goal—remember our discussion of doing your share of the "grunt work." However, also remember that there is much to be learned from such work, and you may want to develop a learning goal or two toward that end.

Assessing Your Progress

The final step in developing your learning contract is to identify ways to measure how well you are doing in meeting the goals. Terms like *measurement, evaluation, outcomes*, and *assessment* are often used to describe the essence of this step. What is important is that you are able to objectively measure your growth using tangible activities or methods. These assessments, though, must connect clearly to the activities you have chosen. For example, you could include in your contract such measurable activities as giving presentations that will be evaluated, creating resource lists that can be reviewed, keeping journals that can be read, developing written documents for review (Kiser, 2008; Rothman, 2000), videotaping yourself, being observed and given feedback, and writing a case study for review by supervisors. If you are stuck, good questions to ask yourself are "What could someone read or observe that would help them tell me how well I am doing in meeting my objective?" or "If I had achieved this objective, how would someone besides me know?" In some cases, particularly if you are using an objective-based approach, you will be encouraged or required to state your assessment approaches in quantifiable terms. For example, instead of saying "present cases to the supervisor," you would say "present at least five cases," and instead of saying "receive feedback," you would say "receive ratings of satisfactory or above on all of my case presentations."

The chart below shows two goals with corresponding activities and assessments:

TABLE 5.2
Sample Learning Contract Sequence

Goal	Activities	Assessment
Function effectively as a team member	Observe a variety of teams and make notes on what seems to work (or not) and why	Present findings to instructor in journal entries
	Read about effective group behaviors	Review understanding with supervisor
	Practice specified behaviors in group	Supervisor observes group, looking for those specific behaviors
Improve my ability to identify personal and community strengths in the clients I work with	Review client files and summarize strengths	Present findings to supervisor in supervision session. Supervisor offers critique and keeps notes on improvement
	Accompany experienced workers on home visits and make notes on strengths visible in the home	Discuss impressions with the experienced worker and get feedback
	Conduct a community assessment	Present written assessment to instructor and site supervisor

SUMMARY

You should expect some anxiety as you work your way through this early phase of the internship. We hope this chapter has helped you clarify some of your anxieties and identify where they come from. We have also addressed the process of setting learning goals, activities, and measures, which should help alleviate some of the anxiety that comes from uncertainty. Many of the concerns discussed earlier, though, have to do with the myriad of relationships you are beginning to form. In the next chapter, we turn to those relationships and the ways you can get the most from them.

For Contemplation

PERSONAL REFLECTION:
SELECT THOSE INQUIRIES THAT ARE MOST MEANINGFUL TO YOU.

1. Look at the "what ifs" listed in your book. How many of them seem familiar to you? What can you add to the list?

2. Have you experienced feelings of fraudulence in your role as an intern? Do you ever feel like an imposter and fear being discovered as less competent than you should be and appear to be? Discuss your experiences with this phenomenon.

SEMINAR SPRINGBOARDS

1. This one involves your field instructor, so be sure she or he is on board before you move ahead with the exercise. Make a list of both your anticipations—what gives you feelings of hope and excitement—and your anxieties—what gives you feelings of apprehension. Then send your copy—anonymously—to your campus instructor so that yours and those of your peers can be copied and distributed in class for a discussion of similarities and differences.

2. As interns approach the internship, they often use metaphors to describe what they think the experience will be like, such as a roller coaster ride, a glider flight, or a trust walk. Think of a metaphor that captures the way you are thinking, feeling, and behaving as you approach your internship. When sharing it with your peers, be sure to explain as well as describe it. Some of our favorites are: a long flight to a long-awaited destination, a roller coaster ride, stone soup (I am the soup, I am not the chef!), planting a tomato garden, and so on.

For Further Exploration

Colby, A., Erlich, T., Beaumont, E., & Stephens, J. (2003). *Educating citizens: Preparing America's undergraduates for lives of moral and civic responsibility.* San Francisco: Jossey-Bass.

Particularly useful for stimulating thinking about civic development.

Crutcher, R. A., Corrigan, R., O'Brien, P., & Schneider, C. G. (2007). *College learning for the new global century: A report from the National Leadership Council for Liberal Education and America's Promise.* Washington, DC: AAC&U.

Excellent, comprehensive statement of student-learning goals in a wide variety of areas.

Garthwait, C. L. (2005). *The social work practicum: A guide and workbook for students.* (3rd ed.). Needham Heights, MA: Allyn & Bacon.

Lots of helpful information on learning contracts.

Howard, J. (Ed.). (2001). *Service-learning course design workbook.* Ann Arbor, MI: OCSL Press.

An invaluable tool that can be applied to developing goals for personal, professional, and civic development. Easy-to-read chapters and worksheets guide you through the process.

Inkster, R. P., & Ross, R. G. (1995). *The internship as partnership: A handbook for campus-based coordinators and advisors.* www.nsee.org.

Inkster, R. P., & Ross, R. G. (1998). *The internship as partnership: A handbook for businesses, nonprofits, and government agencies.* www.nsee.org.

A series of two comprehensive and invaluable handbooks for internship programs and field sites. The first publication focuses on the responsibilities and challenges of the work of the field coordinator; the second does the same for the work of the site supervisor. Both publications have significant value for faculty and other campus supervisors in developing effectiveness in working with placement sites.

References

Brust, P. L. (1986). Student burnout: The clinical instructor can spot it and manage it. *Clinical Management in Physical Therapy, 6*(3), 18–21.

Caine, B. (1994). What can I learn from doing gruntwork? *N.S.E.E. Quarterly* (Winter), 6–7; 22–23.

CAS (2006a). *CAS professional standards for higher education.* Washington, DC: Council for the Advancement of Standards for Higher Education.

CAS (2006b). *CAS self-assessment guides.* Washington, DC: Council for the Advancement of Standards for Higher Education.

CAS (2006c). *Frameworks for assessing learning development outcomes.* Washington, DC: Council for the Advancement of Standards for Higher Education.

Clance, P. R. (1985). *The imposter syndrome.* Atlanta: Peachtree Publishers.

Compact, C. (2003). *Introduction to service-learning toolkit: Readings and resources for Faculty* (2nd ed.). Providence, RI: Campus Compact.

Feldman, D. C., & Weitz, B. A. (1990). Summer interns: Factors contributing to positive developmental experiences. *Journal of Abnormal Behavior, 37,* 267–280.

Garthwait, C. L. (2005). *The social work practicum: A guide and workbook for students* (3rd ed.). Needham Heights, MA: Allyn & Bacon.

Gordon, G. R., McBride, B. R., & Hage, H. H. (2005). *Criminal justice internships: Theory into practice* (5th ed.): Cincinnati, OH: Anderson Publishing.

Gravois, J. (Nov. 9, 2007). You're not fooling anyone. *Chronicle of higher education, 54 (11), 1& A14.*

Hodgson, P. (1999). Making internships well worth the work. *Techniques, September,* 38–39.

Howard, J. (Ed.). (2001). *Service-learning course design workbook.* Ann Arbor, MI: OCSL Press.

Inkster, R. P., & Ross, R. G. (1995). *The internship as partnership: A handbook for campus-based coordinators and advisors.* www.nsee.org.

Inkster, R. P., & Ross, R. G. (1998). *The internship as partnership: A handbook for businesses, nonprofits, and government agencies.* www.nsee.org.

King, M. A., Spencer, R., & Tower, C. C. (1996). *Human services program manual.* Fitchburg, MA: Fitchburg State College.

Nesbitt, S. (1993). The field experience: Identifying false assumptions. *The LINK (Newsletter of the National Organization for Human Service Education), 14*(3).

NSEE. (1998). *Standards of practice: Eight principles of good practice for all experiential learning activities.* Paper presented at the National Society for Experiential Education.

Rogers, G., Collins, D., Barlow, C. A., & Grinnell, R. M. (2000). *Guide to the social work practicum.* Itasca, IL: F. E. Peacock.

Rothman, J. C. (2000). *Stepping out into the field: A field work manual for social work students.* Boston: Allyn & Bacon.

Royse, D., Dhooper, S. S., & Rompf, E. L. (2007). *Field instruction: A guide for social work students* (5th ed.). Boston: Allyn & Bacon.

Wilson, S. J. (1981). *Field Instruction: Techniques for supervisors.* New York: The Free Press.

Yager, G. G., & Beck, T. D. (1985). Beginning practicum: It only hurt until I laughed. *Counselor Education and Supervision, 25*(2), 149–157.

Zlotkowski, E. (Ed.). (2006). *Service-learning in the disciplines series.* Sterling, VA: Stylus Publications.

CHAPTER 6

Getting to Know Your Colleagues

Great discoveries and achievements invariably involve the cooperation of many minds.
ALEXANDER GRAHAM BELL

We are bound together by the task that lies before us.
MARTIN LUTHER KING JR.

A s an intern, the primary work of your organization, whether it is providing direct service, working with communities, improving the environment, or developing marketing plans, is probably your biggest concern. However, your colleagues can make or break your internship. We use the term *colleagues* to describe those with whom you work—other interns, other professionals at the agency, and, of course, your site supervisor. For many interns, their supervisor is the person whose opinion comes to matter most and who has the most impact in either a positive or negative way on learning. Your relationships with your campus supervisor and instructor(s) are also key components of the internship process. Many of the issues you need to consider about your site supervisor also apply to your campus supervisor or instructor. We won't be repeating all of them, but we will be reminding you to consider them as well as the three-way relationship among the campus, your site supervisor, and you. As for co-workers, you will no doubt work closely with some other people besides your supervisors, and you may even work with them in teams. Whatever the case, you are bound to interact with them, and your relationships with them can have a substantial impact on your internship. In this chapter, we focus on the beginning concerns as you meet your supervisor and other colleagues. Like any relationship, your relationship with your colleagues can feel initially like a good match or a poor one. And, like any relationship, they can be worked at, nurtured, cultivated, and improved.

GETTING THE MOST FROM SUPERVISION

Starting in a new place is a blank slate. While this could be positive for some, I felt apprehensive at the thought of starting that relationship of supervisor/ supervisee all over again. I am sure it will become less difficult as time passes, though a sense of apprehension is probably positive because it would keep one attuned.

STUDENT REFLECTION

Certainly, one of the most important relationships at your internship is the one you have with your site supervisor.[1] You will spend a great deal of time with this person, who will also be responsible at least in part for your evaluation. Your relationship with your supervisor is a tremendous source of learning about the work, about yourself, and about the relationship itself (Bogo, 1993; Borders & Leddick, 1987). However, supervisors also indisputably have power over you, and that has a real effect on the relationship.

Regardless of how much experience you have had being supervised, or even as a supervisor, it is normal to be nervous as you approach this relationship. For one thing, internship supervision is different from other kinds of supervision. And even if you have had past internships, this is a new relationship. We find that concerns about supervision focus on both task and process issues (Floyd, 2002).

- **Task concerns** center on the kind of work you will be given to do and how you will perform. You are not unusual if you are worried about what will happen when you make a mistake and your real and imagined shortcomings become visible (Baird, 2007). You may also be concerned about being assessed and evaluated. Your supervisor's perceptions of you are important in and of themselves, but the added prospect of a formal evaluation adds extra weight to those perceptions. Your own issues and personality will determine how important or intense any of these concerns are, and you may have some others we have not mentioned.

- **Process concerns** center on your interpersonal relationship with your supervisor, and perhaps primary among those concerns is acceptance. Interns often wonder whether supervisors will like them. More important, they wonder whether their supervisors will understand them and accept their weaknesses. As in any new relationship, they wonder whether they are going to get along well. You may also wonder how much about yourself you want to share or are expected to divulge and how many of your feelings and reactions you should reveal as you go through the internship. Since you do not yet know these people, you may

---

[1] In some cases, interns have more than one person functioning as an on-site supervisor. When this is the case, one personal typically oversees all aspects of the internship, while the other supervisor, who meets with the overseeing supervisor, handles the on-the-job supervision (direction and support). In other instances, your instructor or supervisor on campus may also be fulfilling some of the functions of a supervisor. In this section, though, we are assuming that you have one on-site supervisor.

wonder how they will react to your disclosures; some interns, especially in counseling settings, worry that their supervisors will take the opportunity to analyze them and expose personal and professional weaknesses (Wilson, 1981).

Again, the cure for some of this is time. As you get to know your supervisor, you may or may not like what you find, but at least it will be known. And a trusting, comfortable relationship takes time to develop. However, clarity can also help, and that is the focus of this section. This is not a section about supervision theory; rather, it is designed to help you become a more informed participant in your supervision. Looking at yourself as a participant, with rights and responsibilities, is a much more empowering stance than looking at yourself as a passive recipient or even a victim. It can help you get the most from your supervision, even in those cases where the supervision you receive is less than adequate.

Being an informed participant involves keeping yourself focused on the learning dimension of the internship. It also means developing realistic expectations of your supervisor and learning to see things from her or his point of view. Being an informed participant means being proactive in assessing the match between you and your supervisor and making the match better if you can. It also means understanding how you will be supervised and evaluated. Finally, it means being mindful of the triadic relationship involved in many internships—the relationship among you, your site supervisor, and your campus instructor.

A Focus on Learning

As your relationship with your supervisor begins and develops, always keep in mind that the main focus of your internship, and of the supervisor-supervisee relationship, is on learning and growth. In a work situation, the primary focus of supervision tends to be on performance of assigned duties, and in exchange, you get paid. Performance is important at an internship, but the primary focus of your supervision is—or should be—educational (Birkenmaier & Berg-Weger, 2007; Royse, Dhooper, & Rompf, 2007). Your supervisor helps you learn and grow, and in return, you provide important service to the site. This distinction can be a bit subtle—after all, it is through performing your duties that you learn—but it is also profound. For example, once you have reached a level of comfort and skill with a certain kind of task, should you continue to perform that task to be a contributor to the agency or should you move on to something else? There is no right or wrong answer, but the way to find the answer is to ask which choice contributes more to achieving the learning goals you worked so hard to formulate.

A focus on learning may also mean a somewhat different role for you as a student. In our experience, students approach their traditional coursework with varying levels of seriousness and focus. Their goals may be largely intrinsic or they may be largely extrinsic, as when the goal is to achieve a certain grade. Here, the goal is to learn as much as you possibly can and to push yourself and your supervisor with that goal in mind.

Developing Realistic Expectations of Your Supervisors

In our experience, interns often bring certain preconceptions about supervisors to the internship (McClam & Puckett, 1991) as well as expectations, hopes, and desires. They expect their supervisors to function in a variety of roles and to do them all well

(Alle-Corliss & Alle-Corliss, 2006; Baird, 2007; Borders & Leddick, 1987; McCarthy, DeBell, Kanuha, & McLeod, 1988; Royse et al., 2007). As *teachers*, interns hope that their supervisors will help them set goals and objectives, help them learn new skills, and be sensitive to their particular learning styles. They hope that their supervisors will be *role models* with great stores of experience and expertise. They also hope that supervisors will use their counseling skills to be a *support* to them and help them through difficult emotional times and decisions. In the role of *consultant*, interns want their supervisors to help assess a problem or situation and generate alternative courses of action. Supervisors are in the role of *sponsor*, *advocate*, or *broker* when they take an active interest in the careers of interns, both at the placement and beyond. And some interns hope that their supervisor will become a *mentor*—someone who takes a deep and special interest in them as a person and a budding professional. The reality, of course, is that no one can do all these things equally well and that your supervisor is a person with concerns, foibles, strengths, and weaknesses just like you.

Try not to make assumptions about your supervisor's level of experience, which may not be a great deal more than yours. Remember, though, that experience is not always the best indicator of quality in a supervisor. Just as brilliant scholars may not make good teachers because students' struggles with the material are incomprehensible to them, so it is that experts in your field may not be the best supervisors. They may have forgotten what it is like to be new to the field, and the gap between your skill levels may be so great that they cannot function as effective role models for you—the expertise seems unattainable. So, a less experienced person is not necessarily a problem. Regardless of her or his level of expertise, your supervisor has a different, more objective vantage point from which to view your experience and your struggles.

Think About It

What Supervisory Roles Are Most Important?

Take a moment now and try to put the roles of a supervisor in priority order. By ranking them from 1 to 6, you indicate which ones are most and least important to you. Yes, they may all be important; that is why we ask you to do this. At times when you are perhaps frustrated with your supervisor for not functioning quite the way you hoped, one thing you can do is come back to this exercise and consider the relative importance of that role.

____ Teacher

____ Role Model

____ Supporter

____ Consultant

____ Advocate

____ Mentor

Your Supervisor as a Person and a Professional

In approaching your relationship with your supervisor, it may be helpful to think about it from her or his point of view. Supervisors have a range of reasons for agreeing to accept an intern. Many supervisors see it as a way to contribute to their profession or perhaps to return the favor that a supervisor did for them when they were students (Birkenmaier & Berg-Weger, 2007). There are some more tangible benefits to working with interns as well. For example, some supervisors have sought out interns because of the positive effects the supervisors observed on the morale, commitment, and proficiency of their staff as a result of working with interns (King & Peterson, 1997). Still others may be motivated by the desire to sharpen their supervisory skills (Birkenmaier & Berg-Weger, 2007). It is also possible that they had no choice in the matter and that the person who agreed to take on the intern is not the person that is assigned to supervise (Milnes, 2001). Also, being an intern supervisor is a complex task, for which many supervisors receive little or no formal training (Baird, 2007). They may have a lot of experience as a supervisor, but everyone has to have their first intern sometime, and it may be you.

Your supervisor also has roles as a worker and perhaps as an administrator or a supervisor of agency personnel in addition to the role of intern supervisor. Your supervisor probably has a supervisor as well or maybe several of them. All of these roles place demands on your supervisor's time, and agreeing to supervise an intern is yet another investment of time (Baird, 2007; Royse et al., 2007). Sometimes, your supervisor has been released from some other responsibilities in order to supervise you but not always. If the investment pays off and the intern does a good job, then everyone benefits, and the intern may even be able to take some of the workload off the supervisor. If it is a difficult internship, on the other hand, and it requires a great deal more time and energy than anticipated, the return for the agency may not be as great. These concerns are probably on the mind of your supervisor. Remember, too, that out of all these roles and responsibilities, your supervisor's top priorities are and should be the primary work of the agency or business. In social service situations, the primary concern is always the clients and communities that the agency serves. All supervisors worry that interns may make a harmful mistake, but in social service situations, such mistakes can harm clients, co-workers, and even the intern.

The Match Between You and Your Supervisors

The goal in developing a supervisor-supervisee relationship is achieving what Birkenmaier and Berg-Werner (2007) refer to as an optimal teacher-learner fit. Of course, this fit is important in any learning situation, but in a one-to-one relationship such as this one, it is both more crucial and more achievable; a classroom teacher, in contrast, is trying to manage multiple relationships. The fit, or match, is not something that is either present or absent; there is lots of room in between. It is also not static; both you and your supervisor can work to improve it. So, your task here is not just to assess the match but to think about how to work within that match for maximum learning.

SUPERVISORY STYLE

In all the roles mentioned earlier, your supervisor has a unique style. Just as learning style is multifaceted, so is supervisory style; many different dimensions are involved. Some aspects of your supervisor's style may be the result of careful study and deliberate choice. Others are the result of attempts to replicate the way she or he learned best or to emulate a favorite supervisor. Still others are matters of personal preference. Some aspects of a supervisor's style may not even be deliberate; she or he may be quite unaware of it or assume that everyone does things in the same way. There are a great many theories about supervisory or leadership style; perhaps you are familiar with some of them. A systematic review is beyond the scope of this chapter, so instead we have selected theoretical orientations to the work, to management, and to supervision that we find especially useful to interns.

Theoretical Orientation to the Work   The work of an individual, and sometimes of an entire organization, is usually guided by a particular theoretical orientation to the work. There are, for example, many different theories about teaching, counseling, management, and community organizing. It is important that you know what theory or theories most inform your supervisor's approach to the work. It is also important to know how your supervisor reacts to other styles. Those reactions may range from an insistence that the intern do things according to the supervisor's orientation to allowing the intern to try and learn from a variety of theories and approaches (Baird, 2007). Costa refers to this continuum as one of collaborative versus hierarchical approaches (1994). A collaborative approach emphasizes mutual discussion between supervisor and supervisee and encourages divergent thinking. In a hierarchical approach, however, the supervisor serves as a communicator of "expert" knowledge. Expertise flows in one direction, as do the questions, except when the supervisee needs clarification on something.

So, for example, if you favor a humanistic or psychodynamic approach to counseling and your supervisor believes in a cognitive behavioral approach, you will need to know to what extent you are expected to use that style with your clients. Or if you tend to focus on systems change and community advocacy and your supervisor focuses on individual work and adjustment, you will need to have a serious conversation about how best to work together.

Expressive vs. Instrumental Styles of Management   The terms *expressive* and *instrumental* describe two general approaches to management or supervision (Russo, 1993). Supervisors who use expressive approaches are more people oriented; the primary concern is for people and relationships. Supervisors who use this style often need to be liked and appreciated, and they cultivate friendships with those they supervise. A supervisor with a more instrumental orientation, on the other hand, is primarily concerned with productivity and task accomplishment. Please note that this is perfectly possible even when the business of the organization is people. A supervisor at a social services agency can be concerned primarily with serving the maximum number of clients and achieving measurable results. For these supervisors, respect is important in the supervisory relationship but not necessarily affection and friendship. They are able to handle conflict and hostility from supervisees effectively.

Focus on THEORY

Situational Leadership and Supervision

Hersey, Blanchard, & Johnson (2008) have identified two dimensions of supervision, which can be combined in four different ways. *Direction* is the first dimension, which involves giving clear, specific directions, close supervision, and frequent feedback. *Support*, the other dimension, is a nondirective approach marked by listening, dialogue, and high levels of emotional support. So, if your supervisor tells you to write a report and says, "There are lots of good ways to do this, and I know you are nervous, but you can handle it. Please ask all the questions you need to and let me know when you are done," this is a high level of support and a low level of direction. A directive approach might tell you exactly what the report should look like and what it should contain, with frequent deadlines for the submission of sections and the final report. These two dimensions can be combined in four ways: high support/low direction, high support/high direction, low support/high direction, and low support/ low direction.

You may think that high support and high direction would be the best all the time. Of course, it is frightening to be given a task and have no idea how it should be done. However, too much direction over a long period of time can discourage independence and deny you the opportunities to try something and learn from it. Sometimes, you need to take risks, and even make mistakes, and that is difficult when you are very closely supervised. Similarly, everyone needs some support and encouragement, but too much can keep you from seeing where you need to improve. It can also make you dependent on your supervisor in a different way; there may be times when you need to function without considerable external support.

Hersey, Blanchard, & Johnson (2008) suggest the most appropriate combination of support and direction depends on the maturity level of the supervisee, and they use the term *maturity* in a particular way. First of all, the term is meant to apply to a particular task or set of task that the supervisee is facing; it is not a global description or a personality trait. They have identified four levels of maturity, which they describe as a combination of *willingness* and *ability*. Each of those levels is best suited to a particular combination of support and direction. For example, suppose an intern is going to be asked to lead a support group. It may be that the intern, at least at first, is neither willing nor able to do it. In that case, the supervisor provides a lot of direction about how to lead the group and monitors the progress carefully. At the second level of maturity, the intern may feel unable but willing to try. In this case, the intern needs high levels of both direction and support. On the other hand, if the intern has the skills but feels uncertain (able but not willing), then high support and low direction are needed. Finally, an intern who has the skills and the willingness does not need much support or direction with regard to the task of leading groups. In other areas and for other tasks, the maturity level and needs may be quite different for the same intern.

Implications of Mind Style for Supervision   In Chapters 2 and 4, we briefly discussed learning style theories and introduced you to two in particular. Another theory that you may find useful is the Mind Styles theory developed by Anthony Gregorc (2004, 2006). This theory describes how people prefer to take in information and how they prefer to order, or sequence, information. People can fall on a continuum from concrete to abstract in their perceptual preferences and from random to sequential in their ordering preferences. Thus, there are four general styles: *concrete random, concrete sequential, abstract random,* and *abstract sequential.* If your supervisor likes to teach you by showing you exactly how to do something, in a structured, sequential process, he or she is using a concrete sequential approach. If your supervisor prefers to speak in general terms about patterns and trends, he or she is more abstract. However, you may also notice that your supervisor, while abstract, approaches everything in an ordered way, from beginning to end, which is an abstract sequential approach. On the other hand, supervisors may not start their sessions with you the same way every time, jump around from topic to topic, and become impatient with you if you take too long to develop a point or a presentation; this is a more random approach.

Other Dimensions of Style   Garthwait (2005) has identified some additional dimensions of supervisory style, including the speed with which decisions are made; how willing the supervisor is to share information; those who like things structured and routinized versus those who are more fluid and flexible; those who focus on details versus those who focus only on the larger picture; and those that like a fast pace versus those that are more "laid back." Keep in mind that each of the terms used here to describe supervisory style defines one end of a continuum; there is plenty of room in the middle. As you get to know your supervisor, you can locate her or him on each of these continua. Remember, though, that supervision is a relationship, and so far, we have only covered one side.

MAKING SENSE OF MISMATCHES

As you read through the last section, you probably found yourself identifying your preferences and tendencies among the styles discussed. What were your reactions and preferences? What factors do you think might have influenced how you reacted and what you prefer? How do you understand the differences you experience between your style and that of your supervisor or co-workers?

If you—or your supervisor—are expecting a certain kind of response and get a very different one, it can be very disconcerting. If, for example, you are looking for a clear sequence and instead get something that is more random or expecting collaboration and get hierarchy, it can be confusing and upsetting. A supervisor expecting a theoretical analysis who instead gets a long discussion of your emotional reactions may think that you either did not understand or are not capable of answering the question. These moments are never easy, but they are easier if they are understood as a *mismatch of styles.* Your supervisor is not necessarily insensitive to your needs any more than you are stubborn or inept. You may each be doing exactly what you think is appropriate and required. Misunderstandings can be avoided by taking a proactive stance toward your supervision. Make note of some of these mismatches that you have experienced with your supervisors or co-workers.

Dealing with Mismatches   When you find an area of mismatch, we encourage you to think about what you can do about it. Most supervisors do not mind being asked about their style. They may or may not be familiar with the specific theories discussed here, but you can find a way to ask about these aspects of style without using the particular jargon described in this chapter. Your supervisor may also be familiar with and use other theories about supervisory styles. If so, you can learn both about that theory and about your supervisor when discussing style. You may also want to consider discussing your style with your supervisor, including your preferences and needs. These needs do not have to be demands. You are merely considering your style, along with your supervisor's style, and looking for potential areas of match, mismatch, and compromise.

Approaches to Supervision and Evaluation

Part of what creates anxiety for anyone is the unknown. Each supervisor and placement site is different, of course, but in our experience, there are some common features of the supervisory process, and you will want to be clear with your supervisor about each of them.

GATHERING INFORMATION

Supervisors need information about interns in order to know how they are progressing and developing. This information guides supervisors in their conferences with you and in the evaluation process. There are several ways for supervisors to gather this information. The approaches all have strengths and weaknesses, and your supervisor may use more than one of them. It should be fairly easy for you to find out which of these approaches will be used.

- **Live Supervision**   Your supervisor either works alongside you or observes as you work. In social service settings, the observation may be done behind a one-way mirror so that neither you nor your clients can see the supervisor, although you know you are being observed (and the client may be told as well). There is no substitute for live supervision; it is the only way the supervisor can see you work firsthand. Your nonverbal behavior and nuances of tone of voice are things that you will not notice, but your supervisor can (Baird, 2007). However, live supervision is often not possible. In social service settings, it can intrude on the relationship with a client. It can also make interns nervous, and, more important, having your supervisor right there and available for consultation may encourage you to be overly dependent on that help.

- **Audio- or Videotaping**   If you are making a presentation, leading a meeting, or working with a client, you can make a tape of yourself and go over it with your supervisor. Some supervisors will go over the tapes privately first, and some may ask you to prepare an analysis of the tape as well. While no one we know (including us) enjoys being taped, it can be of enormous value. Watching or listening to yourself is an excellent way to notice voice patterns, body language, patterns of movements, and other aspects of your performance that you would never be aware of otherwise.

- **Self-report**   A common approach to data collection is the self-report method. Your supervisor will ask you to report, orally or in writing, on your progress with clients or projects. Some agencies use a highly structured and prescribed format for these reports, and others are more flexible. Self-report encourages you to reflect on your own performance and to take responsibility for critiquing your own work, which is something you are going to have to learn eventually. It is also a chance to sharpen your analytical skills. However, it will never uncover the kind of unconscious patterns and blind spots that live supervision and taping can reveal. There is also a natural human tendency to describe yourself in the best possible light, especially when you want to make a good impression (Borders & Leddick, 1987).

- **Peer-report**   A less common approach is that of peer reporting. In cases where your supervisor cannot observe you directly but other people can, the supervisor may request written or verbal reports from these people. In the case of written reports, you may or may not be able to see them. It is important to discuss your placement's policy on gathering information on your performance as early as you can.

SUPERVISORY CONFERENCE STYLES

Kiser (2008) points out that you will have lots of different interactions with your supervisor, ranging from quick conversations in the hall to long talks as you ride to an appointment together. None of these interactions, however, constitute formal supervision. You should expect to have regular meetings with your supervisor throughout the internship. This is a time for you to report on your progress, ask and answer questions, and get feedback. Lots of supervisors will also meet with you spontaneously if there is something important to discuss, and that is fine. However, those meetings allow you no time to prepare, reflect on your experience, and think about issues and questions you may have (Royse et al., 2007). We strongly recommend that some of your meetings be at regularly specified times. Find out from your supervisor how frequent and how long these meetings will be and schedule them ahead of time. It is easy, in the crush of crises and other responsibilities, for interns and supervisors to be tempted to cancel conferences, especially if things are going well (Baird, 2007). Don't do it—few things are more important to your learning than regular and productive supervision.

STRUCTURE OF SUPERVISION MEETINGS

Another important consideration is the structure of the meetings (Baird, 2007; Borders & Leddick, 1987; Garthwait, 2005; Leddick & Dye, 1987). Some supervisors use a highly structured format. They ask the questions, and they ask at least some of the same ones each time. They may also have a structured way that they want you to present your cases or your progress on projects. Still others emphasize more didactic approaches, where they do most of the talking, teaching you about different theories, asking you questions about readings they have given you, and so on. Other supervisors are very loose in structuring the session and respond only to what comes up in the conversation. Some supervisors come into conferences with a set agenda. Others let you ask the questions and raise the issues. Some, of course, do both.

Your supervisor's choice of structure and activities may be based on personal preference, deep beliefs, or assumptions about what will be most helpful to you. Some of them are more suited to particular learning styles than others. So, while you should stay open to all forms of supervision, spend some time thinking about how well suited each of these approaches is to the way you learn best.

PLANNING SUPERVISION MEETINGS

Regardless of the structure, it is crucial that you reserve some time to plan for these conferences (Birkenmaier & Berg-Weger, 2007; Garthwait, 2005; Kiser, 2008; Royse et al., 2007). Keep in mind Kolb's learning cycle, described in Chapter 1. At most placements, there is so much to do that there is a temptation just to work, work, and work some more. Conferences force time for reflection, which is a critical part of completing the cycle, and preparing for them does as well. Planning for conferences can also empower you and give you more control and influence over the session. You should set aside time to think about questions, observations, and problems you want to discuss and to prepare supporting documents if necessary. You may need to put them in some kind of priority order, considering their importance to you and their immediacy in the lives of clients or the completion of projects. You should also be ready to report on your progress with clients and projects.

THE EVALUATION PROCESS

At some point in your internship, you will probably have a formal evaluation. At this time, the supervisor lets you know how you are doing and have done overall. In our experience, most internship programs request two evaluations: one at the midpoint and one at the end. It is natural to be apprehensive about the prospect of reading or hearing an evaluation. But try to remember that the evaluation is another opportunity for growth, empowerment, dialogue, and even assertiveness. Above all, it is an opportunity to learn. Again, the more you know about the process, the less mysterious and threatening it will be and the more you can direct your emotional energy at making the most of these opportunities. As one intern said after reflecting on her evaluation, *"He made me curious about myself and want to work on the areas that need improvement."*

You can take the initiative to find out several things about how you will be evaluated. The structure of your evaluation is an obvious concern. Find out how often you will be evaluated and whether it will be oral, written, or both. Gather information on the specific format. Some written evaluations use scales, where you are given a number from 1 to 5 or 1 to 10, to indicate how well you have met various criteria. Some evaluations let the supervisor write in a more narrative form. Still others use both formats.

It is also important to know what standards are being used in your evaluation (Wilson, 1981). Some supervisors will compare you to other interns, either current ones or those from past semesters. If you are among the best, you receive high ratings. Others evaluate interns individually, assigning ratings based on how much they have grown over the course of the semester. Still others have a standard set of expectations and criteria that you must meet to get a high rating. If the supervisor is using numerical ratings, find out what they mean. What do you have to accomplish to earn a rating of 3?

We knew one student whose supervisor had nothing but praise for her at their weekly conferences. Yet, the midsemester evaluation contained ratings no higher than 3 on a scale of 1 to 5. The supervisor explained that he never gives anyone a 5 because no one is perfect, and if he gave 4s at midsemester, there would be no room for improvement. To him, the 3s were an excellent midsemester evaluation, but to the intern, they felt like an average rating, much like receiving a C on a paper.

You will also want to know the function of the evaluations. Who can see them and for what purpose? Are the midsemester evaluations recorded and counted as part of your grade? What portion of your internship grade is determined by the supervisor's evaluation? You should also clarify when and under what circumstances you will see your evaluation. Many supervisors will make the written evaluation part of a conference. They may discuss the evaluation before filling out any forms or they may go over the written evaluation with you. Regardless of the specific arrangements, you should see the evaluation before it goes to anyone else and have an opportunity to review it and ask questions.

Your Reaction to Supervision

Even under the best of circumstances, the process of supervision often produces at least some anxiety. In your internship, you are being asked to try out new skills and approaches. You should be stretching yourself and taking some risks, and that means there will be times when you don't do as well as you had hoped. Your supervisor may also notice patterns of strengths and weaknesses that you are not aware of. In social service settings, a supervisor who feels you are running into some unresolved personal issues is probably going to let you know that. Under those circumstances, most of us experience two seemingly contradictory emotions. On the one hand, we are there to learn and grow, and supervision is a tremendous opportunity for that. On the other hand, we do not always want to see ourselves clearly or change our ways of doing things. We want to learn and change, yet we resist learning and changing (Borders & Leddick, 1987). Some of us are even resistant to compliments, as in this student's journal entry: "*Positive feedback is something I have a hard time accepting. In fact, I almost cry when I hear it. I am learning that my work in the world is important and well received.*"

These conflicting emotions sometimes manifest themselves in behaviors that help you avoid really looking at yourself and your progress (Borders & Leddick, 1987; Costa, 1994). They include being overly enthusiastic, avoiding certain topics, and going off on tangents. You may also become forgetful, especially about projects, assignments, or issues that make you nervous. Some interns become argumentative, taking issue with every point the supervisor makes. They may or may not do so out loud, but privately and to their friends, they dispute every criticism. Others go to the opposite extreme and agree with the supervisor even when that is not how they feel. If you find yourself exhibiting any of these behaviors, think about the feelings and concerns that may be behind them. The point here is not that any single instance of one of these behaviors indicates that you are avoiding an issue; however, when they become patterns, you should look very carefully at them, discuss them with your instructor, and find a way to work through them.

Managing the Triadic Relationship

Part of the success of your internship depends on the triad of you, your campus instructor, and your site supervisor. This particular triad of you and your two supervisors can be a solid foundation for your learning. A triangle by definition can be a very stable formation; when set on its base, it is hard to knock over. On the other hand, when set on its point, it tips over all on its own. So, too, this triadic relationship can be prone to unproductive alliances. Once you have a clear understanding of the role of the campus instructor in this relationship, you'll have a deeper understanding of why the keys to success in this triadic relationship are mutual understanding and open communication. You can play an important role in making sure each of these keys is in place.

THE CAMPUS INSTRUCTOR OR SUPERVISOR

The role of the campus supervisor in an internship varies from campus to campus, but at the very least, this is the person who will be grading your overall performance. If a campus instructor is also leading a seminar, reading and reacting to your journal, or visiting you at the site, then in some ways that person is supervising you too. And while the relationship is generally less intense and the contact less frequent than the relationship with your site supervisor, many of the same concerns and issues can and do arise. Acceptance is a concern in this relationship, as is assessment. It is important to have a sense of the styles of teaching and supervising used by your campus instructor and see how well they match with your needs.

We have seen situations where the intern and the site supervisor form an alliance against the campus instructor, agreeing not to divulge certain events or to give a glowing report when one is not deserved. We have also seen the opposite situation, where the intern and the campus instructor form an alliance and talk together about their negative impressions of the site supervisor and how best to "get around" that person.

MUTUAL UNDERSTANDING

It is vital for everyone's clarity, and your sanity, that all three of you have the same understanding of the mechanics of the internship. These mechanics include everything from relatively mundane matters, such as required hours, start and end dates, and dress codes, to more profound matters, such as your learning goals, the scope of your activities, the frequency and nature of visits to the site by the campus instructor, and evaluation methods. This last piece is especially important if there are actually two approaches to your evaluation—one used at the site and one used on campus. In some academic programs, mutual understanding about these issues is put in place between the campus and the agency long before any interns are interviewed or accepted. In other programs, due to staffing, inattention, or both, these understandings are not in place, and it will be up to you to make sure they are established.

OPEN COMMUNICATION

Open communication means that issues and concerns are addressed directly by the two or three parties involved. If your supervisor does not seem to like your instructor (or vice versa) or has a problem with something about the program, you can't always avoid talking about it, but you can encourage the supervisor to speak directly to the instructor. If your supervisor seems to want to give you higher ratings than you perhaps deserve or does not want to disclose to your instructor any of your areas for growth and development, it is tempting to let that occur. But to maximize your learning, you need to hear clearly and openly about your strengths and weaknesses. You can help your supervisor not play the "game of school" where the goal is the grade and instead help put your supervisor into the role of a collaborative partner in your learning.

CO-WORKERS

Although your supervisor may be your most important colleague at the placement site, you may actually spend more time with other employees. Sometimes, interns report that a particular co-worker becomes an informal supervisor and that they learn more from and feel more supported by that person than by their assigned supervisor, whom they may only see once a week. Whatever the relative importance of co-workers to you, they are a great opportunity for learning and a potential source of support. They may be mentors, sponsors, and role models offering you knowledge and skills, an insider's and veteran's perspective on the agency and its work, and valuable feedback on your performance.

Just as with your clients and supervisor, you probably have an image of your co-workers as well, and it needs to be tested. And just as with clients and supervisors, another major concern of the Anticipation stage is acceptance from co-workers.

Expectations

You may not have thought much about it, but you probably have an image of what your co-workers will be like. What level of education do you think is needed for working at your site? What level of education do you think most of the staff has? Earlier, we asked you what attitudes, skills, and knowledge were necessary to be successful at your site. If you skipped that section, take a moment now and give some thought to what it takes to be successful at your site, not as an intern but as a full-time professional.

Think About It

What Guides Your Co-Workers?
Most people are guided by some standards or principles in their work. Your organization has a set of rules and policies and may even have a code for professional or ethical behavior. You may have studied ethics and other aspects of professional behavior in your classes as well. To what extent do you think your co-workers are aware of these issues and standards? To what extent do you think they adhere to them? What do you imagine happens to them if they violate those standards?

You can, and should, spend some time learning about your co-workers and finding out information such as:

- What is known about the hiring process at the site?
- What level of education is in fact required?
- What other requirements and preferences does the organization have?
- How many people leave the organization in an average year?
- What do people have to do to get promoted?

You may be quite surprised by the answers you get to these questions.

Acceptance Issues

Most interns are concerned about whether, when, how, and on what terms they will be accepted by co-workers at their placement site. As you ponder this question, there are a number of issues to consider. Obviously, different individuals will react to you differently. Based on what you know now, who are the people at your site whose acceptance is most important to you? Remember, too, that acceptance is not the same as being liked. Most of us want to be liked, but your co-workers do not have to like you in order to accept you. If you need everyone to like you, you may find yourself taking action directed primarily at achieving that goal rather than at what is best for your organization or your learning. On the other hand, here is a reasonable, although not always achievable, desire expressed by a student: *"What I want from them is to remember when they first started out and appreciate what I am going through."*

MUTUALITY

It's important to keep in mind that acceptance is a two-way street. One challenge for you is to accept your co-workers for who and what they are. Again, you do not have to like all of them or even respect everything they do. If you are in a social service program, then in your classes you may have studied and practiced how to be accepting of clients whom you may not like or approve of. Your co-workers are not clients, and they are accountable to standards that your clients are not, but they deserve the same accepting, nonjudgmental treatment that you extend to your clients.

CO-WORKER REACTIONS

Depending on the size of the organization hosting your placement, you will probably get a range of reactions from co-workers, from warmth to indifference and even to hostility. Some interns report feeling like guests, some like employees, and still others like intruders. These two quotes from student journals are a study in contrast:

I felt like an outsider at first. I didn't know anyone at lunch, and they would all sit together.

They are very helpful in that they take time to explain what they are doing and why. They don't seem to be annoyed with my questions and inexperience.

> **In Their Own Words**   Disappointing Reactions from Co-Workers
>
> It seems that my main value to some of them is as a gofer. They give me jobs they don't want so they can take a break.
>
> I was amazed to find out that several of the staff here are brand new in the field.
>
> With my previous field experiences and practicum, I have more experience than they do.
>
> During recreation period, I am out there trying to make contact with the kids, and the other staff members just sit and talk with each other—and they're the ones getting paid!
>
> In private, they make jokes about the residents. They have derogatory labels for almost every one of them.
>
> Some of them don't have degrees in (this field). Some of them don't have degrees at all!

Before you are too harsh on the people mentioned in the first quote, try putting yourself in their place. It is another busy day at work, and they are meeting an intern who they may or may not have known is coming. They may be wondering what your presence means for them, what you want from them, exactly why you are there, and whether they can count on you. Some of them may even resent you. Energetic new workers can be threatening to those who are disenchanted with or exhausted by their work. A new intern's ideas—or even willingness to follow established procedures to the letter—can be threatening or some employees may fear for their jobs.

> *After the visit, I began making notes on my clipboard. The worker I was partnered with asked me what I was doing, and when I told her, she said, "Oh. We don't have to do that." I said that my training manual says that we do, and she just shrugged. Later on, I asked my supervisor, without mentioning what she said. I was right. Later, I found out that she was complaining and making remarks about me behind my back.*

Your co-workers' reactions to you depend on many factors, including their personalities, past experiences with interns, their understanding of (or lack thereof) what an intern is, and their relationship with your site supervisor (Gordon, McBride, & Hage, 2005). For example, agencies sometimes use interns and volunteers as a way to stretch their staffing size and meet their workload needs without hiring additional people (Suelzle & Borzak, 1981). This practice may violate labor laws or ethical practices and often results in a lack of continuity within the agency as well as resentment from frontline staff members who have to deal with the results. Another example is the reactions of co-workers who may or may not have any experience with interns but who have heard and in turn perpetuated stories about interns. You might have been privy to one of those stories at your site. Those stories tend to live on long after the interns leave the site and often take on skeleton qualities; subsequent interns at the site—even years later—may pay the price for the interns who preceded them. As you are learning firsthand, when an intern works out well, it is productive and exciting for everyone involved, but if that is not the case, having interns at the site can be very taxing on many people's time and energy and can be disruptive to the rhythm and flow of the work. You may also find yourself surprised by the variety of approaches to the job taken by your colleagues. Some may be very different from what you expected. You may be disappointed in them, and you may have a right to be.

FEELINGS OF EXCLUSION

Gordon, MacBride, and Hage (2005) have noted that interns often struggle with feelings of intrusion and marginality. If you feel like you really don't belong at the placement, that you don't fit in and can't contribute much, you are struggling with feeling like an intruder. On the other hand, if you are feeling like your co-workers don't want you there and don't see that you can be of much help, then you are struggling with marginality. Of course, as the following student journal entries show, both can occur at once:

> *I feel like my co-workers treat me like an adolescent if I don't present myself properly.*

> *They tell inside jokes and don't clue me in. One of them in particular just looks at me as if I don't belong there, and she is waiting for me to prove it.*

In fact, you may feel like an intruder when you first begin your internship, and you may be treated like one. If so, it will take some time to prove yourself. But, in time, these feelings and reactions usually dissipate, but they sure are troubling while they are present.

For Helping and Service Professionals: Russo's Patterns of Adjustment

Over the course of many years in the field, there are adjustments to be made. There are several ways to cope with the ongoing demands—both physical and emotional—of the work and the day-to-day strains and frustrations that are as much a part of the work as the joys and satisfactions. Russo has described several patterns of adjustment that are found in experienced workers: those who identify with the clients, with their co-workers, and with the organization. Russo emphasizes that these are not static categories; people often move from one to another and back over the course of their careers. It is probably far too early in your own career to determine which category fits you, but knowledge of them may help you make sense of what you are seeing at your placement.

Identifying with Clients   Workers who identify primarily with clients can be further divided into four subcategories. *Reformers* tend to be impatient with anything that they believe interferes with their ability to serve the clients. They often neglect paperwork and will try to change the organization's policies and procedures to better meet the needs of the client (i.e., people before paper). In our experience, this is a very common stance for interns, although they do not actively try to change the placement site. *Innovators* still promote change but are more patient with and understanding of the change process. They listen, they ask questions, and they work with others to try to find the best way to achieve change. *Victims*, on the other hand, are frustrated by systems they see as inadequate and even harmful. They see themselves as the protectors of the clients and are likely to battle the administration openly, sometimes enlisting clients (or interns) in their struggle. Finally, *plodders* identify with the clients and may have some of the same concerns and frustrations as some of the other types described, but they seem to have given up on change. They work quietly with their clients, doing the best job they can. They often have their own way of doing things, and they work without making waves and do not try to influence others or the organization itself to change.

Identifying with Co-Workers   Workers who identify primarily with their co-workers also care about clients, but in addition, they feel a strong allegiance to their profession as a whole. They may be active in unions and/or professional organizations and look to these groups, as opposed to their particular workplace, as their primary guides. Some relatively new workers are attracted to this stance because they are unsure of their own skills and knowledge. They follow the rules of the professional organization rigidly. Other workers in this category are more sure of themselves and regularly consult their union or professional organization for guidance, but consider these groups as one of several sources of wisdom.

Identifying with the Organization   Workers who identify with the organization look to the organization and its policies and procedures as their primary source of guidance. Even though those rules may sometimes work against the needs of a particular client or be in violation of standards or ethical codes issued by professional organizations, these workers believe that following the rules will do the most good for the most people in the long run. Some may be hiding behind the rules so they do not have to think hard or take risks. Others have adopted this stance after careful thought and reflection on their experience. Russo also points out that some of them are conflict avoiders. They realize that the needs of clients, the organization, and the profession can sometimes conflict, but they want to resolve those conflicts quickly, and adherence to the rules is one way to accomplish that.

SUMMARY

Relationships form the primary context for your learning at the internship. As such, beginning them well is an important part of meeting the challenges of the Anticipation stage. Becoming comfortable with a new group of people takes time; how much time depends on the situation you are in, the people that you work with, and the person that you are. Ideally, you will enter your relationships with co-workers with a clear set of expectations and prepared for some of the challenges of acceptance.

   If you prepare well for relationships by questioning your assumptions and gathering as much factual information as possible, you will have a better chance of success. Examining your stereotypes about your supervisor and co-workers will improve your chances of a smooth entry and should reduce your anxiety as well. Acceptance is an issue in all these relationships, as it is in so many areas of life. Examining the issues raised in this chapter, and those raised in Chapter 4, will let you think about and approach the acceptance process from both sides of the relationship.

Although people are of critical importance, they do not tell the whole story of an internship. To complete the picture of your internship and your journey through the Anticipation stage, you must consider your placement site as an organization. That is the subject of the next chapter.

For Contemplation

CHECKING IN

How has your work so far had an impact on your personal development? Your professional development? Your civic development? Comment on changes that have occurred since the last time you reflected on these issues.

PERSONAL REFLECTION:
SELECT THOSE INQUIRIES THAT ARE MOST MEANINGFUL TO YOU.

YOUR SUPERVISOR

1. What is your supervisor's primary theoretical orientation?

2. Characterize your supervisor on the following continua: (a) instrumental vs. expressive, (b) support vs. direction, (c) collaborative vs. hierarchical.

3. In what ways is your supervisor's style well matched to yours? Are there any areas of mismatch?

4. Supervisors have different ways of finding out how you are doing. These methods include direct supervision, self-report, and peer review. Which method or methods does your supervisor use? Which method or methods would you prefer?

5. Does your supervisor follow a structured format? If not, who decides what will be discussed? If you don't have any particular questions or concerns, will your supervisor bring up topics?

6. How are you going to be evaluated by your supervisor? What form will be used? What criteria will you be judged by? What happens to the evaluation after it is written?

7. Here is another exercise for those of you who have had other internships:

   - Compare and contrast the supervision experiences you have had. You should include, but not necessarily limit yourself to, items 1 through 6.

   - Compare your reactions to your supervision experiences. It's not always the case that one is better or worse, but they are always different, and they are bound to affect you differently.

   - What are the things you need from a supervisor? If you were "shopping" for one, what would you be looking for?

   - Do your positive and negative reactions to supervision fit into any patterns in your life? For example, are some of the things that make you uncomfortable in supervision also things that make you uncomfortable in other areas of life? Are you happy with these patterns?

YOUR CO-WORKERS

8. How would you characterize your co-workers? What is their level of skill? Experience? Professionalism? Are they similar or different from what you expected?

9. How do your co-workers seem to be responding to you? How do you feel about that response?

10. What do you hope your co-workers will be able to give you and do for you?

11. Do you recognize any of the patterns of adjustment discussed in this chapter (see pp. 126)?

SEMINAR SPRINGBOARDS

Many interns are concerned about discussing their style and needs with their supervisors. Try using triads to role-play this conversation. Each intern should take a turn being the intern, the supervisor, and the observer. The observer's job is to give feedback to the intern on how effectively and clearly his or her needs were stated. The intern can then discuss what was easy and challenging and perhaps brainstorm some other approaches and try them out.

For Further Exploration

Corey, M. S., & Corey, G. (2006). *Becoming a helper* (5th ed.). Belmont, CA: Brooks/Cole.

Helpful section on supervision.

Gordon, G. R., & McBride, R. B., & Hage, H. H. (2005). *Criminal justice internships: Theory into practice* (5th ed.). Cincinnati, OH: Anderson Publishing.

Excellent discussions of issues relating to supervisors and co-workers. Obviously, this book is aimed at interns in a particular kind of setting, but it has many applications outside criminal justice.

Hersey, P., Blanchard, K., & Johnson, D. E. (2008). *Management of organizational behavior: Utilizing human resources* (9th ed.). Englewood Cliffs, NJ: Prentice Hall.

Situational leadership theory is a very popular approach to management and supervision. Discusses combinations of support and direction and when each combination may be most helpful.

Royse, D., Dhooper, S. S., & Rompf, E. L. (2007). *Field instruction: A guide for social work students* (6th ed.). Boston: Allyn & Bacon.

Excellent section on supervision.

Russo, J. R. (1993). *Serving and surviving as a human service worker* (2nd ed.). Prospect Heights, IL: Waveland Press.

Useful perspectives on supervision and co-workers. Especially thorough treatment of patterns of adaptation found in veteran workers.

Schutz, W. (1967). *Joy.* New York: Grove Press.

In his first stage of group development—inclusion—Schutz talks a great deal about acceptance concerns and the various ways of handling them.

Wilson, S. J. (1981). *Field instruction: Techniques for supervisors.* New York: Free Press.

Useful discussions of supervision and evaluation.

References

Alle-Corliss, L., & Alle-Corliss, R. (2006). *Human service agencies: An orientation to fieldwork* (2nd ed.). Belmont, CA: Brooks/Cole.

Baird, B. N. (2007). *The internship, practicum and field placement handbook: A guide for the helping professions* (5th ed.). Upper Saddle River, NJ: Prentice Hall.

Birkenmaier, J., & Berg-Weger, M. (2007). *The practical companion for social work: Integrating class and field work* (2nd ed.). Boston: Allyn & Bacon.

Bogo, M. (1993). The student/field instructor relationship: The critical factor in field education. *Clinical Supervisor, 11*(2), 23–36.

Borders, L. D., & Leddick, G. R. (1987). *Handbook of counseling supervision.* Alexandria, VA: Association for Counselor Education and Supervision.

Costa, L. (1994). Reducing anxiety in live supervision. *Counselor Education and Supervision, 34*(1), 30–40.

Floyd, C. E. (2002). Preparing for supervision. In L. M. Grubman (Ed.), *The field placement survival guide: What you need to know to get the most out of your social work practicum.* Harrisburg, PA: White Hat Publications (pp. 127–133).

Garthwait, C. L. (2005). *The social work practicum: A guide and workbook for students* (3rd ed.). Needham Heights, MA: Allyn & Bacon.

Gordon, G. R., McBride, B. R., & Hage, H. H. (2005). *Criminal justice internships: Theory into practice* (5th ed.). Cincinnati, OH: Anderson Publishing.

Gregorc, A. F. (2004). *The Gregorc style delineator.* Columbia, CT: AFG.

Gregorc, A. F. (2006). *The mind styles model: Theory, principles and practice.* Columbia, CT: AFG.

Hersey, P., Blanchard, K., & Johnson, D. E. (2008). *Management of organizational behavior: Utilizing human resources* (9th ed.). Englewood Cliffs, NJ: Prentice Hall.

King, M. A., & Peterson, P. (1997). *Working with interns: Management's hidden resource.* Paper presented at the American Probation and Parole Association.

Kiser, P. M. (2008). *Getting the most from your human service internship: Learning from experience* (2nd ed.). Belmont, CA: Brooks/Cole.

Leddick, G. R., & Dye, H. A. (1987). Effective supervision as portrayed by trainee expectations and preferences. *Counselor Education and Supervision, 27*(2), 139–154.

McCarthy, P., DeBell, C., Kanuha, V., & McLeod, J. (1988). Myths of supervision: Identifying the gaps between theory and practice. *Counselor Education and Supervision, 28*(1), 22–28.

McClam, T., & Puckett, K. S. (1991). Pre-field human service majors' ideas about supervisors. *Human Service Education, 11*(1), 23–30.

Milnes, J. (2001). Managing problematic supervision in internships. *NSEE Quarterly, 26*(4), 1, 4–6.

Royse, D., Dhooper, S. S., & Rompf, E. L. (2007). *Field instruction: A guide for social work students* (5th ed.). Boston: Allyn & Bacon.

Suelzle, M., & Borzak, L. (1981). Stages of field work. In L. Borzak (Ed.), *Field study: A sourcebook for experiential learning.* Beverly Hills, CA: Sage Publications.

Wilson, S. J. (1981). *Field instruction: Techniques for supervisors.* New York: The Free Press.

CHAPTER *7*

Getting to Know the Placement Site

DON'T SKIP THIS CHAPTER!

You may be looking at the title of this chapter and wondering what is going on. After all, you have already read about and considered your supervisor and your co-workers. You understand your role, you have a learning contract, and you have read about the stages of an internship. What else is there to know? Although you do know quite a bit about the placement site, you have only part of the picture. Part of your orientation and adjustment to your internship is learning how the agency or company you work for operates and why it operates that way. This knowledge will help you make sense of your experience there, although it may not seem like it right away.

Additionally, if you are not prepared for the organizational dynamics and issues discussed in this chapter, you could be in for some unpleasant surprises. Suppose, for example, that your supervisor, who meets with you once a week, is not around when you are working, but another supervisor is there most of the time that you are. This person seems unusually hostile and appears to resent the time you spend with your supervisor. She also occasionally contradicts something your supervisor has told you. While this situation would not be easy under any circumstances, it will be less mysterious if you understand that your supervisor was recently promoted to that job and the other supervisor was passed over, even though she has been there longer. Whether you know it or not, organizational dynamics are bound to affect you, your colleagues, your supervisor, and even the people you work with if you are in direct service.

LENSES ON YOUR PLACEMENT SITE

We are going to encourage you to look at your placement site through the twin, but related, lenses of systems theory and organizational theory, each of which you may have studied in some of your classes. We have chosen a few concepts from systems and organizational theories that we think are particularly relevant to your work as an intern. We will not be covering all, or even most, of the major concepts in systems or organizational theory, and you may want to investigate them further or even take a course in organizational behavior. In fact, after you read this chapter, you may want to make some sort of organizational analysis part of your learning goals and activities.

Systems Concepts

A system is a group of people with a common purpose who are interconnected such that no one person's actions or reactions can be fully understood without also understanding the influence on that person of everyone else in the system (Egan & Cowan, 1979). Even if you understand each person in the system as an individual, to understand the system, you must understand the way everyone interacts; the whole, in this case, is greater than the sum of its parts (Berger, McBreen, & Rifkin, 1996).

Systems can be analyzed internally or externally (Berger et al., 1996). Internal analysis involves studying the inner workings and components of the system and the way human and material resources are arranged and expended. However, it is important to realize that all systems are hierarchical. This does not mean that they use a hierarchical authority structure; some systems do not. But each system is part of a larger system, and most can be broken down into smaller systems. This hierarchical dimension is easy to see in large organizations, but it is present in smaller ones as well. A family services center, for example, can be broken down into the various programs it runs. However, it is also part of a system of service providers in the city or town in which it operates. An external analysis examines the relationships between a system and other related systems.

Organizational Concepts

In both internal and external analyses, organizational theory can help you look at, make sense of, and navigate your placement site. There are many different organizational theories, but for this chapter, we rely heavily on the integrative work of Lee Bolman and Terrence Deal, whose book *Reframing Organizations* is an excellent resource if you want to explore organizational theory further (2003). We begin with some of the components of an internal analysis. First, we outline some basic and important background information that is important to know about any organization. The mission, goals, and values of an organization tell you in an official and explicit way what that organization is all about and what makes it distinctive. Funding, of course, provides the underpinning for the work and sets crucial parameters. Then we use Bolman and Deal's four lenses or frames—structural, human resources, political, and symbolic—to help you examine how your placement site is organized to carry out its mission and goals and whether it practices the values it espouses.

BACKGROUND INFORMATION

In presenting the concepts in this chapter, we will refer to a variety of kinds of placements, including a fictional human service agency called the Beacon Youth Shelter, an amalgamation of several agencies with which we have worked.

History

Every organization has a history, and it provides an important context for the way things are currently being done. Older organizations have often gone through a number of changes. For example, the Beacon Youth Shelter always had a rule that it would not accept residents with a history of physical violence. However, it has recently agreed to change and allow children with violent histories to come in. That means this population is relatively new to the organization, and there is no doubt an adjustment period will be going on for everyone. Also, both human service and other organizations are sometimes bought by or merged with other organizations. Can you see why this information would be useful to an intern?

Mission

It is easy to gain a general sense of the purpose of your placement; sometimes, the name is all you need, as in The Environmental Collaborative or Child and Family Services. But each organization has an overall mission, and that is a good place to start to see what is distinctive about it. It is likely that your placement site has a written statement of its goals. Here are some examples:

- From a media outlet: "NPR is primarily a news organization. We are always testing and questioning the credibility of others. We have to stand that test ourselves. Whether we are functioning as reporters, hosts, newscasters, writers, editors, directors, photographers, or producers of news, music, or other content, we have to stand that test ourselves. Our news content must meet the highest standards of credibility" (NPR, 2007).

- From a social service agency: "To provide quality and multicultural services to those whose lives have been affected by sexual assault."

- From a large insurance company: "HSB's technical knowledge and expertise is the key that has helped our customers avoid losses, and recover promptly from losses that do occur, for over 135 years. The Hartford Steam Boiler Inspection and Insurance Company benefits businesses and industries worldwide by providing: equipment breakdown insurance and reinsurance; other specialty insurance and reinsurance; risk management consultation; engineering and inspection services" (HSB, 2007).

- From a criminal justice agency: "To serve adults and youth who exhibit or are at risk of criminal or delinquent behavior, substance abuse, or mental illness, as well as other socially disadvantaged persons."

These formal mission statements may be well known to the people who work at the organization and/or to its clients or customers or they may be unknown to either group. A lot depends on who determined the mission and through what process (Caine, 1994). There will most likely be greater investment in and awareness of mission statements that were arrived at collaboratively, as opposed to those created by one or two people and then handed down.

Goals and Objectives

Mission statements are a good place to start, but you will find that agencies that do similar work often have similar mission statements. There is, however, a wide variety in the specific goals and objectives among organizations, even among those with very similar titles. Goals are rather broad and hard to measure. "Putting an end to violence against women" is an example of a goal statement. It is a goal to work toward but not one that is likely to be reached any time soon, and therefore, it is hard for the agency to measure progress toward it. That same agency might have several objectives, including a 25% reduction in the incidence of teen dating violence, the adoption of certain laws and procedures, and the provision of some sort of service and assistance to every woman who requests it. These objectives are much easier to measure.

Values

Organizations also have values or principles that they try to adhere to in doing their work. Again, there are often formal written statements of these values, which may include general principles, rights, and responsibilities. Here are two brief examples:

- "Our coverage must be fair, unbiased, accurate, complete and honest. At NPR we are expected to conduct ourselves in a manner that leaves no question about our independence and fairness. We must treat the people we cover and our audience with respect" (NPR, 2007).

- "The values of Interval House define who we are, what we stand for, and how we do our work. They are principles against which we measure the importance and worth of our decisions and actions. The staff and the board of Interval House hold ourselves accountable for transforming these values into action: empowerment, support, collaboration, safety, diversity, integrity, compassion, confidentiality, dedication, and equity" (Interval House, 2007).

Value Statements from Two Organizations

Interval House
Empowerment

We believe in power with, not power over. We help one another acquire power over their own lives and we respect the decisions that others make. People have the right to make their own decisions and to accept responsibility for them.

continued

Value Statements from Two Organizations *(continued)*

Support

We listen well and we provide support to one another in reaching our goals.

Diversity

We seek and welcome cultural diversity of all kinds as well as diversity of opinion, perspective, talents, and gifts. There can be no excellence without diversity.

Compassion

Although we may disagree, we do not judge. We accept people for who they are, regardless of whether we agree with their behavior, and we believe in the capacity for growth and change.

Dedication

We believe in what we do and we work hard. We hold one another accountable for putting forth the effort and commitment that our work demands.

Collaboration

We believe that the greatest results come from the combined efforts of diverse organizations and individuals. We reach out to work cooperatively with others to promote meaningful and mutually beneficial change.

Safety

We believe that each person has a right to live and work free from fear and safe from harm.

Integrity

We are honest and reliable, open and sincere. We hold ourselves accountable for doing what we say we will do. We evaluate our effectiveness and change as necessary.

Confidentiality

We protect the privacy and confidentiality of those with whom we work, unless doing so would lead to harm to others.

Equity

We believe that all people deserve fairness, impartiality, justice and opportunity.

National Public Radio

"**Fair**" means that we present all important views on a subject. This range of views may be encompassed in a single story on a controversial topic, or it may play out over a body of coverage or series of commentaries. But at all times the commitment to presenting all important views must be conscious and affirmative, and it must be timely if it is being accomplished over the course of more than one story. We also assure that every possible effort is made to reach an individual (or a spokesperson for an entity) that is the subject of criticism, unfavorable allegations or other negative assertions in a story in order to allow them to respond to those assertions.

"**Unbiased**" means that we separate our personal opinions—such as an individual's religious beliefs or political ideology—from the subjects we are covering. We do not approach any coverage with overt or hidden agendas.

continued

> **Value Statements from Two Organizations** *(continued)*
>
> "**Accurate**" means that each day we make rigorous efforts at all levels of the newsgathering and programming process to ensure our facts are not only accurate but also presented in the correct context. We make every possible effort to ensure assertions of fact in commentaries, including facts implied as the basis for an opinion, are correct. We attempt to verify what our sources and the officials we interview tell us when the material involved is argumentative or open to different interpretations. We are skeptical of all facts gathered and report them only when we are reasonably satisfied of their accuracy. We guard against errors of omission that cause a story to misinform our listeners by failing to be complete. We make sure that our language accurately describes the facts and does not imply a fact we have not confirmed, and quotations are both accurate and placed properly in context.
>
> "**Honest**" means we do not deceive the people or institutions we cover about our identity or intentions, and we do not deceive our listeners. We do not deceive our listeners by presenting the work of others as our own (plagiarism), by cutting interviews in ways that distort their meaning, or by manipulating audio in a way that distorts its meaning, how it was obtained, or when it was obtained. The same applies to text and photographs or other visual material used on NPR Online. Honesty also means owning up publicly and quickly to mistakes we make on air or online.
>
> "**Respect**" means treating the people we cover and our audience with respect by approaching subjects in an open-minded, sensitive and civil way and by recognizing the diversity of the country and world on which we report, and the diversity of interests, attitudes and experiences of our audience.

Funding

The organization's budget may not seem very important to you; after all, you probably are not getting paid! However, the financial as well as human resources that an organization has do determine what it can do for clients or customers and often affect the general tone of the workplace. Your placement site certainly has an annual budget. The budget may be part of a larger budget, as in the case of a senior services center that is funded by the city; its budget is part of the social services budget, which is in turn part of the city budget. The budget is usually broken down into categories, such as salaries, benefits, supplies, food, transportation, and so on. It is also important to know how the budget is set, who must approve it, and how changes are negotiated. Of course, budgets can change suddenly; in human services, that usually means they get cut, but occasionally, an agency will be awarded a grant or get a new program approved and funded, which may in turn affect the agency's goals.

FOR HELPING AND SERVICE PROFESSIONALS: SOURCES OF FUNDING

One aspect of human service agencies that is often overlooked by interns is where the money to operate actually comes from. Even if your placement site does not charge for its services, they are not free. The money to pay the staff and administer the agency that provides the "free" services has to come from somewhere. Generally speaking, agencies can be divided into three groups: public, private nonprofit, and for profit.

Public Agencies are funded through local, state, and federal tax revenues. They may have some other sources as well, such as foundation support or grant money, but taxes provide the majority of their funds. Sometimes, the clients are charged a fee on a set scale or one that varies according to their income, but those fees do not begin to cover the cost of operation. What that all means is that these agencies are ultimately accountable to the legislative bodies. That is usually how they were created, and it is typically the body that decides on the agency's budget. Birkenmaier and Berg-Weger (2007) note that public agencies tend to be large, with a vertical authority structure (we will discuss structural issues later on). They further note that public agencies tend to have complex rules and procedures and a lot of accompanying paperwork. These agencies, or their parent organizations, lobby legislators to ensure that their funding is maintained or increased. Examples of public agencies include departments of child welfare, aging, corrections, and education.

Private Nonprofit Agencies get their funding from a variety of sources, including charitable organizations such as the United Way, foundations, corporations, and individual donors. This diversity of funds is a challenge because it means maintaining developing relationships with and being accountable to a number of groups, but it is also a strength in that the organization may be able to avoid being overly dependent on any one source (Birkenmaier and Berg-Weger, 2007). Nonprofit agencies may receive some public funds, usually through contracts with state agencies or though Medicare. Beacon Youth Shelter, for example, contracts with the state's child welfare agency for a certain number of beds each year. Nonprofit agencies are accountable to a board of directors, who are typically volunteers. If your agency has such a board, you may find it interesting to attend a board meeting if the agency will permit it. Nonprofit agencies generally have a lot of paperwork, but unlike public agencies, they tend to go to a variety of funding sources and sometimes to regulatory agencies as well. Fund-raising also takes up a great deal of time and energy, and as an intern, you may have the opportunity to be involved in some fund-raising activities.

For-Profit Agencies get funds to start their work from individuals or private organizations, but they expect to be self-supporting. In fact, they expect more than that—they expect to make a profit. Kiser (2008) notes that this gives these agencies a dual mission to serve their clients and to make a profit. Berg-Weger and Birkenmaier (2007) assert that profit is always the driving factor in these agencies. In any case, they have to pay attention to making money, and they are held accountable for that goal, usually by a board of directors. Keeping client interests primary while still fulfilling their obligations to the board is one of the key challenges for these agencies. Like nonprofit agencies, for-profit agencies may access public funds indirectly by contracting with state agencies. A for-profit agency might, for example, monitor children who are in court-mandated placements and services, and there have been some well-publicized experiments where public schools are operated by private organizations. The growth in for-profit human service organizations is also referred to as privatization and extends into many areas, including nursing homes, group homes for the mentally challenged, and rehabilitation.

This issue of funding is especially important because the source of funding has great power over and influence on the organization. Insurance companies, for example, usually insist on a formal diagnosis before authorizing treatment. Some human service providers are troubled by these diagnostic labels, which tend to stay in clients' files and follow them around. Nevertheless, a diagnosis must be made, and it must be one the insurance company will pay for or the client will have to pay cash or be referred to another agency. Organizations funded by tax dollars are vulnerable to the opinions (informed or not) of the taxpayers and must worry about public relations and influencing the local political process.

ORGANIZATIONAL STRUCTURE

An organization's structure is the way in which it is set up to accomplish its goals (Bolman & Deal, 2003). There are two basic elements of structure: division of responsibilities and coordination of work. There are also many subcategories within each of those two elements. Each feature of an organization's structure can be tight or loose; clearly defined or ambiguous. There are endless configurations, and not one of them is the best in all situations. However, some may work more effectively in certain situations than others. Let us look more closely at your placement site using a structural lens.

Division of Responsibilities and Tasks

Who will do what, and in what configurations, is one of the basic questions facing any organization. Most have developed a formal structure to answer this question, although as we will see later on, sometimes the real operation of the agency differs from the formal structure.

ROLES

Roles describe the positions in the organization and the duties or functions that each one performs on a regular basis. The formal roles in an organization can be found in two places: in job descriptions and in the organizational chart. Understanding these two contexts for roles is important to understanding the integrity of the organization's structure. So, too, is understanding the degree of specialization and the purpose for grouping roles.

- **Job Descriptions**   Most organizations have job descriptions for each position that state, at least in general terms, the responsibilities of the positions. These descriptions may be found in the policy manual or they may be found in the human resources department. They are often generated, or reviewed, when someone is hired. A prospective employee will usually want to see the description, and in a large organization, a higher-level administrator needs to see a job description to approve the hiring. Many placement sites even have written job descriptions for interns.

- **Organizational Chart**   An organizational chart is also valuable when examining formal roles. An organizational chart does not describe the duties of each position,

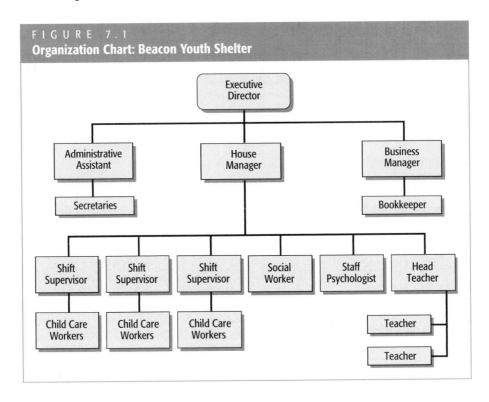

FIGURE 7.1
Organization Chart: Beacon Youth Shelter

but it shows each position relative to one another. It shows who is responsible for whom and to whom each person is accountable. A chart for the Beacon Youth Shelter appears in Figure 7.1.

SPECIALIZATION VS. GENERALIZATION

Another issue in examining roles is the degree of specialization assigned to each role. Some organizations have highly specialized jobs, each assigned to one person (or one category of people). For example, at a child welfare agency there may be one "role" that investigates reports of child abuse. This is an intake worker. Workers who assist victims of abuse are in a different role—they are called treatment workers. If the family needs economic assistance, it is referred to yet another worker. In other agencies, there is less specialization. For example, at a small advertising agency, there are several people who handle accounts and try to handle all the needs of their clients.

Both of these models have advantages and disadvantages. It is a daunting task to know well many functions and expected sets of information, and if someone's role is too generalized, people may get poor advice or miss out on important options. On the other hand, when services are highly specialized, people can get caught in a maze of staff members and may feel overwhelmed and confused. It is frustrating to ask a question and be told "That is not my area." It might have already happened to you!

GROUPINGS OR TEAMS

Roles are often organized into teams. Sometimes, those groupings are functional, and people who have the same or similar tasks are placed in the same department. Other ways to organize groups are according to time (day shift vs. nights shift), clients (as in the case of intake vs. treatment workers), and place (such as a satellite office) (Bolman & Deal, 2003).

Coordination and Control of the Work

Regardless of the particular roles and teams in an agency, there has to be a way to coordinate what they do and guide or control their efforts. How much coordination is necessary depends on the interdependence of the roles involved. In a hospital or a school, for example, the people in various roles are very dependent on one another in order to accomplish their goals. According to Bolman and Deal (2003) there are two dimensions of coordination and control: vertical and lateral. Every organization has them both, but some emphasize one more than the other. And, of course, there is considerable variation within each of these dimensions.

VERTICAL COORDINATION AND CONTROL

Vertical coordination and control refers to the ways in which the work of the agency is assigned, controlled, and supported. This dimension consists of structures for authority (the chain of command), communication, decision making, policies, rules, and evaluation procedures.

Authority Structure The *authority structure* of an organization defines the chain of command, and this chain is visible in the organizational chart. You can see from the chart of the Beacon Youth Shelter that the top administrator is the executive director. The executive director of Beacon Youth Shelter is responsible for all phases of the operation. However, most of the actual running of the shelter is done by other people. The executive director is most closely linked to the house manager, who is in turn responsible for several professional positions. The executive director may or may not attend staff meetings or know any of the child care workers by name.

An organizational chart also gives you an easy view of the length and complexity of the chain of command. Organizations that have many layers and a long chain of command are referred to as *tall* organizations, while organizations with fewer layers and a shorter chain of command are referred to as *flat* organizations (Queralt, 1996). Each has its advantages and disadvantages in carrying out the work and in putting the organizations values into action. Consider collaboration, for example. In a flat organization, more people have input into decisions, but the decision-making process can become quite cumbersome. In tall organizations, fewer people at each level are involved in collaboration, which makes it easier to make decisions but is not as inclusive.

Words to the wise . . . You will notice that there is no intern on this chart. If an organization does not have an intern each semester or year, it may not include one. However, it is important to find out where you fit on the chart. Most interns tell us they fit "on the bottom." While that may or may not be true, it does not help us or the interns to know to whom they are responsible or that person's place in the organization.

You will also notice that there is no place for clients on the chart. While that is not unusual, it is important to know the position occupied by clients in your agency (Queralt, 1996). In some agencies, clients are included in decision making. There may be an advisory board, for example, or a client council. This arrangement is more likely in organizations where clients can choose to use (and pay for) the service or go elsewhere. This arrangement gives clients some power and leverage. However, in agencies where the clients are not there voluntarily or have no other alternatives, clients may have very little influence on decisions.

Communication Flow   Successful communication is, of course, the critical factor in making any organization run smoothly. In a small organization, communication may be very informal. Everyone sees and works with everyone else all day, and important information can be shared relatively easily, although it sometimes is not. Communication in larger organizations is more complicated and often happens through memos, which is short for memoranda. Memos can be sent to just one person or to a whole group of people. Some memos are sent out on paper, but many are sent using e-mail or other web-based technology.

It is possible to look at patterns of communication in an agency. In some organizations, communication flows from the top levels down, with each level playing a role, and from the bottom levels up, with relatively little communication among people or departments at the same level. This arrangement is called a *chain pattern*. In a *wheel pattern*, on the other hand, one person or department is at the hub. Most communication flows directly from various departments to this hub and from the hub to the departments. Again, there is relatively little communication among the departments themselves. In an *all-channel network*, everyone communicates with everyone else (Queralt, 1996). There are many other configurations, and it can take some time to determine what pattern a given agency actually uses.

Decision-Making Mechanisms   Decision making is another important structural consideration. Suppose you are working at Beacon Youth Shelter and you have an idea for a field trip. There is an interesting exhibit at the science museum, and you are successful in interesting some of the residents. But the exhibit is not going to be there much longer. Who must you ask about this trip, and how long will it take to get approval? Your supervisor thinks it is a great idea, but you are surprised to learn that the executive director, who is out of town, must approve all such trips. You also have to obtain permission from each child's parent, guardian, or caseworker, and teachers must approve the absence from school for a day. This story is a bit exaggerated (although, in some cases, not by much), but it illustrates the importance of understanding the decision-making mechanisms at your placement and how they affect your work. What projects or interventions do you have planned that may require multiple approvals?

Policies and Procedures   Every organization has at least some formal written rules that everyone is supposed to follow. These rules can be found in an employee handbook or policy manual, which you may have already seen. Some organizations have very thick manuals; some have more than one. You may need to ask for help in finding the most relevant material. It is also important to know how these rules and procedures were established and by whom (Caine, 1998). Sometimes, it is done by a committee; other times, it is delegated to one person or written by a top administrator. The change process is also significant. Some policies have not been changed, or even reviewed, in many years. Other organizations have a regular method to review and update their policies and procedures.

It is also interesting to see how many people at the agency are familiar with the rules. Both of our colleges have a student handbook; chances are that yours does too. In our experience, students read that book when they have a question or when they think they may have broken a rule or been treated unfairly. Otherwise, they may never open it! At some placements, you will find that this is true for the policy manual as well.

Evaluations   Evaluations are an important structural consideration. Remember that part of the vertical dimension is control, influence, and guidance. Evaluation is one mechanism that can serve those purposes. In most organizations, there is some mechanism for evaluating or assessing performance. Sometimes, a new employee is given an initial evaluation after a few weeks. Periodic evaluations may be done annually, semiannually, or even monthly. Evaluations are usually conducted by the person to whom you are accountable (see the organizational chart).

- **Evaluation Methods**   There are many approaches to evaluation. Some methods allow the employee to set goals and then be evaluated on how successfully these goals have been reached. Other times, the criteria and goals are set by the supervisor, and there are lots of approaches between these two extremes. Some organizations use self-evaluation or peer evaluation in addition to or in conjunction with the supervisor's evaluation. The format of the report can also vary from a simple checklist or set of numerical ratings to more complex narrative formats.

- **Evaluation Criteria**   The ways in which staff members are evaluated say a lot about what is important to the organization. One important issue here is the aspects of the job that are included in the evaluation. Think about your classes in college. You have readings assigned, and many of your instructors may tell you that class participation is important. However, if you are never tested on your understanding of the reading and participation is not a part of your grade, you may be tempted to let these areas go, especially if they are difficult. By evaluating you on certain components of your performance but not on others, the instructor is sending a message, perhaps unintentionally, about what is important in the class.

  Returning to your internship site, employees are supposed to be evaluated based on how well they perform the duties in their job descriptions. In practice, however, some of these duties may not be reviewed, commented upon, or counted toward promotion, pay increases, and other organizational rewards. In addition, a supervisor may comment on aspects of the job that are not in the job description. One human service worker we know was criticized in writing for not being sociable with others at the agency, even though his cases were handled quite well.

- **Performance vs. Outcomes**   Some organizations evaluate performance (also called input), some evaluate outcomes, and some evaluate both (Caine, 1998). Again, an analogy to school may help illustrate this distinction. Teachers of a U.S. history class may be evaluated based on whether they cover all the material by the end of the year and whether students seem to enjoy and be stimulated by the class. Another approach would be to evaluate teachers based on what students are learning. Some schools believe that it is the teacher's responsibility to lay out the information and the students' responsibility to learn. Others believe that the school and its teachers have a responsibility to see to it that every child learns. The evaluation methods chosen reflect these two approaches. Similarly, an organization may evaluate employees based on the number of clients seen or the number of hours worked or it may choose to find ways to measure the change in clients and evaluate its workers in that manner. In private organizations and for-profit human service agencies, the emphasis is almost always on outcomes.

LATERAL COORDINATION AND CONTROL

Although every organization has a vertical structure, those are not always the most effective means of getting the work done. Policies and chains of command are not always followed nor are they always efficient. Lateral methods of coordination and control are those structures that emerge that are less formal and more flexible than vertical ones (Bolman & Deal, 2003). Don't confuse lateral control with a "flat" authority structure. Flat structures are less hierarchical, but lateral methods are still needed. Lateral methods may include meetings, which can occur monthly, weekly, or daily. Some organizations begin each shift with a meeting. Various staff groups then meet once a week for more in-depth discussion, and the whole group meets once a month.

Task forces or temporary work groups are another form of lateral coordination and ones that you may have the opportunity to be involved in as an intern. For example, most human service agencies have to be evaluated by local, state, or even federal licensing organizations from time to time. Generally, the organization has to gather data and write about itself for the visiting team; this is called a self-study. In addition, the team visit, which can last for several days, needs to be organized and coordinated, and large numbers of documents have to be made available. Some of this work can be done on an ongoing basis and built into someone's role. But some of it is temporary and occurs only once every few years. In these cases, a temporary team may be organized to prepare the self-study, arrange documents, and handle the logistics of the team visit.

Task forces usually arise out of some particular problem or challenge. For example, an organization may be concerned at the lack of racial and ethnic diversity in their workforce, feeling that it does not represent the diversity in the surrounding community. A task force may be appointed, consisting of individuals from many segments of the organization, and asked to study this problem and propose a solution.

HUMAN RESOURCES

Another way to look at your placement site is to focus on the people rather than the structure. In doing so, you will see a whole other, and very rich, dimension of life in that organization (Bolman & Deal, 2003). Here, we are concerned not so much with what people are supposed to do but with what they *actually* do as they try to match their own needs to the needs and features of the organization. As with the structural lens, there are a number of components to consider when assessing the human dimension.

Communication Skills

Using the structural lens, you looked at patterns of communication. These patterns emerge over time, and it is almost like you must observe from a distance to see them. Now we turn to the skills that individuals and groups show in communicating with one another. You have more than likely studied communication skills in your academic program, so this will be a review.

One aspect of communication is its clarity. Consider how well people in the organization listen to one another and how they are helped to do that. Another aspect of communication is its openness. In some organizations, people generally say what they think, about issues and about one another, in an honest and clear way. There is no prescription for how to accomplish that goal; specific approaches vary depending on the constellation of individuals and cultural profiles in the organization. Still, every organization needs to find a way to accomplish this goal, and some do so far more effectively than others.

Conflict resolution is another set of communication skills that is visible in various ways and to various degrees in organizations. If there are not some structured, accepted, and supported ways of dealing with conflict, more informal and often less-effective methods will be used. One of us had an intern whose co-workers complained about her in formal and informal meetings, yet no one spoke directly to her nor did anyone try to stop the conversations that were taking place behind her back. When someone finally took her aside and told her of these conversations, she spoke with her supervisor. The supervisor then went to the individuals involved but not to the group and not in the presence of the intern. Needless to say, the situation did not get better.

Norms

As you read the section on the rules of organizations, you may have occasionally wondered how to answer some of the questions about your placement. In some cases, you don't know the answer at all, but in others, you may be wondering, "Do you want to know what's written down or do you want to know what's really going on?" Often, the formal rules, roles, goals, and values only tell part of the story. Consider the speed limit on the interstate. The formal limit may be 55 or 65 mph, but how fast you can go before risking a ticket is the informal rule that most people concern themselves with. If you live in a college dormitory, there is probably a time after which you are supposed to be quiet. But the unwritten rule usually is that you can make all the noise you want as long as no one complains. These are examples of informal rules or norms that contradict or modify the formal rules. Other norms are not products of the formal rules. Most schoolchildren can tell you that it's wrong to "rat" or "tattle," often to the chagrin of adults. This is a very powerful rule for many children, yet it is not written anywhere.

Every organization has unwritten rules. Think back on jobs you have had and on your current placement. There is an official start time, but there is usually a grace period of at least a few minutes. In some places, you are expected to work late—for no extra pay—and in other places, you are not supposed to do that even if you want to. Many organizations have an informal dress code to go along with the formal one. Sometimes the informal rules come about because the formal rules have eroded or became obsolete but have never been changed (Gordon, McBride, & Hage, 2005).

Here are some examples of unwritten rules that interns have encountered:

- Go to lunch with your co-workers or they will think you are a snob.
- Never go out to lunch; work at your desk.
- Don't miss the boss's holiday open house.
- You can date another employee, but keep it quiet.

And here is one observed by one of our interns: "*...the more time out of the office the better. The assumption is that when you are out in the field, you are visiting directly with clients . . . which is how the work is billed.*"

The written rules are usually easy to find, but how do you go about finding the unwritten ones? First of all, you will usually know when you have broken one. Just like in school, when you wore the "wrong" thing, used an expression that was suddenly "out," or sat at the "wrong" table in the cafeteria, people will react strangely; perhaps someone will even take you aside and clue you in when you violate an unwritten rule. Another way to discover the unwritten rules is to know the written rules and observe what happens when they are broken (Caine, 1998). In some cases, nothing happens.

A Case in Point

Unwritten Rules Rule

At one family services agency, the hours were 9 A.M. to 7:30 P.M., Monday through Friday. However, in recent years, the agency had been doing more and more outreach programming, working with groups in the community, meeting during the evening in church halls or people's homes, and developing a lot of weekend programs. Fridays are an especially busy day, as the finishing touches are put on the weekend activities. As a result, Mondays have become an unofficial day of "downtime." Few appointments or meetings are scheduled, and many staff members do not come in until noon. If you were an intern there and wanted to work on Monday or tried to launch a new group that meets on Mondays, you would receive a chilly reception, and you might wonder why.

Either the rule has been bent in a way you don't know about, or it has been ignored altogether. As one of our students said: "I learned the rules by observing my surroundings and conversations."

Informal Roles

There is also a network of informal roles and relationships. It will not take you long to notice that some people perform duties that are not in their job descriptions. It may be that they were asked to do it and they didn't feel that they could say "no." Consider, for example, the employee whose supervisor discovers that her route home takes her past where his son has piano lessons. One day, the supervisor asks if she would mind picking up his son, since he will be in a meeting. The employee is happy to help, but it soon becomes an expectation, and that may become problematic.

Another way job descriptions expand is when something is not in anyone's job description; someone just decides to do it, and after a while everyone relies on this individual. There may be one person, for instance, who remembers everyone's birthday, takes up a collection for a gift when someone gets married or retires, or makes the arrangements for parties. Other roles in organizations might include the jester, who makes everyone laugh, and the go-between, who seems to have access to just about everyone.

Sometimes people expand their job description to make their jobs more interesting, especially if they have had the same job for a while (Gordon et al., 2005). At the Beacon Youth Shelter, a child care worker became interested in grant writing, volunteered to work on some grants, and eventually became quite good at it. Of course, his job description expanded and his paycheck did not, but it did make him a more influential person in the organization, which is another reason people sometimes try to expand their roles.

Cliques

There are informal subgroups or cliques in many organizations. At Beacon Youth Shelter, for example, there is a group of child care workers who have been there a long time; they are referred to as "the veterans." Few of them had any formal education to prepare them for their work; they learned on the job. Some of them rose in the organization to become shift supervisors, and until recently, the house manager position was held by a former member of this group. In recent years, though, there have been more and more workers hired who were graduates of programs in human services or social work. They are often younger and full of new ideas. Some of these ideas work, but others don't. The veterans sometimes shake their heads at these "whiz kids," and the whiz kids sometimes get impatient with the veterans.

Of course, both groups have a valuable perspective to offer the organization, and most of the time, they work well together. Still, there is some tension and competitiveness between them. It doesn't help that the new house manager did not come from within, is a graduate of a human services program, and has a master's degree. The veterans miss the easy relationship and open access they had with the old house manager; they could just approach her privately and accomplish a lot. Not only do the veterans not have this relationship with the new manager, but some of the whiz kids seem to be developing this informal access.

Interns Encounter Cliques

I have noticed that there are several subgroups: a sports one, a political one, and a friendship outside work one.

I've noticed that there is a core group that seems closest to the director . . . I would term this as the "in crowd." I don't believe the director purposely treats anyone with less respect professionally, but this core group is afforded more personal time, where lingering joking and storytelling take place.

Management Style

Just as your supervisor has a style with you, people in supervisory positions have their own styles with the people or groups they supervise. There is considerable variation among individuals, of course, but there is often a prevailing style in the organization itself. If you go back and reread the section on supervisory styles, you can think about how this applies to your placement site.

Staff Development

Most organizations put some time and energy into helping their employees grow as professionals and do their jobs better. These activities are often referred to as *staff development.* Staff development may take place during a portion of regular staff meetings or special meetings may be held exclusively for that purpose. Some organizations or units take their staff away on retreats. Others send them to workshops offered by colleges or training organizations. Still others support their employees in continuing their education.

The connection to values is not just in how much time and money is put into staff development but in the purpose of the activities (Caine, 1998). Some activities are directed at helping people learn to do their current jobs better. Others are directed at learning a new set of regulations, policies, or procedures. Still others are aimed at helping the employees learn more about theories so that they can come up with and share applications to the organization's work. Finally, some staff development activities are directed at career development. These activities are undertaken with the understanding that most of the employees will not and should not stay in their current jobs. They may move up in the organization or they may move on to a parallel, or better, position somewhere else. Staff development of this kind is an investment in an individual and the profession as well as in the current functioning of the organization.

ORGANIZATIONAL POLITICS

Your age, your interests, and your work experience affect the associations you have with the word *politics.* In our experience, many interns associate that word only with elections for political office. Others apply the term more broadly but find that the word has negative connotations of wheeling, dealing, and backbiting. In fact, as professors of human services, we have had a number of students tell us that they want to go into the helping professions to "get away from all the politics" that they imagine exist in the corporate world.

In reality, you cannot escape politics. Politics are everywhere, and they affect you whether you see them or not. The politics of an organization can be difficult for an intern to see, especially if you are not there for very many hours each week or for very many weeks. Still, the more you can learn how to look at your placement and organizations in this way, the more successful you are likely to be.

Bolman and Deal (2003) call this lens the political frame. When you look through this lens, you see organizations as coalitions, made up of different individuals and groups. They may be formal groups or the informal groups and cliques described earlier. Those individuals and groups have substantial differences in perspective, beliefs, and priorities. That diversity can and should lead to a richness in the decisions made in the organization, but it is a mistake to think that the differences can be resolved or integrated all the time.

According to the political frame, when resources are scarce, decisions about them are critical. Many organizations of every kind experience periods of shrinking budgets, due to factors such as funding cuts, differing corporate priorities, or poor profits. In most nonprofit agencies, resources are *always* scarce; scrambling to raise money and adjusting to reduced allocations from legislatures, shortfalls in fund-raising goals, and escalating costs are just facts of life. In these circumstances, conflict, bargaining, and negotiation are inevitable and not necessarily a bad thing to do. So, the first thing to look for when you examine your placement site from this perspective is the various groups and the resource issues over which they compete. The next thing to look at is how each group accesses and uses power and influence.

Power and Influence in an Organization

Formal authority and power are not the same thing. Sometimes in organizations, individuals or groups who are "lower" on the organizational chart have more power and influence than some of those above them. Secretaries often appear at the lower levels of the chart; in theory, they have little power and influence. However, it is often secretaries who know where everything is and how to navigate the paperwork maze. They overhear a lot of conversations, and they often control important resources, such as supplies. Often, if you want to know how to get something done, you are better off consulting an experienced secretary than the policies and procedures manual.

There are a number of ways to have power and influence in an organization (Bolman & Deal, 2003; Homan, 2007). Position power is one of them, but power also comes from information and expertise. If your organization is using a new set of software programs or a new computer system and you have someone who is very knowledgeable about them, that person has enormous power and influence. Pity the people who have insulted, slighted, or ignored the tech expert; their computers will have a problem sooner or later, and they will get fixed later, not sooner. Control of rewards, resources, and "inside" information is another source of power, as we noted in the case of secretaries. People who are well connected to those who control these aspects of agency life generally have what they need and know what they need to know, often in advance of everyone else. Finally, individuals can have personal power. This power may come from high levels of charisma, from natural charm and sociability, or by carefully building networks of friends and attending to each of them.

Power in organizations can also be found in the informal networks of influence. In some instances, people on the staff are related (or were related) to each other. In other instances, they have shared decades of connection through schooling, neighborhoods, or churches. In still others, outside activities such as softball may unite them. Determining these informal influence networks takes time and reflection. You might try arranging everyone in order according to the influence they seem to have on decisions or on daily operations. Russo suggests that you interview people in the organization and ask them to whom they give advice, to whom they give orders, and from whom they accept advice and orders (1993). This may be impossible in practice, and not necessarily well received by the staff, but if you could do it, you could then construct a web of connections that would tell you a lot about how power and influence flow in the organization.

ORGANIZATIONS AS CULTURES

A final way to look at your placement site is as a culture that, like any culture, is understood at least in part by its symbols and rituals (Bolman & Deal, 2003). When you look at behavior in organizations this way, you are looking not so much at the function of certain practices but at their *meaning*. Commencement exercises are really not very functional if you think about it. Diplomas could be mailed to graduates, and it would be a lot easier and cheaper. But the ceremony means a great deal to the graduates and to the faculty, staff, and parents. It is a celebration of the values of the information—a way to say, "Look, here is what we do well." It is an important annual reminder, in the face of the day-to-day frustrations and aggravations, of why faculty members, administrators, counselors, coaches, and others in schools do what they do. Ceremonies marking thirty days—or five years—of sobriety in Alcoholics Anonymous are another example.

Less dramatic events, such as staff meetings, can be looked at in this way as well. They may not be the most efficient way of communicating critical information or even the most productive forum to discuss difficult issues, but they are a way for people in the organization to see each other visibly interacting and inhabiting the same space.

Like other organizations, your placement site may have smaller, less formal rituals as well. Going for pizza on Thursday, or stopping for coffee on the way back from an off-site meeting, or gathering in the lounge during a break may seem silly to you, but ignore them or decline an invitation to participate and you are at your peril. These rituals symbolize collegiality, concern, and cohesion.

For Helping and Service Professionals: The Importance of Language

Language is another aspect to the symbolic or cultural frame and another key to operational values. For example, consider the way staff members talk about clients and the labels they apply to them. Sometimes, staff members will refer to their clients in openly derogatory terms, like "retard," "dummy," or "geezer." Other times, the labeling is more subtle. For example, sometimes certain clients are referred to as "difficult" or "hard cases." When we hear this, we are always tempted to ask: difficult for whom? This comment is made using the experience of the staff as a base for description, not the experience of the client. The same client could be described as "having a hard time accepting authority" or "unwilling to accept personal responsibility." These phrases are much more descriptive of the *client's* experience. Of course, people are going to become frustrated and say negative things from time to time, although preferably not to the client. A pattern, however, of one sort of comment or label can be a clue to an unconscious organizational value.

Organizations also use rituals or symbols to make their values and mission visible, both to the public and to the employees, so that they do not become irrelevant to daily work. Posters, letterhead, pens, even e-mail signatures can reflect the mission and values.

The internal workings of any organization are a complex and fascinating thing. It will take you some time and energy to learn to look at organizations in the ways discussed here. What we hope for is that you learn to do it better than you do now and that you come to see the value of taking time for these analyses. Still, the picture is not complete. Just as it is impossible to understand a system without understanding all its component parts and relationships, it is impossible to understand that system without understanding its relationship to other systems and to larger societal forces and pressures.

THE EXTERNAL ENVIRONMENT

You may already be beginning to see how the world outside the agency has an impact on its functioning. The number and nature of the organization's relationships with other organizations and agencies, sometimes called the *task environment* (Queralt, 1996), will affect your work and the ability of your placement site to do its job. In a more indirect but no less important way, the surrounding community, the economic climate, and some political issues affect you, your clients, and your agency (Queralt, 1996).

The Task Environment

Most organizations cannot function without smooth relationships with other organizations. This is certainly true in human services. For example, the Beacon Youth Shelter works with two state agencies that refer and pay for residents. A local food bank helps supply their food. A large clinic in the area contracts with Beacon for certain psychological services. Some of the students attend school off-grounds. Occasionally, a resident runs away or causes some problem in the community; good relationships with neighbors and the local police are essential. In addition, residents do not stay at Beacon forever; sooner or later, a new placement must be found for those who cannot return to their families or live on their own. This aftercare placement process involves detailed knowledge of and good relationships with a variety of other placement sites.

Of course, other organizations can be sources of competition as well as collaboration. That is particularly true in the world of for-profit organizations, but it is not limited to them. If you are interning with a nonprofit, other organizations in your area that provide similar services may compete for grants, state contracts, corporate sponsors, and so on.

Furthermore, many organizations (and most human service agencies) have one or more agencies that monitor them in some way. Some agencies are part of a larger parent organization. That is not the case with Beacon Youth Shelter, but other shelters in the city are funded by organizations like Catholic Charities or the Salvation Army.

The Sociopolitical Environment

There are a number of political issues that affect placement sites; we will just mention a few of them that seem to be common across many different placements. Local politics can have a considerable impact. For example, in some states, there is a cap on property taxes, which make up a large portion of each town's budget. Towns can vote to exceed the limit, and these votes are often taken to provide a special service, such as a new gymnasium or road repairs. If your placement is heavily funded by the town, these are important issues for your agency.

The attitudes of the people in the town toward your organization, its clients, and the work you do are important. Some social service agencies, like group homes, seem to inspire at least apprehension, if not fear, among neighbors. People agree on the need for them but want them located somewhere else. The same is often true of certain kinds of manufacturing plants, such as biodiesel, or large installations, such as cell phone towers. If your organization has just opened up or launched a venture in a hostile or unwelcoming community, it has a large public relations task ahead of it. Some agencies make deliberate efforts to involve people in the community in their work. This is especially important when the clients are residents of that community, as they are in a Planned Parenthood center, for example. If these residents are suspicious of and hostile toward the agency, chances are that fewer people will come for services.

State and federal politics can have an impact as well. A new governor or mayor often means new people in charge of government agencies and sometimes means a change in philosophy, along with new regulations and changes in the way funds are distributed. Federal issues such as welfare reform, abortion rights, or changes in the tax code may also have a large impact on the mission, goals, and tactics of your placement site.

THE ORGANIZATION AND YOUR CIVIC DEVELOPMENT

In Chapter 1, we made the case that an internship can be a vehicle for civic development. In Chapter 5, when we discussed the learning contract, we invited you to develop learning goals and activities that were aimed at this aspect of your development.

Using the lenses of organization and systems theories we have presented here, you can further explore the civic dimension of your internship by examining how your placement site recognizes and fulfills its civic mission.

You may recall that in Chapter 1, we pointed out that even private organizations can embrace a commitment to the identification and solution of societal problems. Does your placement site make that commitment? Consider this statement from an organization we discussed briefly earlier in the chapter, Hartford Steam Boiler (2007):

Corporate Citizenship

The Hartford Steam Boiler Inspection and Insurance Company is committed to making a contribution to the quality of life in the community. We support the community by:
- Direct corporate donations;
- Our employees' participation in fundraisers and volunteer opportunities;
- Matching our employees' charitable gifts.

Our corporate-giving program follows some general guidelines:
- Our corporate contributions program focuses primarily on the Greater Hartford community, the home of our headquarters for over 135 years.
- We support organizations that are focused on several categories—education, civic, housing, health and human services, the arts, and culture.

Corporations also execute their civic mission by inviting community leaders to serve on their boards of directors or on subcommittees of their boards. They may also invite community leaders in for more informal conversations about the activities of the company.

For Helping and Service Professionals: The Civic Dimension

If you are in a social service setting, you may think that the civic dimension of your agency's work is obvious, but we invite you to think about it more deeply. Remember that part of the concept of a civic professional is one who employs the art of Practical Reasoning (Sullivan, 2005), which goes far beyond the acquisition and application of skills. Does your agency promote this sort of development in its staff?

Another question to ask is how well your agency recognizes the community context of the work it does. We will be discussing the community context in depth in the next chapter, but the question here is whether your agency chooses to be aware of and work with the constellation of assets and challenges in the surrounding community, or whether it provides services to individuals and families in ignorance of these assets and challenges.

Finally, just as in a corporate setting, you can ask whether your agency actively seeks input and involvement from the community in determining its priorities and assessing its effectiveness. And again, this can be done by inviting community leaders or community members to sit on committees or participate in agency evaluations.

SUMMARY

In this chapter, we have tried to introduce you to some ways of thinking about organizations. Although you may feel somewhat overwhelmed by the information, we have really only scratched the surface. Systems theory, organizational dynamics, organizational development, values, and staff development are enormous topics. In our experience, many interns find these topics fascinating. Others do not find them so interesting. However, we believe that the concepts we have introduced here, at a minimum, will help you understand both your placement site and what is happening to you there. At the very least, this chapter should stimulate some thought now and be a resource if you begin to have problems at the placement.

For Contemplation

CHECKING IN
How has your work so far had an impact on your personal development? Professional development? Civic development? Comment on changes that have occurred since the last time you reflected on this part of your internship experience.

PERSONAL REFLECTION
As you can see, there are many aspects to an organization. The questions below are broken into categories. Some of them will require that you ask questions and do some research. Choose the questions within each category that seem most meaningful to you.

BACKGROUND INFORMATION
1. How many years has your agency existed? Have there been any major changes during that time? Describe them.
2. What is the agency's overall goal and philosophy concerning the clients? Working with clients? Managing employees and interns?
3. What is your agency's annual budget? What categories is it broken into?
4. Where does the money come from? Do clients have to pay? If not, does some other agency pay? If the services are "free," where does the agency get the money for payroll and operating expenses?

STRUCTURE
5. Obtain or create an organizational chart of the agency. Are interns on the chart? If not, place them in an appropriate location.

6. Where are the formal rules of the agency located? Have you looked at this resource?

7. How are decisions supposed to be made at your site? What evidence do you have that decisions are or are not actually made that way?

HUMAN RESOURCES

8. Have you observed any informal rules? Do any of them conflict with the formal rules?

9. Have you observed instances of people performing tasks and functions that are not in their job description? What do you know about how this happened? How about informal roles such as "office clown," which aren't in anyone's job description? Do there seem to be informal or unwritten expectations of you as an intern? How do you feel about those expectations?

10. Are you aware of any cliques or informal subgroups that exist at your agency?

POLITICS

11. What are the various subgroups within your agency that may be vying for power or resources?

12. Which of the sources of power identified in this chapter can you see being used in your agency?

ORGANIZATIONAL CULTURE

13. What are some important symbols and rituals at your agency?

THE EXTERNAL ENVIRONMENT

14. What are some of the agencies with which your site has important relationships? What do they do?

15. Are there any agencies that monitor or control your agency? Are there periodic site visits?

16. What have you been able to learn about the relationship between your agency and the surrounding community?

17. What local, state, or national political issues are affecting your agency and its clients?

THE CIVIC DIMENSION

18. To what extent and in what ways does the organization that you intern for fulfill its civic mission? Does it foster a culture of civic professionalism? Can you see additional ways in which this might happen?

SEMINAR SPRINGBOARDS

Analyzing organizations is not easy, and some of the concepts in this chapter may be abstract or difficult to understand. If so, one thing you could do as a class is to analyze the school you go to using these various frames. It will be fun, and it will help you learn.

For Further Exploration

Berger, R. L., & Federico, R. C. (1985). *Human behavior: A perspective for the helping professions* (2nd ed.). New York: Longman.

A thorough discussion of systems theory, including many concepts not discussed in this chapter.

Egan, G. (1984). People in systems: A comprehensive model for psychosocial education and training. In D. Larson (Ed.), *Teaching psychological skills: Models for giving psychology away*. Belmont, CA: Brooks/Cole (pp. 21–43).

Egan, G., & Cowan, M. A. (1979). *People in systems: A comprehensive model for psychosocial education and training*. Belmont, CA: Brooks/Cole.

A wonderful book that integrates a systems and individual development approach to human services. Excellent chapters on systems. Unfortunately, the book is out of print. If your library does not have a copy, try the preceding Egan reference.

Gordon, G. R., & McBride, R. B., & Hage, H. H. (2005). *Criminal justice internships: Theory into practice* (5th ed.). Cincinnati, OH: Anderson Publishing.

An excellent chapter on organizations from a criminal justice perspective. Covers both public and private settings.

Queralt, M. (1996). *The social environment and human behavior: A diversity perspective*. Needham Heights, MA: Allyn & Bacon.

A social work text with an excellent chapter on organizations.

Russo, J. R. (1993). *Serving and surviving as a human service worker* (2nd ed.). Prospect Heights, IL: Waveland Press.

Interesting chapter on organizations, with examples from criminal justice and mental health settings.

Senge, P. M. (1990). *The fifth discipline: The art and practice of the learning organization*. New York: Doubleday/Currency.

Very interesting and accessible discussion of systems theory and organizations. The applications are business oriented, but the concepts are very clearly explained.

References

Berger, R. L., McBreen, J. T., & Rifkin, M. T. (1996). *Human behavior: A perspective for the helping professions* (4th ed.). Boston: Allyn & Bacon.

Birkenmaier, J., & Berg-Weger, M. (2007). *The practical companion for social work: Integrating class and field work* (2nd ed.). Boston: Allyn & Bacon.

Bolman, L. G., & Deal, T. E. (2003). *Reframing organizations: Artistry, choice and leadership* (3rd ed.). San Francisco: Jossey-Bass.

Caine, B. (1994). What can I learn from doing gruntwork? *NSEE Quarterly* (Winter), 6–7, 22–23.

Egan, G., & Cowan, M. A. (1979). *People in systems: A comprehensive model for psychosocial education and training*. Belmont, CA: Brooks/Cole.

Gordon, G. R., McBride, B. R., & Hage, H. H. (2005). *Criminal justice internships: Theory into practice* (5th ed.). Cincinnati, OH: Anderson Publishing.

Hartford Steam Boiler. (2007). From www.hsb.com. 11-17-07

Homan, M. S. (2007). *Promoting community change: Making it happen in the real world* (4th ed.). Belmont, CA: Wadsworth.

Interval House (2007). *Proudly facing the future: Strategic plan 2007–2010.* Hartford, CT: Interval House.

Kiser, P. M. (2008). *Getting the most from your human service internship: Learning from experience* (2nd ed.). Belmont, CA: Brooks/Cole.

National Public Radio. (2007). From www.npr.org. 11-17-07

Queralt, M. (1996). *The social environment and human behavior: A diversity perspective.* Needham Heights, MA: Allyn & Bacon.

Russo, J. R. (1993). *Serving and surviving as a human service worker* (2nd ed.). Prospect Heights, IL: Waveland Press.

Sullivan, W. M. (2005). *Work and integrity: The crisis and promise of professionalism in America* (2nd ed.). San Francisco: Jossey-Bass.

CHAPTER *8*

Getting to Know the Community

I think that every community in someone's life plays a role in who they are and how they interact with others. Because there are so many different communities in my agency, it is hard to try and understand every community that every client belongs to or is from. Trying to understand is helping me learn how diverse the agency really is.

STUDENT REFLECTION

INTRODUCTION

In the last chapter, we encouraged you to widen the circle of your thinking to stretch beyond your colleagues and clients and consider the placement site itself. In this chapter, we encourage you to widen the circle of your thinking even further and consider the community context of your work. In our experience, some interns are naturally drawn to the community context, others are led to it by their experience with clients and their struggles, and still others never really feel drawn to learning about the community. However, all interns and all internships are affected by the community and its dynamics.

In this chapter, we will spend some time helping you understand why the community context is important. We will also spend a little time on definitions and categories of communities. Finally, we will help you consider some of the important aspects of any community and how you might go about learning this information in your particular situation.

THE COMMUNITY CONTEXT
AND THE CIVIC PROFESSIONAL

Regardless of where you are interning and what profession you hope to enter, learning about the community is an important part of your personal, professional, and civic development. We have commented several times in this book that every profession has a social obligation and its work has public relevance. Sometimes that social obligation is enacted nationally, or even globally, but it is also enacted locally (as we mentioned in the last chapter) in its relationship with the community where it is located. Effective local enactment is not possible without a deep knowledge of the community itself.

By now you are becoming more familiar with the concept of a Civic Professional (Sullivan, 2005). Part of being a civic professional is understanding the community context of your work. The art of Practical Reasoning calls for you to move back and forth between the theoretical and human contexts of your profession, and the community forms an important part of the human context.

For Helping and Service Professionals:
But I Don't Want to Work in the Community!

For some of you in social service settings, the community *is* your client. If your agency helps to organize community discussions of neighborhood or municipal issues or if you are helping a group of people mobilize for political advocacy, you are, of course, working with some individuals and groups, but the *main* target of your work is the community itself. If that describes your work, you probably don't need much convincing about studying the community; for you, this is as fundamental to your work as thinking about individual clients is to other interns.

However, even if your work does not focus on the community directly, it is important to take some time to think about the community. It is all part of learning to think in systemic terms—something you have probably studied in your program and something that we encouraged you do to in considering your family and the agency. It is a hard lens to acquire for many people, and it takes practice and persistence. But once you do acquire it, it is equally hard to stop looking through it! Once you learn to see the world in this way, things will never look the same.

For one thing, your clients live their lives and struggle with their issues in a larger context. Communities shape the context of people's lives in many ways. You should not ignore some of the larger reasons why your clients have the problems that they do (Homan, 1999; Rogers, Collins, Barlow, & Grinnell, 2000) nor larger arenas in which assistance or solutions may lie. You are probably accustomed to thinking at least a little bit about your clients' families, even if they are not the main focus of your work, but the context goes beyond that. One of our students, for example, worked with adults who were developmentally disabled. After a while, she became quite frustrated at how few of them lived independently and had jobs, believing that they would feel much better about themselves and probably function at a higher level that way. She learned that attempts to build group homes or get employers to accept adults with developmental disabilities were often delayed or scuttled by the perception, whether based in reality or not, of the attitudes of the community at large. She also learned that her agency had done very little to try to educate or influence that community. Some of her clients were frustrated too, and she wondered whether her energy should be spent trying to help them feel better about and adjusting to something she believed was unfair.

For another thing, your agency does not exist in a vacuum any more than your clients (Garthwait, 2005). Many agencies grew directly out of community needs. The range of agencies and programs that have expanded to meet the needs of persons with HIV/AIDS is a perfect example. Regardless of their origins, though, all agencies are affected by the rhythms of the community. For example, even though a homeless shelter that admits clients who are actively using drugs and alcohol meets needs in the community, there are some people in the community and in the neighborhood who are adamantly opposed to it. Some of them are philosophically opposed to the approach that these agencies use; more are concerned with the safety and general appearance of the surrounding neighborhood. People concerned about discouraging tourists worry about the effect of a homeless population. When factions like these are able to gain power and influence, or when a high-profile crime or death occurs, efforts are renewed to close these shelters. In other communities, efforts are made to close programs for the homeless and conduct sweeps of people living on the streets in hopes that the homeless will simply go away.

Finally, communities are powerful. They can be powerful allies or powerful opponents. One of us once worked with a local school to create a series of parent workshops. The topics were chosen based on input from school counselors and social workers, who in turn were listening to the parents. Transportation was arranged and child care was provided. Flyers were sent home to parents. Almost no one came. What we had not done was to find the people who were influential in the parent network in the community and enlist their help or, better yet, their input from the beginning. They did not oppose our work, but they did not support it, either. We later learned that our university, like many others, had a long history of offering programs to the community and a long history of not delivering what was promised or pursuing its own as opposed to the community's agenda. The parents did not know us, and their general instinct was not to trust us.

Communities are important factors in your work for all these, and many other, reasons. So, let us help you identify a community to study and suggest some of the things you may want to find out.

In Chapter 1, we also discussed the internship as an opportunity for civic development, for you to acquire and strengthen attitudes, skills, and knowledge that will make you better prepared to contribute to a healthy democracy. We revisited this concept in Chapter 5 when we encouraged you to set civic learning goals for yourself. Wherever you end up living, you will be part of a community. Being an active citizen of that community means understanding it as well as you can. For some students, the dynamics of communities become clearer when they leave their home communities and go somewhere very different, as when a student from a suburban background does a service-learning project or internship in a poor urban area. However, even though the elements of community that those students see may look different from those they are used to, they are really the same elements just manifested differently. Fred traveled to New Orleans early in 2007 and visited the Ninth Ward, one of the areas most devastated by Hurricane Katrina and one of the slowest to recover. In a speech at the conference he was attending, a speaker reminded the attendees that while there may not have been a hurricane to point it out and draw national attention, there are neighborhoods like the Ninth Ward in every city. Where, she asked, is your Ninth Ward (Hughes, 2007)?

WHAT IS A COMMUNITY?

The first step in the learning process is to identify the community, and this is sometimes not easy. For one thing, we use the word *community* to refer to different kinds of groups (Garthwait, 2005; Homan, 2008). As common a term as it is, it is also most elusive because of its inherent variability (Kempers, 2002). Sometimes, we are referring to a geographical area—a city, a town, a neighborhood, or even a block. Other times, we are referring to an age group, as when we refer to the community of senior citizens; or an interest group, such as local environmentalists; or even a cultural group, such as the Caribbean community. Members of these groups may live in different parts of the geographical community. To complicate matters even further, most communities of every sort can be further divided into smaller communities.

For another thing, the community where your organization or agency is located may be a different community from the one in which its clients or employees live. In that case, both communities are probably important to know about. In some social service settings, the clients who come to an agency come from a wide range of geographical locations and are members of many different communities of interest, age, or culture as well. So, if you work in a clinic that provides services to persons with HIV/AIDS, the community where the clinic is located, the communities where your clients live, and the community of people with HIV/AIDS are all communities that have an impact on your work and your clients.

We say all this not to confuse you or to undertake an abstract theoretical exercise. The point is to get you to think about all the communities there are for you to consider and to acknowledge the difficulty in choosing one or two. Chances are, though, that one or two is all you will have time for. We encourage you to consult your instructor and site supervisor in identifying the community or communities that are most important in helping you and your organization achieve your goals.

A COMMUNITY INVENTORY

Communities are complex and fascinating, and it takes time, experience, and some expertise to get to know them. We encourage you to invest some time in some kind of community inventory and to make it part of your learning goals. We cannot begin to cover all the important aspects of communities or working in them in this short chapter, but we will suggest some aspects of communities for you to consider and explore. We chose facets of the community that seem especially relevant to you as an intern, with ideas drawn from our experience and from the work of Mark Homan, author of two seminal texts on community development (1999, 2008), as well as Garthwait (2005) and Rogers et al. (2000). We begin with basic information that you need to know, leading off with the concepts of assets and needs and moving on to ideas about analyzing communities. Just as we did in Chapter 7, we have organized this discussion into Bolman and Deal's categories of structural considerations, human resources, symbols, and political considerations.

Basic Information About Communities

If you have chosen a geographic community to study, begin by learning its dimensions. How large is it, and what are its recognized boundaries? Boundaries can be a town line, a street, a river, or anything that people generally accept as marking the edge of a community. Next, consider the general appearance of the community. What sort of impression does it give at first glance? What catches your eye? Are there visible local landmarks? How about landmarks you might miss at first but that every resident knows about (like a controversial cell phone tower)?

Next, consider some basic information about the people in the community. How many people live there? Is the population dense or spread out? Is it stable or does it fluctuate? Communities with seasonal workers, summer residents, or colleges are examples of communities where the population changes over time. Finally, what is the average income of the people who live there? How does it compare to state or national averages and to the cost of living in the community?

Assets and Needs

When considering any community, it is easy to think in terms of needs. And to some extent, that makes sense; all communities do have needs, and it is important to look at how well and in what way the needs are met. It is easy, though, to move from "needs" to "needy"—to look only at what the community does not have or does not do well. That is especially true if your interest in the community is stimulated by the desire to understand problems being experienced by individual clients. Working with youth in a Boys and Girls Club in the inner city, for example, it is easy to look at crime rates, poverty, and poor transportation as factors contributing to the problems your clients have.

When you think of communities this way, you are equating needs with deficits. It is all too easy, then, to miss the strengths or assets of the community. And that is especially easy if the community meets its needs in ways that are unfamiliar to you. Some writers in the field of community development are trying to reframe discussions of communities to focus on assets or capital (Hodgson, 1999; Kretzmann & McKnight, 1997). As we proceed with our discussion of community inventories, we are going to try to use the language of strengths and assets, and we encourage you to do the same.

Structural Considerations

This category of assets concerns the formal and informal structures that allow people in the community to meet their needs and how well those structures work. We begin with basic human needs, such as food, shelter, clothing, and medical care, and the issue here is *access* to these assets across a variety of incomes and circumstances. What provisions has the community made to assure that these needs are met? How successful are they? For example, how does the community handle those who are temporarily or permanently displaced from their homes? Some people will have insurance for such circumstances, but some will not. What then? Are there shelters? Perhaps the community relies on networks of family and friends to handle situations like this. Does that work? Another example is grocery and other retail stores. How easy are they to find and get to?

Another set of assets has to do with education. What is the quality of the schools in the community? What about adult education and training opportunities? Is accessible and affordable postsecondary education available? And don't forget about the informal learning, mentoring, and educational networks that exist in many communities.

Employment is another basic need, and opportunities for employment are a community asset. Some communities have a variety of employment opportunities available, while others have almost none. Sometimes, people in the community have to travel a long way to get work. Also, sometimes, the jobs that are available require special skills, but no training is available. As you learn about these features of the community you are studying, the unemployment statistics (which you should also determine) will have more meaning.

Communication is also a basic need; people need to know what is going on in their community and in the world around them. Most communities have both formal and informal means of communicating necessary information, and these are powerful community assets. Identify the major newspapers and TV and radio stations available in the community. Do those media outlets provide adequate coverage of news and events in the community you are studying? Large metropolitan newspapers often overlook issues in outlying areas or in certain segments of the city. Some communities have their own means of communication; some have newspapers or newsletters of their own, and many have gathering places where people go to catch up on what is going on. Finding these local resources is crucial.

There are also basic systems in communities that meet needs or provide access to the services described above. Adequate public transportation to, from, and within the community makes an enormous difference in people's abilities to meet their needs, as do the quality of roads, sidewalks, crosswalks, and walk lights. In some communities, clean air and water are taken for granted. Others struggle with pollution or even with a lack of potable water. Similarly, you probably don't think very much or very often about waste removal and drainage systems, unless you have been in places where sewers back up, garbage overflows, and water sits in stagnant pools. Not only do these problems detract from the general attractiveness and desirability of the community, but they each pose serious health risks.

Human Resources

People, of course, are an enormous asset, and the idea here is to get to know something about the people in the community, their social and emotional needs, and their strengths as well as their liabilities. In some communities, it is easy to see that people feel safe and secure. They move about freely and easily; they congregate in a variety of places. In other communities, people are afraid to go outside or to gather in certain places. And, of course, there is lots of room between these two extremes. Pride is another social and emotional issue. Find out what people who live in the community are proud of about where they live or about the members of their community.

In the chapter on organizational issues, we mentioned operational values and told you that you can learn a lot about an organization by how it spends its time and money. The same is true for people in the community you are studying. When people have free time, where do they go and what do they do? There will, of course, be a range of answers to that question in any community, but you can probably discern some patterns. Some communities, for example, have a strong recreational focus; people gather in large and small groups to play sports, play cards, bowl, and so on. In others, there is a large number of volunteers and a large number of volunteer opportunities. In still others, though, people tend to come home from work, lock their doors, and spend time only with themselves and their immediate family.

As you learn about the people in the community, remember to look for their strengths. In human services, since the focus is often on problems, it is easy to look at people and communities in terms of what they do not have and what they do not offer. This can also be true of corporations as they consider the relevant issues in a community. And it is an important part of the picture. Communities also have strengths, however, and those strengths are often found in the skills, talents, and values of the people who live there.

Community Symbols

Communities, just like organizations, can also be looked at as cultures. Among the assets of any culture are its symbols and traditions. Earlier, we mentioned landmarks. Some community landmarks have symbolic importance beyond what is immediately obvious. A church, for example, can be a center of social life in a community. As such, it is a symbol of unity in the community. Communities also often have special celebrations, like town fairs, carnivals, dances, or parades. They also sometimes have their own traditions around holidays.

Political Considerations

The issue here is who has power and how decisions are made, and there are both formal and informal aspects to these dynamics. Find out the status and strength of the major political parties in the community. In some communities, there are parties in addition to Republicans and Democrats that have a strong following. Another part of the formal structure in a community is its local decision-making structure. There are many of these, ranging from small towns that are governed by small committees and a town meeting to large municipal governance structures.

There is more to the political life of a community than decision making, and many more sources of political power and influence than the formal mechanisms. At issue is the control of vital resources, such as money, goods and services, and information. In some cases, the institutions that control money, energy, natural resources, and information are located outside the community and operate without any input from or regard for community interests. And don't forget the informal means of access; there are both legal and illegal avenues to loans and credit, for example, and to goods and services.

Sometimes, power is conferred formally, by election or appointment or by occupying a position of authority in a major institution. Other sources of power in the community are less obvious from the outside. People in communities derive power and influence from their families, for example. In some communities, there are families who have lived there and been powerful for generations. Other people derive power from their connections, their personality, or both.

The following chart shows all the dimensions of a community that we have discussed and should serve as a summary as well as a guide for your investigation:

Elements of a Community Inventory

Basic Characteristics	*Structural Issues*	*Human Resources*
Geographic	**Basic Needs**	**People**
Landmarks	*Food*	*Sense of safety*
General appearance	*Shelter*	*Sense of belonging*
Size & boundaries	*Clothing*	*Sources of pride*
	Medical care	*Core values*
People		*Skills*
Number	**Employment**	*Talents*
Size	*Training Needs*	*Other strengths*
Density	*Opportunities*	
Stability	*Location*	*Political Considerations*
Income	*Unemployment %*	**Formal Power**
		Major political parties
Cultural Issues	**Education**	*Other parties*
Symbols	*Schools*	*Governance structure*
Landmarks	*Adult education & training*	
Gathering places		**Formal Control**
	Communication	*Money*
Rituals	*Media*	*Energy*
Fairs	*Informal communication*	*Natural resources*
Celebrations		*Goods & services*
Parades	**Transportation**	*Information*
	Air & Water	**Informal Sources of Power & Influence**
	Waste & Drainage	

HOW DO I FIND ALL THIS OUT?

First of all, you are probably *not* going to find out all this information, depending on the length of your internship and the importance you, your instructor, and your supervisor assign to these issues. However, everything you learn about a particular community and about how to do this sort of research will be valuable, and there are several places to look. Some of this information you can find by yourself on the Internet or by visiting local or college libraries; most libraries have reference librarians who can help you as well. Homan (2008) suggests several resources that you might find in the community and a library, including:

- Census data
- City directories
- Newspaper files
- Local magazines
- State yearbooks
- Town reports
- Political directories

Another source of information is needs assessments of the community, which are undertaken by charitable organizations, such as the United Way, or by local hospitals, colleges, or universities (Homan, 2008; Kiser, 2008).

For other sources of information, you will have to go out into the community itself. Local businesses, or large businesses that operate in your community, are a good resource for information. So are chambers of commerce, local human service agencies, and other groups that either service or advocate for your community (Homan, 2008). If you can find someone at your agency—or anywhere else—who can connect you with some of the informal networks we have discussed, you will learn things about your community that you cannot learn anywhere else. Be prepared, though, to learn that no one in your placement site has these connections. Homan notes that too often agencies and other organizations sit in neighborhoods, but not *with* the neighborhood, and conduct their business in isolation. In that case, you begin with whomever you can find—clients, merchants, faith community staff—and ask them who else you should be talking to in order to get a better feel for the community.

SUMMARY

We hope that we have aroused your interest and curiosity about the community or communities that form the context for your work. Even if we have not, we hope you will go out and look for some of the information discussed here and see if you can connect it to the lives and struggles of your clients.

For Contemplation

CHECKING IN
How has your recent work affected your personal development? Professional development? Civic development? What kinds of changes have occurred since you last thought about them?

PERSONAL REFLECTION:
SELECT THOSE INQUIRIES THAT ARE MOST MEANINGFUL TO YOU.

1. Think about the clients you work with or the agency that you work for. What sorts of communities have an impact on them? Include communities of place as well as communities of interest, age, or culture.

2. Which of these communities do you think is most important for you to learn about and why?

3. Conduct a community assessment. First, plan with your instructor the scope of the assessment and the amount of time you have to devote to it. Next, choose from the aspects of communities listed in this chapter a set that fits the scope of your assessment. Keep track of *where* you find information, not just of what you find.

SEMINAR SPRINGBOARDS
If your classmates are all working in the same community, or if several of you are, you may want to consider taking a quick tour of it, by car or public transportation. Plan ahead for what you will be observing, and divide up responsibilities. One person might keep track of the number and nature of small businesses, another of churches, another of bars, and so on.[1]

For Further Exploration

Homan, M. S. (1999). *Rules of the game: Lessons from the field of community change.* Belmont, CA: Wadsworth.

Packed with practical suggestions and accumulated wisdom.

Homan, M. S. (2008). *Promoting community change: Making it happen in the real world* (4th ed.). Belmont, CA: Wadsworth.

Clear, cogent, and thorough introduction to principles of community development.

Kretzmann, J. P., & McKnight, J. L. (1997). *Building communities from the inside out* (2nd ed.). Evanston, IL: ACTA Publications.

Good resource for adopting an assets-based approach to communities.

---

[1] Thanks to Mark Homan at Pima Community College for suggesting this exercise.

References

Garthwait, C. L. (2005). *The social work practicum: A guide and workbook for students* (3rd ed.). Needham Heights, MA: Allyn & Bacon.

Hodgson, P. (1999). Making internships well worth the work. *Techniques*, September, 38–39.

Homan, M. S. (1999). *Rules of the game: Lessons from the field of community change.* Belmont, CA: Brooks/Cole.

Homan, M. S. (2008). *Promoting community change: Making it happen in the real world* (4th ed.). Belmont, CA: Wadsworth.

Hughes, M. (2007). *Fulfilling the promise of a just democracy: New Orleans after Katrina.* Paper presented at the Association of American Colleges and Universities.

Kempers, M. (2002). *Community matters.* Chicago: Burnhams, Inc.

Kiser, P. M. (2008). *Getting the most from your human service internship: Learning from experience* (2nd ed.). Belmont, CA: Brooks/Cole.

Kretzmann, J. P., & McKnight, J. L. (1997). *Building communities from the inside out.* Evanston, IL: ACTA Publications.

Rogers, G., Collins, D., Barlow, C. A., & Grinnell, R. M. (2000). *Guide to the social work practicum.* Itasca, IL: F. E. Peacock.

Sullivan, W. M. (2005). *Work and integrity: The crisis and promise of professionalism in America* (2nd ed.). San Francisco: Jossey-Bass.

CHAPTER 9

Getting to Know the Clients:
A Chapter of Special Relevance for
Helping and Service Professionals

*The interpersonal work is the critical aspect of any internship—
graduate or undergraduate. It is also the most difficult
for most people.*

STUDENT REFLECTION

I f you have chosen an internship providing direct service, it is natural that many of
your thoughts and much of your excitement and anxiety are directed toward the
population you are working with, who may be referred to as customers, consumers, patients, students, or clients, depending on the setting. Whether you are working
in a government agency or a business, a human service agency, some other service
or nonprofit organization, it is easy to become preoccupied with paperwork, policies
and procedures, and office politics. As an intern, you will have all these concerns
to deal with plus the additional concern of being evaluated by at least two supervisors. However, as J. Robert Russo reminds us in his book *Serving and Surviving as a
Human Service Worker* (1993), the people your organization serves are the reason for
your job. We have found that concerns about those who are served generally fall into
two categories: the nature of the population served (which involves your assumptions)
and your relationship with them. An area of emerging concern that is related to both
of these categories is the personal safety of the helping professional.

RECOGNIZING THE TRAPS:
ASSUMPTIONS AND STEREOTYPES

Since I started my internship, I can honestly say that my impressions have changed. When I first started, I did have stereotypes. I was wrong, and I am not ashamed to admit it. When I was exposed to the clients and what their situations were like, I started to change.

I'm embarrassed to say so, but the media taught me all my impressions of the clients.

STUDENT REFLECTIONS

As you begin your work, you no doubt have some expectations and assumptions about the people you will meet. You probably have some knowledge of the population as a group from the placement process and your first couple of days at the site. You know in general terms who the site does and does not serve and some of the needs the organization can and cannot meet. For example, populations can be described as homogeneous or heterogeneous with regard to various characteristics. The more *homogeneous* a population is, the more alike its members are; the more *heterogeneous* the population, the more diversity is found within it. So, for instance, a high school has a population that is relatively homogeneous in age but may be heterogeneous with regard to characteristics such as race, social class, and intellectual acuity. Beyond this factual information, though, it is important to be aware of the image you have of your prospective clients. You are bound to have one, and it is not all based on facts.

Think About It

Checking Out Your Assumptions

Imagine for a moment the people you will be helping or serving. Each one will be somewhat different, but you can describe them within certain parameters. Think about their backgrounds, their reasons for coming to the agency, their personalities, life histories, and typical behaviors.

Think about their race, ethnicity, social class, religion, and sexual orientation. How heterogeneous do you imagine the population to be? In our experience, interns often imagine that the clients will be very similar to one another or they go to the other extreme and see each one as totally different, missing some of the important commonalities among them.

Think about how similar and/or different they are from you. What do you have in common with them? What aspects of their lives and experience are totally unfamiliar? Here, we often find that interns imagine they have almost nothing in common with their clients. That is, of course, not entirely true, but it is easy to feel

continued

Think About It *(continued)*

that way. On the other hand, interns who have struggled with the same problem that brings people to the agency (such as alcoholism or other addictions) may assume that they know just how it is for them. A little thought, and some reflection on classes you have taken, will tell you that is not true either, but it is another tempting assumption.

Think also about the source of that image. Probably very few of you have had extensive and varied experience with the people you will be working with. If you've not worked with this population before, what might be generating those images for you?

In spite, or perhaps because, of their inexperience, many interns make unconscious generalizations; they form stereotypes. The word *stereotype* has some pretty negative connotations, and you may be reluctant to consider that you have some. However, whenever you make a judgment about someone based on little or no information, you engage in stereotyping. And that is a major trap in being effective in the helping and service professions and in being a civic professional.

Focus on SKILLS

Recognizing Your Stereotypes

Try imagining two people. One is a slim, slightly pale man in a tweed jacket wearing wire-rimmed glasses and carrying a beat-up briefcase. The other is a tall, heavy man with a large belly, long hair, and a big beard, dressed in a T-shirt, sunglasses, dirty jeans, black boots, and a black leather vest with a Harley-Davidson insignia on it. In spite of yourself, are you making assumptions about what each of them does for a living? About how educated they are? Their personalities? Which one is more likely to have read existential philosophy? To have been in a barroom brawl? It seems to be a human tendency to generalize, and the fewer people in any group we actually know, the more we are likely to generalize from the few that we do know or have read about.

This example is based on physical appearance, and this is one way people make assumptions. Another way of getting trapped by your assumptions is by becoming too focused on a client's behavior. Some clients served by human service agencies display fairly unusual or even bizarre behavior. It is easy to jump from the behavior you see to all kinds of conclusions, especially if this is your first time working with these behaviors. Until you are more informed about the nature of the behavior, it's best to reserve judgment about what you are actually seeing. However, you may find that in spite of yourself, the reality of what some clients have done, especially those who have committed violent acts, makes it hard to see past their behavior to the unique features of each person. What's important here is that you are aware that you are making assumptions and that you consciously make a commitment to objectivity in spite of the assumptions that live or linger in your mind.

Uncovering the Roots

Your assumptions and stereotypes can come from several places. Perhaps you have met a person with the same needs or heard one speak. Perhaps you have been such a client yourself or know someone who has. The media are another powerful source of stereotypical images and assumptions. There are scholarly books and documentaries on various client groups, but a more powerful source of images is the mainstream media, especially television. Think, for example, about how many mentally challenged adults you have seen depicted on prime-time television. Were any of them leading anything like normal lives? How many of them had committed a crime? Couple the images of this population portrayed in the media with your own lack of direct experience, and you can begin to see where your image and stereotypes may originate. As one student said, *"Given all the media attention to crime, in all the genres, it would be difficult not to form an impression of the criminal population."*

Getting Beyond the Traps

The first step in getting beyond stereotypes and assumptions is to admit you have them. Check with other interns; they have them too. Theirs may not be the same as yours, but they have them. The next step is to gather as much factual information as you can and hold your assumptions up to the light of objectivity. At the beginning of this section, we asked you several questions about your clients or customers and what you were assuming about them. Now may be a good time to go back and try to find factual answers to those questions. Don't be discouraged if the answers don't come easily; you have held on to some of your assumptions for a long time.

Rethinking Client Success

When you think about being successful in your work with clients (whether they are called clients, patients, students, or customers), what does that actually mean? Be as specific as you can. Think about the kinds of goals that make sense for the population you serve. Think about the attitudes, skills, values, and knowledge that you need to help them empower themselves. This may be a good time to talk with your co-workers and site supervisor about this idea of being successful in your role as helper. Their experience will serve you well in understanding your assumptions. In the sections that follow, we consider many aspects of the relationship the intern develops with those she or he serves. For much of the discussion, we use the term *client* without specifying the various names that clients are called in different internship settings. If you are in the helping professions, then all of what follows applies to your work. However, if you are placed in other kinds of service professions, you'll need to read the following sections with your population in mind.

ACCEPTANCE — THE FIRST STEP

I want clients to respect me. It would be nice if they liked me, but it's not necessary. Most important is that they trust me to know that I will do whatever is in my power to help make their lives easier.

STUDENT REFLECTION

As interns think about the kind of relationships they want to have with those they serve, several concerns often arise. One of the most common, especially for those in the helping professions, is how clients will react to them (Baird, 2007). In talking with undergraduate students who were going to perform community service with homeless individuals, Ostrow (1995) reported that one of their major anxieties was how they, as relatively affluent and fortunate young people, would be perceived by the homeless population. You may be wondering, "What kind of reception am I going to get from these people? Will they respect me? Will they listen to me? Or will they just write me off?" The theme in these concerns is acceptance, which is a crucial part of the foundation for any relationship. You have probably read about how important it is for you to accept those you serve, but they need to accept you too. You need to find ways to have them accept you, and that's not always easy.

Being Accepted by Clients

Acceptance can mean different things to different client populations. Some clients show their acceptance simply by being willing to talk with you; until they accept you, they simply ignore you. Other client groups may show their acceptance by including you in their conversations, confiding in you, considering your suggestions, following your directions, or accepting the limits you set. How do you imagine you will know whether your clients have accepted you? Be as concrete and specific as you can.

Think About It

What Does Acceptance Mean to You?
Think for a moment about what the word *acceptance* means. Don't try to think of a definition for it; instead, think what the word *means* to you personally. Ask a peer to do the same, and you quickly realize that you aren't necessarily in agreement. Most people, though, know what it feels like when they are accepted and what it feels like when they are not. Think about how acceptance *feels* to you as well as what it means to you. Being aware of how you think about acceptance allows you to better understand your reactions to acceptance issues.

TUNING IN TO THE CLIENTS' WORLD

I couldn't survive a minute in their world, and they know it. I didn't grow up in the city or in the street culture. I haven't been in a fight since grade school. And I'm going to give these guys advice?

STUDENT REFLECTION

Getting those you serve to accept you can be a real challenge. You will have an easier time with it if you think about the situation from their perspective. The main challenge is to enter, but not become lost in, their world. Schmidt (2002) writes of the power of perception in shaping our day-to-day experience and the importance of entering the perceptual world of your clients—another dimension of the human context of your work that is so important to a civic professional. Entering that world means going beyond facts and figures, reports, testing profiles, and school grades found in client files, even beyond catalogues of experiences that clients may tell you about. It means trying to understand how their experiences have shaped the way clients see the world. In the book *Crossing the Waters*, Daniel Robb writes about his experience working in a residential setting with troubled adolescent males (Robb, 2001). Here, he recounts an insight gained while working with one of the boys to fix a car: "*I saw as we went that Louis had never had a man show him how to do anything. So, he expected to do it wrong, expected to be shamed for it, because unknown territory is not filled with angels*" (p. 173). Here is another example from a student journal: "*Most of these kids build a stone wall as a façade because it is the only way they have learned to survive. Behind that wall is an individual with feelings and emotions.*"

MEETING WITH RESISTANCE

Your clients have probably had many frustrating and stressful experiences. In addition, they may be for the first time in a position where they cannot meet their own needs, where they have to let strangers into their business, and where their welfare is not in their hands (Royse, Dhooper, & Rompf, 2007). They may be expecting a miracle or a quick solution to their problems, and they are not usually going to find one. Or maybe they have just figured that out. If they are clients with a long history of services that did not help them, they may be expecting more of the same (Royse, et al., 2007).

In some cases, clients may be particularly wary of interns. Some respond better to volunteers, reasoning that they are there because they want to be, as opposed to interns, who are fulfilling a requirement for school (Royse, et al., 2007). In addition, interns generally do not stay very long. Some may be there for one academic year, and some may even be hired, but many are gone after one semester. Many clients have had unfortunate experiences with people leaving them or letting them down, and they are reluctant to invest in a relationship that will end soon. One of our students who was working at a group home for adolescents reported that one of the residents was quite hostile and rebuffed all attempts at contact. On further inquiry, the intern learned that this resident had gotten very close emotionally to an intern the prior semester and, although she knew the intern would leave, was heartbroken when it happened.

EXPLORING CLIENTS' CULTURAL PROFILES

Part of their perceptual world, of course, is shaped by your clients' cultural profiles. These profiles must, to a large extent, be understood one client at a time, just as your own profile, explored in Chapter 4, is different from almost anyone else's. Try to find out, then, not just what subgroups your client may belong to but how that has shaped her or his behavior and perceptions. And try to avoid the trap of only looking for the weaknesses or liabilities in a client's cultural background. See the strengths as well. For example, a client whose ethnic background encourages her or him to seek help from family and friends, but never from strangers, may be reluctant to let you in. But remember that this same client has a wonderful resource in family and community—one that can be used to help. Many of you have probably studied this topic already, and this book is not the place for a lengthy discussion of multicultural issues. However, we provide some references for further reading at the end of this chapter.

PUTTING CLIENT BEHAVIOR IN CONTEXT

Another advantage to understanding the perceptual world of your clients is that you can try to understand their behavior in that context (Baird, 2007). This will help you not to be overly flattered by early compliments or expressions of trust. It will also help you not to be overly flattened by clients who turn away from you and your attempts to connect. Remember that clients often have an agenda they are not going to tell you about but one that you can understand once you come to know them. A client who approaches an intern and says "You are the only decent person in this place—I can trust you" is either trying to gain an advantage, or rushing into closeness on very little data, or both. Similarly, a client who gives you a hard time may be doing it precisely because you are new to see whether you will back away as so many others have done.

Focus on THEORY

Transference and Overidentification

In counseling and psychotherapy, the client's behavior just described is referred to as *transference*. Clients transfer to you feelings and reactions that are rooted in their past experiences. Marianne and Gerald Corey (2005) emphasize that transference can take many forms, including wanting you to be the parent, husband, friend, or partner they never had; making you into some sort of superhelper who can fix anything; refusing to accept boundaries that you set; and easily displacing anger or love onto you.

While you are working to enter the world of your clients, take care not to get caught in it. If their experience is similar to yours, it is easy to have old feelings of yours come to the surface and you react to those feelings. It is also easy to become so absorbed in the experience of the clients that you take on the feelings and reactions that they have. This is called *overidentification*. Once that happens, it is very easy to cross the line from wanting your clients to do well to *needing* them to do well. Maintaining distance, even while developing rapport with clients, will let you navigate the overidentification trap. It will also help you in confronting clients when necessary and helping them see how their perspectives may be creating problems for them.

SEEKING COMMON GROUND

Part of developing acceptance with clients is finding some common ground. A frequent concern for interns is their belief that they have nothing in common with their clients. A young woman interning in a shelter for single women and their children said:

> I don't have any kids. I have no idea what it's like to be a single parent, or to grow up with one parent. I have no idea how to survive on the streets or what it's like to be abused by a husband. What do I have to offer these women? Why should they listen to me?

These are real concerns, but remember that even having the exact same experience as your clients is not always a guarantee of success. Common experience can be a help in establishing acceptance, but it can also be a hindrance. You may have had some of the same things happen to you as have happened to your clients, but assuming or saying you know how your clients feel can be alienating for a client whose experience was quite different from yours.

If you don't have much common experience, what can you do? You can't manufacture experience you haven't had, and pretending you know how it is will only make things worse. However, common ground does not always mean common experience, and showing respect for your clients' experiences is one way to gain their acceptance. Furthermore, even if you have not had the particular experience, you may be able to think of something in your background that gives you a hint of what your clients may be feeling. Again, we turn to the words of Daniel Robb (2001), who recalls being cut from a baseball team despite having superior skills:

> I turned and walked away from him, my good, dusty glove tucked under my arm, cursing the whole corrupt system. . . . It was clear to me at that moment that there was no trusting the men in charge. And I felt different. . . . I the kid who wasn't part of the team, the kid without a father. I had been cut. . . . Which all sounds maudlin and sappy until you remember the leverage of those days, the ferocity of your little peers, their cruelties and name callings. . . . The dangers of trust, the betrayals, the sweetness of being accepted and the tangles of rejection. These boys out here on the island, they were given no acceptance in the deeply psychological sense. . . . They're here because they were cut (p. 56).

Learning to Accept Clients

Your ability to accept your clients is just as important to the relationship as their ability to accept you. In fact, it is more important. If you cannot accept your clients, they will know it in some way and on some level, and they will be far less likely to accept or trust you. Really, we are talking about a process of mutual acceptance here, and much of the work you did to get your clients to accept you will help you to accept them as well, especially the work you did to understand their perceptual world.

WORKING WITH TROUBLING BEHAVIORS

As we mentioned earlier, one important part of acceptance is to be able to see clients as more than their behaviors. Listed next are some concerns drawn from our experience and that of interns and beginning helpers (Cherniss, 1980; M. S. Corey and Corey, 2005; Russo, 1993). As you review these behaviors, think about how you might react to clients who:

- lie to you
- manipulate you to get something they want but cannot have
- are never satisfied with what you have to give and always seem to need more
- become verbally abusive and physically threatening
- blame everyone else for their problems
- are sullen and give, at most, one-word answers or responses
- ask again and again for suggestions and then reject every one
- refuse to see their behavior as a problem
- make it clear they don't like you
- refuse to work with you

Can you see these behaviors as understandable, although not helpful or praiseworthy, given what you know about your clients? Can you stay open to clients who do these things and see their behavior as only one part of who they are? If not, take some time to consider why.

UNDERSTANDING YOUR REACTIONS

The most important thing you can do is try to understand your reactions. Remember, just as clients bring their past experiences and emotional tendencies to the relationship, so do you. *Countertransference* is the term counselors and therapists in the helping professions use to describe situations in which your perceptions of and reactions to clients are distorted by your own past experiences and hurts (M.S. Corey & Corey, 2005). Look back to Chapter 4 and reconsider some of the things you discovered or affirmed about yourself. Do any of those things help explain your reactions? For example, think about the reaction patterns you may have. If you have trouble confronting others, then clients who challenge you overtly will be especially difficult for you. Knowing this will help you avoid assigning all the blame to the client. If your psychosocial identity includes a shaky sense of competence, then clients who are not making progress may actually anger you more than they otherwise would.

KNOWING YOUR IDENTITIES

> *Some clients evoked feelings of mistrust and prejudice as well as feelings of sorrow. I am most shameful of the feelings of prejudice.*
>
> STUDENT REFLECTION

Your cultural identities may help explain some of the reactions you are having as well. Consider the preceding quote from a student's journal. This quote shows an intern who is willing to look to herself, as well as to the client, to discover the source of her reaction. The willingness to admit prejudice is especially impressive. In time, this intern may well move past feelings of shame and be able to work on ways to overcome the prejudice and mistrust she lives with. Admitting these feelings, instead of pretending not to have them, is the first step.

DEALING WITH SELF-DISCLOSURE

Another important issue in developing relationships, and a frequent concern for interns, is self-disclosure. How much about yourself should you reveal? There are actually two kinds of self-disclosure at issue: personal information and personal reactions.

Disclosure of Personal Information It is natural for clients to want to get to know you, and they may ask you questions about your life and relationships. Some of these questions will seem quite comfortable and easy to answer. Others may not. You may feel that the questions are too personal or that they concern matters you would rather keep private. You may feel that certain information is inappropriate for certain clients.

A word to the wise . . . These are issues you should think about but not decide on your own. Your agency may have strict policies, and very good reasons for them, about how much you can disclose, and they may not always seem obvious to you. In some placements, interns are told not to have personal photos on display or divulge their last names. Matters of personal safety (which we will discuss later) and client boundaries sometimes dictate limits such as these. Be sure to discuss these issues with your supervisor.

Disclosure of Personal Reactions The second kind of self-disclosure is more immediate. You will no doubt have opinions about and emotional reactions to things your clients say or events at the site and wonder whether it is appropriate to share your thoughts with clients. The question "What should I share with clients?" is not possible to answer and, in fact, is not a helpful question. It fails to consider the wide variety of clients, situations, and goals that form the context of your work. It is better to ask a "three-dimensional" question (Hunt & Sullivan, 1974), such as "What sort of self-disclosure is appropriate with which clients and for what purpose?" If you are working with battered women, for example, it may be appropriate to share some of your relationship struggles as a way of establishing an empathic connection; clients sometimes think their counselors have no problems of their own. However, if you are working with a heterosexual teen of the opposite sex, such disclosure is probably inappropriate, as it can blur an important boundary and create confusion about your intentions. Asking three-dimensional questions will help you, in collaboration with your supervisor, to find answers that work for you and your clients.

Managing Value Differences

In Chapter 4, we stressed the importance of being aware of your values. In their book *Issues and Ethics in the Helping Professions*, G. Corey, Corey, and Callanan (2006) stress the importance of values awareness in relationships with clients.

There will be times when you are dealing with a client whose values are very different from yours, and that can be perplexing and stressful for both of you. Let's suppose that honesty and straightforwardness are strong values for you. Of course, you are dishonest occasionally; you know it is wrong when you do it, but it makes sense to you under the circumstances. Let's also suppose that you are dealing with a client whose values differ from yours. In her experience, being honest, especially with human service workers and others in authority, means being taken advantage of by a service delivery system she perceives as unfair. For example, in some states, a woman on welfare will have her benefits reduced if she is married, even if her husband is not employed. So, she lies to you as a way of solving an anticipated problem.

One of the challenging aspects of situations in which it is obvious that there are differences between you and your client is deciding how to respond. And important to deciding on how to respond is understanding the options that you have. G. Corey et al. (2006) make a distinction between *exposing* your values to clients and *imposing* your values on them. One choice you do have is exposing differences by telling clients about them. Another choice you have is imposing your values by attempting to influence clients to change theirs. And a third choice is to do neither, but to work with your differences.

Think About It

How Will You Respond to Value Differences?

Let's take the example of sexuality and family values—hardly light topics, but certainly ones you will encounter in your work with clients. You certainly may encounter clients whose values are at odds with yours in these areas. Think about how you might respond to each of the following case situations and why you would respond in that way.

- You are working with an adolescent client who is sexually promiscuous and thinks it is just fine. Furthermore, he uses no birth control and says a pregnancy would not be his responsibility.
- You are working with another client who thinks that a marriage must stay together at all costs, and you are in favor of divorce in some cases.
- You are working with a client who thinks that unmarried couples should not have children.

Specific Client Issues

As you negotiate acceptance with your clients, there are a number of issues that seem to arise frequently as you launch your relationships with them. These issues tend to permeate the helping relationship and thus can be formidable companions in the helping relationship. Being aware of them will help you prepare and not be surprised when they appear.

AUTHORITY ISSUES

One particularly challenging issue concerns authority. You have certainly been in a position where you were subject to authority and so have those you serve. Authority is often an important issue in forming a relationship with them. Shulman (1983) has pointed out that clients tend to perceive you as an authority figure. This is especially true if you or your organization has some kind of legal authority over clients, such as the authority a benefits worker has to deny benefits or a probation officer has to surrender a client to the court. However, even when the authority is not explicit, you should be prepared for clients who see and react to you as an authority figure (Shulman, 1983).

Very often, clients are in a "one-down" position. They need something they cannot get for themselves. It could be something tangible, such as food or shelter, or something less concrete, such as control over a substance abuse problem or help in understanding some bureaucratic procedure. You and your agency have what they need; you are holding the cards. Think about when you have been in a similar situation. How did it feel? Your clients will bring their fears about past experiences with authority to their relationship with you, and those factors will shape how they respond.

FINDING EQUALIZERS

One way clients try to reduce their one-down feelings is to assess your background and experience, often with a goal of finding a flaw or an equalizer. We refer to this phenomenon as *credentialing,* and it takes place in nearly every internship. Sometimes, clients will literally ask about your credentials, as when they ask about your education and training or experience with clients. Other times, the credentialing is more subtle, such as, "Do you have children?" or "Have you ever been arrested?" It may be as simple as asking about your age. Sometimes, clients are just trying to get to know you, but you may be surprised at the persistence with which these questions are asked. You may also be surprised by their reaction when they find the information they are looking for, and you may feel dismissed. Knowing that these assessments are a normal part of building a relationship will help you decide how to respond.

LIMIT TESTING

Clients also will often test your limits. They want to see where your personal boundaries are and whether and how you enforce agency rules. Because you are an intern, they may be genuinely unsure of your role and what you can offer them (Shulman, 1983). Some clients have experienced many workers, not to mention other people in authority, and have been treated in many different ways. They need to know what to expect from you, and although they may not like it when you set a limit, when you do so with firmness, compassion, and consistency, it ultimately helps them trust you. Other reasons for testing you may include a need for recognition and attention or an attempt to gain status with their peers (Shulman, 1983).

You may think of this phenomenon as occurring more with children, as they try repeatedly to pick up something they have been told to leave alone or try to poke or bite you. You may also associate testing with people who are in an involuntary situation, such as juveniles in a detention center or runaway shelter, who refuse to do chores or curse in front of you to see what you are going to do. And, in fact, it is a very real issue in some criminal justice internships, where interns are often referred to as "officers in training" precisely so they are not seen as easy targets for limit testing. However, other kinds of clients can test you too. An elderly client can press you to stay longer than you are able, a client at a soup kitchen may try to go through the line twice, a parent may ask you to stay late and watch the children until they can be picked up.

Interns are sometimes tested even more because they are initially seen as "not real staff" (Gordon, McBride, & Hage, 2005). These behaviors can be exasperating, especially if you are not expecting them. Try not to imagine that really effective workers never have these challenges; of course they do. They may have learned to handle them a bit more smoothly and quickly—and you will too someday. Taking these challenges personally only makes it harder to meet them effectively. Remember that you once did this kind of testing, and you may still do it occasionally (Russo, 1993). Think back for a moment on your behavior with substitute teachers or babysitters. Perhaps you have even tested some of your college faculty in this way.

PERSONAL SAFETY OF THE PROFESSIONAL

Safety is definitely a concern, since I work in a courthouse with metal detectors and my office is locked at all times. I know my level of risk is higher, but then again, I sometimes feel safer knowing that there are detectors, guards, locked doors, etc. I guess how safe I feel really depends on how I choose to look at things.

STUDENT REFLECTION

There are a lot of people interested in your personal safety in the internship. Your safety is foremost in the minds of your campus and site supervisors. It is the concern of your co-workers. And it is the concern of your campus administrators. Increasingly, we hear interns themselves express concerns about their personal safety, and we are not alone. Birkenmaier & Berg-Weger (2007, p. 57) cite several studies of real and perceived risk to social work interns and note that such concern is warranted because, among other reasons, (1) social workers are second to police officers in the risk of work-related violence directed at them and (2) the "number and lethality of safety risk incidents on the job has increased for social workers."

Certainly, the potential for risk in the helping professions can be high. Human Services is such a broad field of professions, though, that it is impossible to make general statements about risk. Here, however, are some examples, drawn from our experience, of internships with various levels of risk.

- Steven interns at a men's shelter in the city. It is the only shelter in the city that allows men who are actively abusing substances to be sheltered, and many men come in under the influence. Others congregate outside, seeking services only in a medical emergency. Furthermore, Steven is asked to ride along with an outreach worker, who travels to abandoned buildings to try and help addicts be more safe in their behavior and perhaps to come for treatment.

A Matter of Law

Campus Liability for Internships

Student safety is the issue getting the most publicity when it comes to off-campus programs. Why? Because students are injured or killed while enrolled in such off-site programs as internships and study abroad. If an internship is required of a student for completion of a degree requirement, as it often is in the academic programs of the helping professions, then the college may face liability if the student is injured or killed while involved in the internship.

Case in Point: A 2000 ruling by the Florida Supreme Court has implications for all campuses with such a requirement (Nova Southeastern University Inc. v. Gross, 758 SO. 2d 86 (Fla. 2000)). NSU offers a doctoral program in psychology, and the internship sites for that program are identified by the college. The doctoral student was assaulted on-site and in turn sued the NSU for negligence. The court ruled in the student's favor, noting "the student-institution relationship created a special duty on the part of the university to use ordinary care in assigning students to internship sites. There was evidence that there had been earlier assaults at the internship site and that the university had been aware of the fact but had not warned the student." (Connell, Franke, & Lee, 2001, pp. 2–3).

- Carlos interns for an agency that monitors adjudicated youth in the community. He travels with a staff member all over the city, to schools, playgrounds, and apartments, to check on these adolescents and talk with them.

- Su-Je interns at a Student Assistance Center at a huge, urban high school. The school has students of every race and from dozens of ethnic backgrounds, and intergroup tensions run high. Confrontations and fistfights are not uncommon. The center offers peer mediation, among other services, but sometimes, these sessions erupt.

- Kavita interns at a women's shelter. She helps a staff member facilitate a support group for victims of domestic violence. One night, one of the clients tells the group that she is scared. Her abuser found out she was coming to the group and got very angry. He found her journal, and in it were names of clients and staff members. She snuck out to group that night but is frightened about what he will do when he gets home and finds her gone.

Give some thought to these examples, which we will return to as we continue our discussion of safety issues.

Assessing and Minimizing Levels of Risk

The personal safety chapter couldn't have come at a better time. A social worker in my office was attacked this week by one of her clients. She was hit on the head and has permanent damage to her eyesight. I hadn't thought much about my personal safety until this point.

STUDENT REFLECTION

Your level of risk depends on several variables, some of which are evident in the preceding examples. Although you should be made aware of any and all potential risks to your safety before you start working with clients, safety issues do arise that cannot be anticipated. What is important here is that you are as fully informed as possible about your safety.

ASSESSING CLIENT RISK LEVELS

Before beginning your internship, you should have a good sense of the likelihood of being exposed to violence. If you are interning in a residential setting, a locked facility, or in the field of criminal justice, the likelihood is higher than if you are interning in the maternity wing of a hospital. And you certainly should have been informed of that by the person who placed you in your agency. As we mentioned in Chapter 2, when it comes to clients, three factors should be considered: the client's developmental stage, motivations, and immediate situational factors and conditions (Baird, 2007, p. 172). Those factors include access to weapons, mental status, medication/controlled substances, stress level and precipitators, and history of violence (pp. 172–173). It's important to keep in mind that some clients are habitually violent, whereas other clients may become violent only under certain circumstances, such as the influence of drugs or the lack thereof. And unfortunately, some clients are in terribly frustrating and even desperate circumstances (Garthwait, 2005) and are not necessarily predictable when it comes to their behaviors. For example, if one of the adjudicated youths that Carlos is monitoring has gotten into more trouble and believes that he will be turned in, he may go to great lengths to stop that from happening. Nonclients are sometimes a risk as well. Parents whose children have been removed from the home, partners of victims of domestic violence, and friends of an adjudicated youth are all examples.

MINIMIZING CLIENT RISK LEVELS

As we emphasized in Chapter 2, if you suspect the possibility of violence by clients or their family/friends, it is critical that you consult with others immediately, starting with the campus placement coordinator who arranged your placement, both of your supervisors, and your co-workers as to the history of violence at the field site. You may have ignored this suggestion the first time we made it; if so, we remind you that history remains the most dependable predictor of future behaviors. If there is a history of violence, it is time to meet with your supervisors and develop a safety plan. Because it is a safety factor, it needs to be noted on the learning contract with specific safeguards described.

ASSESSING SITE AND COMMUNITY RISK LEVELS

There are organizational variables as well that can pose safety risks. As we said in Chapter 2, agencies vary considerably in the extent to which they have assessed risk to interns and workers as well as implemented appropriate procedures. For example, the high school where Su-Je interns gave her and all the staff careful instructions about what to do in case of a violent incident, when to respond, and how to get backup when needed. Agencies have the responsibility to be thorough in their assessment of risk factors and levels and respond accordingly with policies and procedures for staying safe. Your responsibility is to determine how well your site meets these criteria for safety.

Location and hours of operation are another set of variables. Some human service agencies—and schools—are located in risky neighborhoods (i.e., at risk for violence or crime), especially for someone who does not live in that community. The neighborhoods that Steven travels to are more dangerous after dark. If your work takes you out into the community, as Carlos's and Steven's does, you may find yourself in such neighborhoods as well.

MINIMIZING SITE AND COMMUNITY RISK LEVELS

If you in any way feel unsafe or otherwise not comfortable traveling to and from your site due to location, hours of operation, or security measures in or near the building, consult with the site supervisor and co-workers immediately to determine how they ensure their personal safety working at the site. Be sure to inquire about the history of safety as it relates to your concern, requesting specific data about crime in the area at night and past crimes committed in the parking area or on the premises of the site. Of course, you can always contact the local police department and request such information if the staff is too new to the site to have it. If your concerns prove to be a safety factor, they need to be noted on the learning contract with specific safeguards described.

Facing the Fears

In order for you to embrace your field experience as the learning experience that it is, you will need to face your fears and assess the level of risk at your internship— even if your agency has not done that or not discussed it with you. It becomes your responsibility by default because it is *your* internship. And we know that may not be easy to do. Many interns are reluctant to discuss their concerns about safety for fear of being seen as overly timid, not ready, or not committed to the field (Birkenmaier & Berg-Weger, 2007).

Fear is a powerful emotion. In this instance, it can keep you from confronting your prejudices, if you have or are aware of them, and from making realistic assessments of your safety. It may well be that some of your fears are unfounded, or exaggerated, because of prejudices and stereotypes. For example, some persons with mental illness or mental retardation exhibit some very odd behaviors, but that does not necessarily make them dangerous (Baird, 2007). A Caucasian intern in a predominantly Hispanic high school is not necessarily at greater risk for violence, but he may *feel* like he is. The only way to deal with these stereotypes is to acknowledge and confront them. If you are not comfortable discussing these issues with your supervisor, you may want to talk with your campus instructor instead or with your classmates. What's important is that you do discuss them in supervision.

The likelihood is that once you have thought through your fears and gathered the factual information you need, you will feel ready to face what challenges there may be. If that is not true, if you feel like you are in over your head, you must discuss that with your campus instructor as soon as possible.

SUMMARY

This section on clients may have raised more questions than answers for you. If so, we accomplished our goal, for that was the point of the chapter. It is very important that you are aware of and prepared for some of the challenges in the early stages of working with clients. You have materials for reflection and for discussion with peers, and you have unanswered questions to explore. The exact shape and pace of your experiences with clients we can not know, but you can.

This chapter ends Section Two of the book. We have encouraged you to get oriented, to explore the basic dynamics of your clients, colleagues, agency, and community. In so doing, you have addressed a number of the concerns and issues of the Anticipation stage. With effort, and a little luck, you feel more settled in your placement and more committed to the work. In subsequent sections we consider the challenges that await you as the rest of the stages of your internship unfold.

For Contemplation

CHECKING IN

How has your work so far had an impact on your personal development? Your professional development? Your civic development? Comment on changes that have occurred since the last time you thought about these issues.

PERSONAL REFLECTION:
SELECT THOSE INQUIRIES THAT ARE MOST MEANINGFUL TO YOU.

1. What did you know before you started the internship about the population you serve, particularly the clients or customers you work with? What informed you about them? Did you have experience with the population before?

2. Now that you have gotten to know those you serve a little bit, how have your initial impressions changed? Or have they remained the same?

3. How would you characterize the population your internship site serves? In what ways are they similar to and different from each other (race, gender, ethnicity, personality, social class, etc.)? In what ways are they similar to and different from you?

4. Have your clients challenged your credentials? If so, how? If not, how do you think they might do so?

5. How easy do you think it will be for you to set limits with those you serve? What issues will you need to set limits on? If the prospect seems difficult, how much of that difficulty is because of what you know about the people you serve and how much is because of what you know about yourself?

6. What aspects of yourself are you willing to share with your patients, students, or clients? What aspects are you unwilling to share and/or seem inappropriate to share with this particular population? How will you respond if you are asked about these areas?

7. What other kinds of challenges are you expecting from those you serve? Include general as well as specific challenges (i.e., those that pertain to one particular client). Why do you think these issues and behaviors will challenge you? Think about yourself. Is there something about you that makes these issues and behaviors especially troublesome?

8. If you have had other internships or field experiences, compare the current group of people you serve and your reactions to them with previous client groups.

SEMINAR SPRINGBOARDS

1. As a class, make a list of the populations that you are working with. Brainstorm a list of the more common stereotypes that society has about each group. Remember that you are not being asked whether you subscribe to these stereotypes, just what they are. Where do you think these stereotypes come from? You may want to revisit this list later in the internship to see how these stereotypes have held up to the light of experience.

2. As a group, put aside some time to talk about the safety issues that each of you faces. Frame your discussion by focusing on risk factors, risk levels, and safeguards that need to be in place. Remember, all risk levels that warrant safeguards need to be brought to the attention of your campus supervisor. An important piece of that discussion is acknowledging the feelings that the risk levels evoke and how you will manage them so you thrive instead of survive each day in placement.

For Further Exploration

Axelson, J. A. (1999). *Counseling and development in a multicultural society* (3rd ed.) Belmont, CA: Brooks/Cole.

A comprehensive and informative text on multicultural issues.

Birkenmaier, J., & Berg-Weger, M. (2007). *The practical companion for social work: Integrating class and field work*. Needham Heights, MA: Allyn & Bacon.

Thorough chapter on personal safety, with helpful exercises and inventories.

Corey, M. S., & Corey, G. (2006). *Becoming a helper* (4th ed.). Belmont, CA: Brooks/Cole.

Helpful sections on client issues.

Corey, G., Corey, M. S., & Callanan, P. (2006). *Issues and ethics in the helping professions* (6th ed.). Belmont, CA: Brooks/Cole.

Excellent section on the ethics of imposing values on clients.

Gordon, G. R., McBride, R. B., & Hage, H. H. (2005). *Criminal justice internships: Theory into practice* (5th ed.). Cincinnati, OH: Anderson Publishing.

Excellent discussions of issues relating to clients. Although this book is aimed at interns in a particular setting, it has many applications outside criminal justice.

Pedersen, P. (Ed.). (1985). *Handbook of cross cultural counseling and therapy.* Westport, CT: Greenwood Press.

In this: Another good resource on multicultural issues. Triad model is helpful.

Pedersen, P. (1988). *A handbook for developing multicultural awareness.* Alexandria, VA: American Association for Counseling and Development.

A very "hands-on" short book with lots of exercises to help you.

Schutz, W. (1967). *Joy.* New York: Grove Press.

In his first stage of group development—inclusion—Schutz talks a great deal about acceptance concerns and the various ways of handling them.

Shulman, L. (1983). *Teaching the helping skills: A field instructor's guide.* Itasca, IL: F. E. Peacock.

Especially helpful regarding authority issues.

Sue, D. W. (1981). *Counseling the culturally different.* New York: John Wiley & Sons.

Classic text by one of the leaders in the field; informative and provocative.

References

Baird, B. N. (2007). *The internship, practicum and field placement handbook: A guide for the helping professions* (5th ed.). Upper Saddle River, NJ: Prentice Hall.

Birkenmaier, J., & Berg-Weger, M. (2007). *The practical companion for social work: Integrating class and field work* (2nd ed.). Needham Heights, MA: Allyn & Bacon.

Connell, M. A., Franke, A. H., & Lee, B. A. (February 2001) Issues related to off-campus programs. *Nobody said this was going to be easy: Legal and managerial challenges for department chairs and other academic administrators.* American Council on Education. Originally prepared for Stetson University Law School conference on law and higher education.

Corey, M. S., & Corey, G. (2005). *Becoming a helper* (6th ed.). Belmont, CA: Brooks/Cole.

Corey, G., Corey, M. S., & Callanan, P. (2006). *Issues and ethics in the helping professions* (7th ed.). Belmont, CA: Brooks/Cole.

Garthwait, C. L. (2005). *The social work practicum: A guide and workbook for students* (3rd ed.). Needham Heights, MA: Allyn & Bacon.

Gordon, G. R., McBride, R. B. & Hage, H. H. (2005). *Criminal justice internships: Theory into practice* (5th ed.). Cincinnati, OH: Anderson Publishing.

Hunt, D., & Sullivan, E. (1974). *Between psychology and education.* Hinsdale, IL: Dryden.

Robb, D. E. (2001). *Crossing the water: Eighteen months on an island working with troubled boys—A teacher's memoir.* New York: Simon and Schuster.

Rogers, G., Collins, D., Barlow, C. A., & Grinnell, R. M. (2000). *Guide to the social work practicum.* Itasca, IL: F. E. Peacock.

Royse, D., Dhooper, S. S., & Rompf, E. L. (2007). *Field instruction: A guide for social work students* (6th ed.). Boston: Allyn & Bacon.

Russo, J. R. (1993). *Serving and surviving as a human service worker* (2nd ed.). Prospect Heights, IL: Waveland Press.

Schmidt, J. J. (2002). *Intentional helping: A philosophy for proficient caring relationships.* Upper Saddle River, NJ: Prentice Hall.

Shulman, L. (1983). *Teaching the helping skills: A field instructor's guide.* Itasca, IL: F. E. Peacock.

Suelzle, M., & Borzak, L. (1981). Stages of fieldwork. In L. Borzak (Ed.), *Field study: A source book for experiential learning* (pp. 136–150). Beverly Hills, CA: Sage Publications.

FACING NEW FRONTIERS

In the preceding section, we helped you prepare for, examine, and work through many different issues that tend to arise early on in the internship. Once you have resolved those issues—or at least once they are not foremost in your mind— you can turn your attention to other matters that take on importance as the internship matures. This section of the book is designed to keep your progress moving and your internship alive and vital.

In Chapter 10, we identify common traps that can await you at this point in your internship and encourage you to take a careful and thoughtful inventory of your personal, professional, and civic growth and development. We also explore a common, although not universal, phenomenon: the feeling that your internship is falling apart, which is indicative of the Disillusionment stage. In Chapter 11, you will study a model for thinking about and dealing with the obstacles and difficulties that you have identified. This section of the text is less theoretical and considerably more applied than the previous one. Our primary objective is for you to examine your own experience. We believe you already have the theoretical tools with which to conduct such an examination.

Although you may find the reading in this section of the book a little less dense, the issues you will read about are every bit as difficult as those faced in the Anticipation stage. However, if you have worked through the concerns you had in Stage 1, then you know the sense of satisfaction and empowerment that comes from dealing with issues effectively. In the previous section, you achieved an informed, realistic commitment to your work. In this section, you achieve a clearer vision of that work and an increased sense of confidence that through reflection, effort, and reaching out for guidance and support, you can handle whatever comes your way.

CHAPTER 10

Taking Stock and Facing Reality: The Disillusionment Stage

I knew I would learn an incredible amount because of the actual hands-on experience that you cannot get from a book. But I was surprised at the impact the internship had on me and to learn how much I need to work on.

STUDENT REFLECTION

You have dealt with the issues and concerns that accompany the first weeks of the internship and moved successfully through the Anticipation stage. Now what? Do you relax and coast through to the end? Well, you might be able to do that (although it's not likely), but you will miss many learning opportunities if you do. Do you wait, apprehensively, for the problems to develop? After all, you read about the stages of an internship and that the Disillusionment stage is often the next stage. The problem with that stance is that it is passive and reactive, and one focus of this book is to help you become an *active* and *proactive* learner, shaping your own learning experience as much as possible. So, what we want you to think about now is how you can keep yourself growing and moving forward. There are three major components to this effort, and you will continue to need and use them throughout your placement. Each of them, though, can involve and evoke powerful feelings, so learning how to do them and how to deal with their effects is very important to your success in the internship.

- Take inventory of where you are and where you want to be (assessment).
- Identify problem areas that need attention (problem identification).
- Develop and implement plans to take action in both of these areas (problem solving).

In this chapter, we begin with some thoughts about growing and learning. We then present some ways for you to take stock of your progress so far. We pose some general questions about your internship and ask you to think about issues and challenges in the major arenas of your internship: the work, the people, the site, community, and yourself. It is also at this time in the internship when many people enter the Disillusionment stage. If that is happening to you, or in case it happens later, we will help you make sense of that experience and think about ways to move through the challenges. The third step in the process, problem solving, is the subject of Chapter 11.

THINKING ABOUT GROWTH

One of my anticipations was "growth." This I can feel every day. I can feel myself growing as a person as well as a professional.

STUDENT REFLECTION

As time goes on, many interns become absorbed in the knowledge and skills they are learning and lose sight of the larger picture of them in relation to their overall development. Personal, professional, and civic growth and development are integral to each stage and phase of an internship, not just during transitions from one to the next. We focus your thinking on that growth and development at this time so that you are more tuned into each of these processes as they occur during your internship. We hope that you will face challenges so that those processes allow you to flourish.

The Human Side of the Internship

It is important to remember that your internship is fundamentally a *human* experience. It is precisely this human element that makes the internship one of mutual learning and vulnerability for you and for the others who are part of your experience in the field (Georges & Jones, 1980; Wagner, 1981). When the focus of the internship is on collecting data, analyzing information, working on projects, providing services, or even acquiring skills rather than on *living the experience* of the internship, the effects of the human side of the experience on you and your work are easily ignored or only mentioned in passing, as if they were of little importance to the real work of the internship. So, how do you make sure you are attending to the *human* side of the internship? The answer is in the concept of "remaining open."

- Remaining open to individuals by listening, reflecting, paying attention, resisting temptation to categorize and label people (including you!) into boxes.
- Remaining open to the relationships in your internship and the joys and challenges they bring.
- Remaining open to and accepting of your own emotional experiences and the opportunities they present. The more you are open to your feelings, even those you wish you didn't have or think you're not supposed to have, the more you will be open to the emotional experiences of others.

Experiencing Change

The setting may change, but I continue to be who I am for the most part, while I can sense come change within myself. Others are who they are. The field experience is rich because what may seem familiar experiences are reviewed with a different perspective. . . . They are getting a twist and that is good.

Conventional wisdom says work smarter, not harder. Word of the day — resiliency. . . . Feel the emotion, bounce, and continue. Change is happening.

STUDENT REFLECTIONS

Moving on and changing are integral parts of the internship process. You expect them to happen, as do your site supervisor and campus instructor(s), so much so that if they do not, it can be cause for consternation. *Moving on* suggests taking on additional responsibilities that further test and develop your knowledge and skills as well as shape and test your attitudes and values. You also expect your workload to increase and your performance to improve. Essentially, you seek out greater challenges and strive to meet them, often with considerable satisfaction. *Changing*, on the other hand, suggests that an internship affects you in more noticeable and pervasive ways. During the first seminar class, just after the internship begins, we often ask our interns a question that we will pose to you here:

> What would you say if we told you that the person you are today, reading this passage, is probably not the person you will be when you complete your internship?

Our interns tend to look at us very strangely and quietly when we make this prediction. How about you? What is your reaction? Would your reaction be different if the question were asked at the very beginning of your placement? Maybe the question strikes you as a little dramatic, and perhaps it is, but change involves something more substantial, more far-reaching, and more challenging than just taking on new responsibilities, learning new theories, or improving your skills. It involves looking at yourself, your work, the placement site, other organizations, civic responsibility, and even society in new ways. This integration and reintegration of attitudes, skills, and knowledge across the personal, professional, and civic domains are the hallmarks of a civic professional, and it is not solely an intellectual process. Change of this sort can be enormously exciting and frightening at the same time.

In a time of change, how can you keep yourself moving forward? For one thing, you can remember the experiential learning cycle described in Chapter 1 and mobilize your own resources and those around you so you can seek and have new experiences. Remember, though, that action and experience are only part of the picture. The need to *do* is very high for most interns, as you have probably discovered, and the needs of placement sites can also be high due to the sheer volume of work and sometimes to reductions in the workforce. However, the activities of processing and reflecting are equally critical to your professional development and learning, and they need to be structured parts of your internship experience in activities such as journal time and discussion groups with peers as well as designated personal time for thought and reflection. When the balance between action and reflection is compromised, so too is the overall quality of your field experience. How are you doing so far in achieving this balance in your internship?

> ## Focus on THEORY
>
> ### The Essentials of the Change Process
>
> There are natural tensions involved in change, and you are bound to feel them. According to Jon Wagner (1981), you deal with two sets of perspectives during your internship: the perspectives you bring to the field and the ones you are developing as a result of your work. On the one hand, you are drawn to the reality of the work, its vitality, its unpredictability, and its dynamic yet problematic character; on the other hand, you seek the comfort of "home," where you know the situations and the problems and where you are known to others. Your challenge is to bring these two sets of perspectives together in such a way that learning goes on. If you keep both your feet safely planted at home—concentrating only on how you currently think and what you already know—you underextend yourself in the experience and risk not growing; however, if you bury yourself in unfamiliar soil, you overextend yourself and risk sinking in the experience.
>
> Resolving these differences requires that you embrace both perspectives at the same time, "straddling them" in Wagner's (1981) words, so as to create a perspective or vision that incorporates the past as well as the emerging ways of looking at things. Sounds great, right? It can be, but William Perry (1970) reminds us of two opposing human urges: the urge to progress toward maturity and the countering urge to conserve. These competing emotions accompany the two perspectives just discussed and can interfere with their integration. Similarly, Robert Kegan (1982) believes that all change involves letting go of old familiar ways, and that can be frightening. So, expect some excitement as well as some trepidation, and remember that both these emotions are normal to have and can help you grow.

There is another balance that you need to pay attention to as well: the balance of support and challenge. You will need support as you experience the emotional, human side of the internship, and you will need challenge in the tasks you are given. You need a balance of these dimensions from those around you as well as from yourself. Yes, from yourself. More importantly, you need support for who you are now and the way you think and feel at this time, and you need to be challenged to move forward and take risks (Kegan, 1982).

TAKING STOCK OF YOUR PROGRESS

With thoughts about learning and growth in mind, we turn to assessing your progress. This process involves considering some general questions as well as some specific issues. One theme you will notice throughout this section is the reaction you have when something you have read or heard about appears in front of you, or within you, in real human terms.

Keeping the Contract Alive

If your learning contract is to serve as a valuable guide and point of mutual clarity for you, your site supervisor, and your campus supervisor or instructor(s), you all need to work together to keep it from becoming obsolete.

Think About It

How Viable Is Your Learning Contract?

Give some thought to these questions as you consider the viability of your learning contract:

- When was the last time you looked at the contract or thought about it? In Chapter 5, we described the learning contract as a living, changing document, but we also know how easy it is to set it aside and forget it once it is handed in. Now is the time to take stock of this document.

- Are there differences between what you planned and what you are doing?

- Have some of the goals been reached and now seem uninteresting?

- Are there activities you know that you will not get to do?

- Have changes in personnel, needs, or priorities affected the contract?

- Have new and interesting opportunities presented themselves? If so, they may need to be incorporated into the contract.

EXPANDING YOUR KNOWLEDGE

Some of your goals are knowledge goals, and you want to be sure you are challenging yourself, and being challenged, to acquire more concepts and think in new ways and at deeper levels. The exact nature of these knowledge challenges is, of course, an individual matter, but there are two general issues to think about.

Integration of Theory and Practice   You probably have studied lots of theories in your classes and even given some thought to their application. However, as you leave the role of observer and become a more active participant, your theoretical knowledge changes as the result of trying to apply it (Benner, 1984; Garvey & Vorsteg, 1992). This process can be troubling; you may find that your theories don't seem to work. Could all of your professors and textbooks have been wrong? Not at all! You have started down the road from *theoretical* reasoning to *practical* reasoning, as we discussed in Chapter 1. The integration of theory and practice is a normal challenge in a number of professions because not all theories work equally well in all situations. Furthermore, it may take you awhile to recognize your theories in action because they might look different when applied to real problems. Remember that both theory and practice are transformed when the two meet in a human context.

Developing Guiding Principles for Your Work   These principles are derived from your experiences in class, in the workplace, and in other field and life experiences

(Brenner, 1984). Often, interns can tell us about what they did in a certain situation and why but struggle to answer questions such as:

- Can you formulate some general ideas about what works for you and what works in an organization like yours?
- Can you identify accurately and in specific ways your strengths and weaknesses in given situations?
- What are you thinking about as you approach a client, customer, or co-worker in crisis?
- How do you know when to press for a response and when to let the individual be silent?

You might be able to quote textbook answers but do not feel a personal connection to those answers because they are not coming from your experiences.

EXPANDING YOUR SKILLS

When we discussed learning contracts, we mentioned that there are three levels of learning. You are moving into the second phase or level of learning now, where the focus shifts from learning *about* the work to learning *how to do* the work. When you

---

Focus on THEORY

Skills Acquisition

Dreyfus and Dreyfus (1980) studied how health care professionals acquired and developed the skills to carry out their work effectively. They identified five levels of skills acquisition: novice, advanced beginner, competent, proficient, and expert. These levels of proficiency reflect changes in three aspects of skills performance:

- A shift from reliance on abstract principles to past experiences
- A change in perceptions from seeing a series of parts to seeing things as wholes with relevant parts
- Movement from being a detached observer to being an involved learner

These findings and Brenner's guiding principles are particularly useful when thinking about this new phase in your internship. As a novice, you had little understanding of the situations in which the work must be carried out. The basic skills that you brought to the field were generic and in need of *contextual meaning*; that is, knowledge of how to use your skills effectively in a particular situation. For example, in the field of the helping professions, knowing the names of the resources in a community and being familiar with the indicators of potential suicide do not necessarily mean that you know how to interpret a client's behavior and prioritize your responses to deal with a life-threatening crisis. By the end of the internship, though, the student intern may be expected to develop the skills to respond effectively to just such a crisis. Certainly, it would have been helpful to have taken a course to prepare for understanding the contextual meaning of the work, but that may be neither practical nor available.

wrote your learning contract, you probably wrote about skills in fairly general terms. As you begin to acquire new skills, you can think about skill development in more specific terms and specify smaller steps for yourself.

Also, remember that the skills you are developing or can develop at your internship may be professional, personal, civic, or some combination of the three.

TESTING YOUR ATTITUDES AND VALUES

The practice of every profession rests on a foundation of values and ethics; without them, skills can be used to all kinds of ends. And in most professions, attitudes are as important to success as skills and knowledge. It is no longer enough to know *about* these values and attitudes or be able to recite them; now you need to begin to focus on how to live them. As you look back at your learning contract, did you set some goals for yourself in the area of attitudes and values? If so, how are you doing with living those values and acquiring those attitudes? If not, is it time to do so?

CONSIDERING THE ISSUES: PREDICTABLE CHALLENGES

> *I had arrived to class feeling okay about my internship. I did have some doubts, but I felt that with time, things could work out. As I listened to my classmates talk about their experiences, I felt my insecurities grow. It seemed that I was the only one in the group with any doubts about their internship and what was happening to it.*
>
> STUDENT REFLECTION

Just as there were issues that most interns face in the Anticipation stage of the internship, there are new problems and challenges that emerge as you move beyond that stage. And while challenges can be both exciting and unnerving, problems are never enjoyable. In our experience, nearly all interns experience some problems, but how many, how severe, and when they occur are difficult to predict. We are not talking about something minor that is quickly resolved. Nor are we talking about something more significant that is manageable but trying. Rather, we are referring to situations that are more substantial, troubling, and that stay with you for a while. Dealing with them is part, but not all, of the challenges of the next stage of your internship. Keep in mind that it is through challenges and problem solving that growth occurs. Without them, you may have learned *how* to do the work but not necessarily how to do it when faced with complexity, uncertainty, and setbacks.

There are two broad categories of challenges and problems. The first set are those that are more predictable in that they are likely to occur right at this stage of the internship. They may look different in different internship contexts, and you may not have them all, but the sources of these problems are a normal progression of issues and concerns.

Hitting the First Bump

Problems can be of importance or can be inconsequential, and as an intern, you probably won't know how to assess that early on. The first problem, though, usually has an added impact, precisely because it is the first one. Once you have encountered—and resolved—the first problem or two, subsequent problems may be harder to resolve, but the shock of the first few will be behind you. You will also have the added confidence of knowing that you are capable of addressing problems and taking a proactive role in your internship.

Rising Expectations

After you have settled into the internship, people often expect more of you, and that can be hard. Take, for example, the case of the business intern who has stellar performance early on. Because her work was so outstanding in the first five weeks of her placement, she ended up being well over her head with the volume of responsibilities she was willing to take on. As her peers observed, she needed to put the brakes on and regain control of the internship: *"I feel as if they will look down upon me (if I slow down), and I will get a bad grade. Plus, I don't know what my priorities are. . . . It's making me stressed. There are too many things, and I have no focus."*

Anticipation and Actuality

One common source of problems is the gap between expectations and reality. There is almost always a difference between what you anticipated about the internship and what you actually experience in the field. Fortunately, if you acknowledged and dealt with your concerns in the Anticipation stage, you will be less likely to have a wildly different reality from what you expected. However, there will be some discrepancies

---

Focus on THEORY

Difficulties and Problems

An interesting and useful perspective on problems and their resolution is offered by Watzlawick, Weaklund, and Fisch (1974). The terms *problem* and *difficulty* are often used interchangeably, and indeed we have done so in this book, but these researchers draw a distinction between them. Difficulties, they say, are normal, if unpleasant, events that either have a fairly simple solution or no solution at all. However, in your attempts to solve these difficulties, you may make them worse or bring on additional troubles. It is then that you have created a problem. Many of the issues you face as an intern are normal and common. Knowing that they are part of the internship may make it a little easier for you, but probably not easy enough. However, if you choose to ignore or misdirect your energies in trying to address difficulties, they can snowball into full-blown problems.

between your expectations and the realities of the placement, some of which may be troubling. It may be that the issues and personalities are not what you expected or that you are reacting differently than you thought you would. In addition, issues will arise in placement that you simply never considered or never knew existed because you had no way to anticipate them. These issues derive from that dimension of knowledge about which you don't even know you don't know . . . if you know what we mean! One student said, *"What I don't know I don't know changes all the questions."*

Reactions and Responses

Just as there is a range of problems, there is a range of emotional reactions to them. Some interns seem to take the problems in stride; others are really thrown. As you read each sentence, try not to fall into the trap of saying to yourself, *"Well, I am going to be the first type."* Perhaps you'll be very accepting and supportive of those who do seem to be thrown, but you won't allow yourself that experience. Remember, allowing yourself to *feel* your problems, as well as to catalogue and analyze them, is one way of remaining open to the human experience of the internship. There is no right way to react. Your responses and reactions—and those of your peers—will depend on your emotional styles, the particular intrapersonal issues touched by a problem or situation, your willingness or unwillingness to be open with yourself and others about your feelings, how you perceive moving beyond the problems, and, of course, the nature of the problems themselves.

Sometimes students disclose their feelings of disappointment to the faculty, whether online, by e-mail, by conversation in person, or by telephone; other times, they disclose them in the seminar class to their peers. Regardless of how or when the student shares feelings, it is usually difficult because of all the expectations he or she had of the field experience and how the student now feels.

In spite of our efforts to reassure them, interns often believe that if they talk about their difficulties or problems or discuss their concerns or mistakes, then their grade will reflect these shortcomings. It is true that the grade you earn in your internship is influenced by your strengths as well as by the competencies that need continued development. However, your campus instructor or supervisor recognizes that all interns have shortcomings, as do practitioners and faculty members. Your willingness to recognize and face the issues helps the faculty instructor gauge your growth in placement. In fact, not doing so may give the faculty reason to question how much you are gaining from your placement (Gordon, McBride, & Hage, 2005).

Most importantly, whenever you or your peers are discussing feelings of disappointment, which in our experience tends to happen in the seminar class, it is very important that effective communication and feedback skills be used to provide support. Knowing when to clarify, paraphrase, and reflect feelings is critical to you and your peers being willing to come forward again with similar concerns. Failing to use "I" statements, being vague and general, and being interpretive when giving feedback is sure to disappoint all who are part of the discussion. You or a peer may soon feel disillusioned with your internship, and it is important that you are prepared to support your peers and be supported by them. This is a good time to revisit the principles of effective communication and feedback that we discussed in Chapter 2.

CONSIDERING THE ISSUES: VARIABLE CHALLENGES

The second set of issues, challenges, and problems is less predictable. We cannot say when they will occur or even whether they will. What we can do is tell you that they fall into several broad areas: the work, the people, the internship site, your civic development, and your personal issues, and we invite you to think about each of these areas as you read.

Issues with the Work

Work issues are those that center on the demands being placed on you, the responsibilities you are undertaking, and the opportunities you are being given. There are three issues for you to consider: the level of challenge, the content of the work itself, and the balance of needs between you and the placement site.

THE CHALLENGE OF THE WORK

Are your responsibilities overchallenging, underchallenging, or right on target? Think about the volume, the depth, and the scope of what you are asked to do. Too much responsibility can be overwhelming and leave you feeling ill equipped to do the work; too few demands result in being underchallenged, which can be humiliating, frustrating, and disappointing (Burnham, 1981; Royse, Dhooper, & Rompf, 2007) and leave you feeling bored and not connected to the energy of the work or the field site.

THE CONTENT OF THE WORK

Are you engaged in the activities called for in your contract? If not, why not? If this question raises a flag for you, then it is a supervision issue, and it is your responsibility to bring it to the attention of your supervisors on campus and in the field. Whether it's a contract between the campus and the internship site or your individual learning contract, the tasks identified on it were deemed academically worthy at one point; if you are not being guided in learning those tasks, then it's time to reevaluate the tasks and agree upon a viable contract for learning. Why you are not learning the tasks is an important question to think about before such a discussion with your supervisors.

Are the activities you *are* engaged in the ones in your learning contract? If not, should they be? Exciting opportunities do arise, such as when you fill in for someone in an emergency and discover you like what you are doing, and it's not in your learning contract. There is nothing wrong with such a change as long as you keep your goals in mind. If those goals are changing as well, then a shift in tasks makes sense. If not, then adding new tasks should be discussed in supervision.

THE BALANCE OF NEEDS

Balancing your needs as a learner with the site's needs for an intern can be a challenge. Your supervisors are responsible for working with you to ensure that the balance is being maintained. If it isn't, the issue should be raised as a point of discussion in supervision with both campus and site supervisors. The balancing act becomes even more complicated when you are employed at your field site or offered employment while interning. Potential problems can be managed if you clarify your needs in both roles (intern and employee) by discussing them with your supervisors. It is wise to keep supervision separate for these two roles. This usually means two different supervisors in the field—one for the internship and one for employment. Sometimes, internship responsibilities shift because the site needs help and you are there and can help. Of course, you will want to help, and you will be nervous about saying "no" to your supervisor. But are these responsibilities a good match with what you most need or want to learn? If so, they need to be incorporated into the learning contract. If not, they need to be discussed in supervision before you take them on.

Issues with the People

I wanted the work, not the system. I realize now how significantly affected I am by the negativity of the staff. I want to separate myself from them so I don't have to deal with it. How do I do that? Can I do that?

Before beginning the internship, I had these ideas about what the people would be like; what the procedures would be like. Some of those expectations were met, and some blew my mind. It was the reality of fair-does-not-equal-justice and the judge-does-not-equal-unprejudiced. . . ."

STUDENT REFLECTIONS

For most interns, the internship unfolds in the context of a web of relationships, with supervisors, co-workers, peers, and for some interns, clients or customers. Once the acceptance concerns have been addressed, the relationships continue to develop and new issues will undoubtedly present themselves.

CONCERNS WITH SUPERVISORS

You may be having concerns with your site supervisor, whether it's your impression of the supervisor as a professional, the supervisor's style of supervision, or aspects of the supervisory process. In many ways, the same issues that arise with site supervisors arise with the campus staff or faculty who oversee your placement, perhaps conduct the seminar class, grade your papers, and perhaps issue your grade. Although you do not spend anywhere near as much time with these people as you do with your site supervisor, they are important relationships nonetheless. Now is a good time to reflect on all of these relationships because issues with supervisors tend to threaten the foundations of trust and acceptance you are working to develop. In turn, your concerns about your supervisors can leave you feeling vulnerable in dealing with them. If you are facing these challenges, the following sections may prove helpful as you work your way through them.

The Perils of Idealizing Some interns initially are so impressed with their supervisors, both on campus and on site, that they idolize them. If you find yourself feeling this way, remember that the bubble eventually bursts. The supervisors will make mistakes, might even snap at you, miss an appointment, or exhibit one or more of any number of foibles that all of us can fall prey to at one time or another. Realizing that supervisors are human, yet still professional, with substantial responsibilities can be an eye-opener.

Think About It

How Are You Reacting?
- Have you found yourself idealizing your supervisor, putting her or him on a pedestal?
- Has your supervisor displayed consistent behaviors that are confusing for you and that she or he seems to see no need to explain?

In the helping professions in particular, interns can become confused by callous and cold behavior on a supervisor's part, which might anger and disappoint them. It's important to know that professionals who have to deal with heinous acts of violence as part of their workday use outward appearances of insensitivity and callousness as ways of coping. Knowing this beforehand may prevent confusion and frustration on your part.

Reacting to Supervision Styles   By now, you know a good deal about your supervisor's style and your own supervision needs. You have a sense of what is suited to your needs, what is not, and what needs have changed since the internship began. All these factors can affect the supervisory relationship for better or worse.

This is a good time to ask yourself how your emerging needs match up with different styles of supervision and what you need to do to get your needs met, as did this intern: "*I have learned a lot about her style. I have had to change my style in some ways to meet hers. I know that she runs things in a certain way, and I have finally learned to make my needs match her style. I make sure all of my needs are met, but I make sure that they are on her grounds, so to speak.*"

Even if you and your supervisor are well matched, you may have differences of opinion about strategies, projects, or other issues. Acknowledging and discussing differences are expected practices in many professions. However, you may feel anxious about doing so with your supervisors, especially at this stage of your internship (Wilson, 1981). Remember the section in Chapter 2 about the power of discussion? In most cases, open discussion, not confrontation, strengthens the supervisory relationships, as long as your supervisors respond empathically and educationally to your concerns. If you are not ready for an open discussion, you may find this intern's comments helpful: "*I wouldn't say that we have disagreed, but I would say that I have disagreed with her and not shown it. I dealt with it in my own way by means of reflection and discussion with peers. Then I realized how to fix it without stepping on her toes.*"

Think About It

How Do You Match Up?
- How do your needs match up to your supervisor's style?
- How do you address differences between you and your supervisor?

Reacting to the Supervisory Process: Feedback and Evaluation This is the part of supervision that can be the most anxiety producing yet can yield the most valuable information for your success. You will be getting feedback on a regular basis from both campus and site supervisors, and you will be having some type of evaluations of your performance at the site. Keep in mind that you have a right to a high-quality evaluation. Suanna Wilson (1981) suggests that a high-quality evaluation contains both positive comments and comments on areas for improvement. Consistent with the principles of effective feedback mentioned several times in this book, Wilson further suggests that a high-quality evaluation is concise, specific, and describes behavior (as opposed to using labels). If you do not receive an evaluation that is guided by these qualities, you should consult with your campus supervisor. This individual is a rich resource for you. Whether faculty or staff on campus, this supervisor is there for you. It may be consultation that you need or it may be a discussion about the supervision you are receiving and the feedback you are getting about your performance on-site. Remember, if you don't understand why you received a comment or rating, it will be very difficult to learn from it. Working with your campus supervisor, you can learn the most by learning how to question feedback as well as question evaluations in productive ways.

Some interns accept both positive and negative feedback gracefully. Others are uncomfortable with it; they become defensive about negative feedback and think of it as criticism and/or they give someone else the credit for their work when they are complimented. If you fall into the latter group, it makes sense that you talk about this in supervision or in your seminar class to learn how to deal with it in effective ways. Thinking of feedback in terms of a tool that allows you to grow will serve you well; you will focus on the comments rather than only on the grade, and you will know the specific areas to focus your energy for improvements.

Interns have a variety of reactions to and experiences of evaluations. For some, the feelings are those of satisfaction, joy, and relief; for others, the feelings are those of anger, disappointment, and frustration. For both, there may be tears and disbelief. Evaluations can be a painful experience because of the introspection, self-doubt, and self-confrontation that can arise from them. Not everyone is comfortable scrutinizing their skills, abilities, and commitment to the field of study or perhaps questioning the value of the work itself and their choice of profession (Lamb, Baker, Jennings, & Yarris, 1982). For some interns, the experience can be disorienting. Some find out that they excel in the classroom but not

---

Think About It

How Are You Responding?

- What opportunities have you had for feedback?
- How have you responded to the feedback you have received?
- What has been your experience with evaluations of your performance, and how do you respond to them?

in the field, at least at this point in their internships; others find out that an excellent grade in one internship does not guarantee an excellent grade in a subsequent internship. Some interns experience competitiveness during evaluations and have trouble staying focused on learning from their own evaluations. This is especially true if other interns are at the site and there are differences in how the evaluations are conducted, what criteria are used, and the supervisor's style of "grading."

CONCERNS WITH CO-WORKERS

Perhaps your relationships with co-workers are healthy and productive; we hope so. You should be on the lookout, though, for some common problems. For one thing, if you find yourself being given work by anyone other than your site supervisor, and you were not informed about this possibility up to this point, then it's time to discuss the situation with your supervisor so you are clear on what to expect. If you find co-workers seeking out your friendship, know that they may have hidden motives for befriending you, and it may take you some time to figure out what is going on. Often, these employees are mostly marginal and most likely to reach out for alliances. In addition, if there are factions at the site, people may become friendly in order to recruit you to their "side" or clique. If you are not feeling welcomed as a team member at the internship site, or if you have had an interpersonal conflict with a co-worker, they can be major sources of stress (Yuen, 1990) and need to be discussed in supervision with your campus instructor or your site supervisor.

If you have a concern about the kind of example your co-workers are setting for you, this is another issue to raise in supervision. For example, you may have seen some behavior for which there is no good explanation. In any field, there are professionals who are lazy, jaded, harsh, or unethical. Experiencing such circumstances can leave you feeling quite alone in your reactions. In addition, the support you need may not be forthcoming from your co-workers. These issues can arise in an internship and should be addressed in supervision as early as possible.

It is also the case that what behaviors look like may not be what the behaviors are about. For example, perhaps you are interning at a state child welfare agency and you overheard some workers coolly discussing the murder of a child, which you recall from the morning paper. As the conversation went on, you realized that the child was

---

Think About It

How Are You Being Treated?
- Do co-workers give work to you or otherwise exert influence over you and tend to do so without your supervisor's awareness?
- Do co-workers enthusiastically seek your friendship and support?
- Do you feel that you are not a member of the team?
- Do co-workers model or exhibit behaviors that give you cause to pause?
- Do you have feelings of being "just an intern"?

a former client of the agency. Other than this conversation, it was business as usual, and few outward emotions were evident. If you did not know that such an outward appearance of insensitivity and callousness is one way workers cope with heinous acts of violence that are part of their workday, then you might have become very angry with the situation and very disillusioned about the people who work in the field.

CONCERNS WITH PEERS

Your peers are your classmates and other students who are interning at your site. These are the people who know best what you are experiencing, and they can be an enormous source of support throughout your internship. On the other hand, they can also let you down by talking too much (or too little), offering unwanted advice, focusing only on their own issues, or simply appearing indifferent. You'll need to decide what it is you expect from your peers in terms of support and assess whether or not it is realistic.

Another concern is that of competition and rivalry as it relates to same-site interns. The rivalry can emanate from either the staff or the interns and usually results in interns being labeled winners or losers. Competition in and of itself is not unhealthy nor necessarily should be discouraged, as all may benefit from it. However, when competition becomes the focus of your energy and each of you is watching the other's assignments, level of support, progress, and involvement with staff, difficulties can develop. If that is the case, then it's important to bring the matter to supervision for guidance.

Think About It

How Are You Connecting with Peers?
- Can you depend on your peers for support?
- Do you feel competitive and contentious with your peers?
- Do you think that your progress is lagging behind that of other interns, either at the site or in your seminar class?

A third issue that creates challenges for same-site interns as well as interns in the same seminar group is differences in progress. There are lots of reasons why placements turn out as they do. However, when there are two interns at the same agency and they are having very different experiences, the peer relationship can be affected. No longer are the interns sharing similar experiences, and it is difficult to keep both involved in supporting each other. Such differences in progress need to be addressed as potential issues by you, your supervisor, and your instructor.

For Those in Direct Service: Concerns with Clientele

When you are engaged in direct service work, a great deal of your psychological, intellectual, and emotional energy is directed into this aspect of your internship. As you move from an intellectual way of understanding the clientele to emotionally connecting with them, you are bound to have reactions to them and their situations. The challenge here is to understand your reactions and learn from them rather than letting them interfere with your work and your general well-being.

REACTING TO CIRCUMSTANCES

As you get to know those you are helping or serving, learning about their individual and collective life situations can be emotionally demanding and possibly damaging to you. For example, if you have already had your first encounter with issues of disease, unnatural death, or abuse, you already know that the work is never easy with those who are so socially vulnerable (Brill & Levine, 2004; Wilson, 1981). So, too, is dealing with and understanding the behaviors of those among us who are abusive, violent, or mentally ill. Destructive, irrational, or violent behavior directed at peers, family, or loved ones can be difficult to hear about or witness. If the behavior is directed at you, as it might well be if you are interning in the criminal justice system or in a psychiatric setting, the behavior can be most unsettling (see Chapter 9 for a discussion of personal safety issues). In all of these instances, supervision is critical to your understanding of how these situations can and are affecting you emotionally and how to ensure your mental and physical safety.

As you come to know your clientele and their situations better, the issue of empathy can surface again. In Chapter 9, we discussed the problem that some interns in the helping professions have finding common ground and establishing empathy with clients. If you are successful in establishing empathy, you can face a new set of problems. Understanding others' viewpoints about and experiences in the world challenges your own perspectives, and such challenges can be threatening at times. For example, one intern from a middle-class background was working with juvenile offenders. She could not understand how anyone could commit crimes and think the behavior was right, although she understood that people make mistakes or act impulsively. As she listened to clients, though, and came to understand how hopeless many of them felt and how angry they were at a system that never seemed to benefit them, she began to understand their contempt for the legal system and the reasons for their own code of ethics. She was quite rattled to discover that something she thought she knew for sure was not as certain anymore.

Think About It

How Do You React to Life Situations?
- Have you thought about the emotional effects on your personal life of direct service work with those who are violent, mentally ill, abusive, victimized, or otherwise socially vulnerable?
- What happens when you try to understand others' viewpoints and experiences of the world? Are you able to do so and empathize with their situations?
- How have you responded in the past to emotionally stressful situations? Do you find yourself preoccupied with them and unable to think of anything else?

Your emotional reactions to this work can be problematic when they interfere with your learning or when you are unable to carry out your field responsibilities. If you find that you cannot get the clients out of your mind and cannot fully concentrate on other tasks, a problem exists. *"I'm overwhelmed by the sad stories of their lives; I can't get them out of my head. This is too much for me."* We typically hear comments like this from students who work with violence and other forms of extreme emotional content. These interns tend to develop needs different from those of their peers early in their field experiences. For example, they may need more "air time" in seminar class. They also may be more cynical in their perspectives and preoccupied with loss, pain, and violence (King & Uzan, 1990). We help these interns understand how the work is affecting them. If you find yourself experiencing these challenges, it is very important that you discuss them in supervision and seek guidance in handling them.

Focus on THEORY

Burn Out, Compassion Fatigue, and Vicarious Traumatization

The work of helping professionals can be replete with varied stressors, including heavy workloads, a nonsupportive workplace, and the challenges of severely complex and chronic client problems. As a future professional in this field, it is important that you are aware of the ways in which your life's work can negatively affect you. If you believe that you are experiencing any of these reactions to your work, inform your supervisors immediately so that you can learn how to prevent ongoing effects and engage the most effective interventions for you. Self-care is a primary goal for the helping professionals. Without self-care, the professional is quite vulnerable to all three of these clinical reactions to the work.

Burnout (BO) reflects general psychological stress that progresses gradually as a result of emotional exhaustion. Interns tend to key into their co-workers who appear to be experiencing burnout—almost as if interns had an intuitive sense of when helping professionals are in trouble with their work. Symptoms occur in any profession and are related to client problems of complexity and chronicity. There are emotional, physical, and behavioral issues, work-related matters, and interpersonal problems (Trippany et al., 2004).

Compassion Fatigue (CF) is a form of burnout. It can develop when the worker is exposed to a secondary stressful event(s) and in turn becomes emotionally depleted and unable to provide the empathy needed for effective client care. Physical and spiritual exhaustion result as well. Physicians experiencing compassion fatigue have described it as being pulled into a vortex and unable to stop the downward spin (Pfifferling & Gilley, 2000). Symptoms, which develop quickly, are usually associated with a specific event and include generalized fear, sleeping problems, flashbacks of

continued

the event, and avoidance of event-related matters (Dubi, Webber, & Mascari, 2005). Other symptoms include (Pfifferling & Gilley, April 2000):

- Abusing drugs, alcohol, or food
- Anger
- Blaming
- Chronic lateness
- Depression
- Diminished sense of personal accomplishment
- Exhaustion (physical or emotional)
- Frequent headaches
- Gastrointestinal complaints

- High self-expectations
- Hopelessness
- Hypertension
- Inability to maintain balance of empathy and objectivity
- Increased irritability
- Less ability to feel joy
- Low self-esteem
- Sleep disturbances
- Workaholism

Vicarious Traumatization (VT) occurs only among those helping professionals who work specifically with traumatized clients (Trippany et al., 2004, p. 32) because of their personal reactions to their clients' trauma. If you are working with victims of physical or sexual assaults or domestic violence, victims of natural disasters such as Hurricanes Katrina and Rita, victims of catastrophes such as the Minneapolis bridge collapse, or victims of terrorist attacks such as 9/11 and the Oklahoma City bombing, then you are working with clients who survived the trauma, and that makes you vulnerable to VT. Not all who survive or are affected by trauma develop Posttraumatic Stress Disorder symptoms, and not all who work with people who have experienced or been exposed to trauma develop VT (Harvard Mental Health Letter, 2007; Trippany et al., 2004, p. 32).

Trauma is described as exposure to a situation in which a person is confronted with an event that involves actual or threatened death, serious injury, or damage to physical integrity and inspires intense fear, helplessness, or horror. The victim may experience the event directly, witness it, or be confronted with it in some other way (DSM-IV-TR, APA, 2000; Harvard Mental Health Letter, 2007). Not only can the victims be dramatically affected; the helping professionals who work with the victims can be significantly affected as well, including profound changes in the core aspects of their beings (Trippany et al., 2004). It is the traumatic histories of the clients, not the demanding nature of their cases, that contribute to developing VT. The onset of symptoms is acute and often includes the symptoms of burnout in addition to changes in trust, safety issues, feelings of control, issues of intimacy, esteem needs, and intrusive imagery (Trippany et al., 2004, p. 32). There are disruptions in the cognitive schema of the helping professional, including identity, memory, and the belief systems as well as changes in one's sense of spirituality and one's perceptions of the self, others, and the world. These effects are not limited to the work with clients but rather permeate the professional's private life.

The first step in preventing these reactions is basic awareness of them. The second step is a self-care plan to ensure that you are safe from the devastating effects of the work.

REACTING TO PROGRESS

When I try to talk about the jam he got into the night before, a difficulty at home, or his drinking problem, he changes the subject. If I keep it up, he shuts down, or just gets up and leaves.

STUDENT REFLECTION

If you have developed a reasonably comfortable relationship with your clients, you may be wondering how and when you can move beyond what many interns call "chitchat" into some "real" work. This concern is not limited to those who work one-on-one with clients. Interns at sites such as shelters, soup kitchens, or drop-in centers often report that clients are more accepting of them at this point in the internship. They have stopped ignoring the interns and even sit and talk with them—about everything, that is, except their problems. Does that mean that they are not progressing? Not necessarily so.

As you spend more time with and around those you are helping or serving, you may find them engaging in all sorts of behaviors that seem unproductive or even counterproductive. For example, they may refuse to talk; they may talk on and on about their problems but show no interest in solving them; they may agree to try new behaviors and then not follow through. Initially, interns who are not making progress with their clients that think they should be making are often frustrated with such behaviors. If you fall into this group of interns, supervision will help you understand that upon examination, there are many ways to understand such actions, which may or may not turn out to be unproductive actions.

REACTING TO BOUNDARY ISSUES

The clinical area of setting and maintaining boundaries can be a very challenging one for interns. Appropriate boundaries are very important in a helping relationship, although clients do not necessarily accept or appreciate them. When you set boundaries, you are making your clients aware of the limits of your relationship with them and what is and is not permissible. Clear boundaries will help your clients understand your role and what to expect of you. They also let the clients know where in your relationship there is flexibility. Ultimately, they help the clients feel safe.

Clients often test boundaries, and that can be hard on the intern. Some boundaries are very clear and unquestionable, such as sexual relationships with clients. If a client is attracted to you, even if the attraction is mutual, there must be no sexual relationship. This is a clear boundary. Other boundaries are more flexible and determined by situation. For example, the issue of socializing with clients can be a complex one, especially if services are delivered in the community (as opposed to in an agency) or clients are from a culture that expects social contact with helpers.

By now, you should be aware of the policies at your field site; if you are not, it is time to ask about them. You do need to know what is expected when clients ask you for personal information that seems both irrelevant and inappropriate, ask to call you at home or stay beyond their time limit, or invite you to dinner, to their homes, or to their weddings. Without realizing it, you could find yourself involved in any of these situations. If so, they might prove to be very difficult to manage and extract yourself from. As one of our students pointed out: *"I realized this week that the staff and I are not at this facility to be friends with the clients. . . . Our main goal is to assist them in their recovery. . . . We must draw the line at a professional level. . . . We must be careful not to get too close. . . . (I'm) aware of (the) trap before I get sucked into it. I imagine that once a staff member gets sucked in, it is difficult to get out.*" This intern had a keen awareness of the importance of boundaries; that's impressive because even seasoned professionals find themselves involved in situations that can be hard to escape at times. If this discussion raises issues for you, it makes sense to bring those issues to your supervisor's attention.

NEGATIVE REACTIONS TO CLIENTS

You may be surprised to discover that there are clients and behaviors you just do not like. You may even feel frustrated and refer to these clients as "difficult." The term *difficult client* implies that the difficulty rests solely with the client. While it is true that certain behaviors, behavior patterns, and issues would be troubling to anyone working with the clients, the depth and breadth of the emotional reactions to the client vary greatly from person to person. The question is not which clients are difficult but which clients *you* find to be difficult. Two studies conducted in the 1980s found similar results in surveying helpers about their perceptions of stressful client behaviors. Suicidal statements, depression, anger/aggression/hostility, apathy or lack of motivation, and the client's premature termination were most stressful to helping professionals (Deutsch, 1984; Faber, 1983).

Perhaps, too, you are working with clients whose race, ethnicity, age, social class, lifestyle, stage of identity development, or any other aspect of their cultural identity is new and very different from your own. You know how you *want* to feel about your differences, but it is important to be open, at least with yourself, about how you *actually* feel. What you probably did not expect to deal with are your reactions as you learn about people who are quite different from what you expected. For example, there are the sex-abusing clergy, the substance-abusing physicians, and the embezzling attorneys. We have met them in our work, and you will meet them in your work as well.

Issues with the Site: Values, Systems, and Philosophy

As with other aspects of your work, seeing the dimensions of organizational life unfold can be quite different from studying about them or discussing them in the abstract. Three areas that particularly challenge interns in this regard are the organization's values, the formal and informal systems, and the operational philosophy. Once you appreciate how discrepancies can develop, you are ready to be part of the solution.

ORGANIZATIONAL VALUES

What an agency values is evident in how the work gets done. What you see, though, may not always reflect what the agency says it values in its mission statement, and you may find yourself having a value conflict with the agency and the system. This is often the case with the paper-versus-people issue. Because there is never enough time in the helping professions to give clients all that they need and still attend to the required documentation of the work, tension develops. "People before paper" may be what is valued by the intern, and it may be what the organization wants to value, but ideals get put aside because of the paperwork they create. Value conflicts such as this can be disheartening (Sparr, Gordon, Hickham, & Girard, 1988).

SYSTEMS OF INFLUENCE

The site and system have found ways to organize their resources to get the work done, both formally and informally. The informal organizational structure, or the structure of influence, can be a source of issues for an intern because it is this structure that influences what *really* happens in the course of the agency's work (Stanton, 1981). Sometimes, these informal networks function quite smoothly and support the overall goals and work of the agency. Other times, they seem to undermine those purposes. Although you have read about such discrepancies, actually seeing them operate and feeling their effects can leave you feeling confused at best and frustrated at worst. Ignoring this reality is sure to interfere with your work.

OPERATIONAL PHILOSOPHY

You may be starting to feel the effects of the agency's *operational* philosophy, not the one described in the policy manual but the one that has evolved to get the work done. Many interns are impressed with their organization's operational philosophy, although there is also room for philosophical and political debate. You might find yourself surprised by the pace of the system and the realities of dealing with bureaucracies, underfunded programs, or reductions in staff.

Issues in Civic Development

Your internship, whether in journalism, sociology, the helping professions, or political science, provides you with an "authentic occasion" to discuss with members of the internship community issues concerning the public good (Newmann, 1989) and what your responsibility is as an intern at that site. Each profession has its contract with society as well as its ethical and moral commitments. These obligations create the web that connects the work to a larger social purpose. Most importantly, you get to think about the public relevance of the profession and see in practice how its social obligations to society are or are not carried out at your field site (Sullivan, 2005).

If you are interning in a government agency, the helping professions, or in a non-profit organization, you will find these obligations evident in its mission statement and policies. But they are just as important in the private sector in a business or corporate setting. Remember, too, that as we pointed out in Chapter 7, formal statements of mission are only part of the picture. Just as you need to focus on how your own values are lived, as opposed to merely stated, this is a good time in the internship to assess the operational values of the site, whether it is a business, human services, or a government agency.

Think About It

The Social Contract in the Private Sector

If you are interning in the private sector, it's important to consider these questions:

- What responsibilities do they have beyond maximizing profits and developing wealth?
- Should they be contributing to the solutions of social problems in their communities or in the communities of their employees?
- In this green age, what is their responsibility to eco-stability and sustainable resources?

As you become familiar with the knowledge, skills, and values of the profession, you become aware as well that the professionals you work with demonstrate what Sullivan (2005) calls "practical reasoning," referring to the fluidity with which they move back and forth between the theory they studied years before and the application of that theory in the present day. You are also becoming aware of the values of that profession. Perhaps you witness how the professionals expect themselves to invest in caring for the future public world by being part of advisory groups and think tanks today, maintaining memberships in civic associations, being "good neighbors," using an ethics of care, and being actively involved in the community by sitting on oversight boards (Battistoni, 2006; Howard, 2001).

As an aspiring professional, you are already seeing the challenge of understanding working directly with individuals, families, or citizen groups. Developing solid problem-solving skills, using critical thinking skills, engaging in coalition and community building, being politically aware of the issues, and using advocacy skills are just a fraction of the civic skills that are and will be expected of you in your role as a professional. And your internship will probably give you opportunities to polish those skills. If they are not already part of the skill set you are developing, you might want to ask about developing them.

The issues of civic development that many interns tend to deal with at this point in the internships are:

- Values and differences in values between what the profession expects and your personal as well as emerging professional values
- The meaning of the concept of "public good" and what responsibility you see yourself as having to it as an intern
- The responsibility to resolve differences through discussion and reasoning
- Understanding your civic responsibility in the profession you have chosen

Issues with Yourself

We have been encouraging you to take stock of your evolving relationships with your work, the people, and the system. However, we have not paid much attention to another important factor in the relationship equation: YOU! Finally, it is time to take a look at what you are discovering about yourself. Interns consistently tell us that they learned a tremendous amount about themselves during their placements: "... *It's a power trip in understanding me*," as one of our interns recently said.

THINKING ABOUT YOUR MEANING MAKING SYSTEM

Your meaning making system consists in part of your attitudes, values, behaviors, unresolved issues, and psychosocial and cultural identities. Combined, they create the filters or lenses through which you look at your life and make sense of your experiences. Typically, these become more alive as an internship progresses. For example, if experiential learning is new for you, you may have discovered that your learning styles and preferences differ from what you experienced in traditionally taught classes.

THINKING ABOUT BUMPING INTO YOUR OWN ISSUES

Unfortunately, you cannot predict with accuracy how you are going to think, feel, or react until you get into a given situation. Inevitably, there will be differences between what you thought would happen and what actually happens. For example, it is one thing to think about your values and what it may be like to have them tested or be involved in a value conflict. It is another thing to experience the often unidentifiable feelings that conflicts in values create. This intern's reflections are dealing with just that: *"Once I got somewhat settled in . . . I began to realize that it's not all 'up,' not everyone is nice, and working this type of schedule makes me very tired."*

Your experience at the internship can also teach you about aspects of yourself that you have missed up to now. You may not know you hold a value strongly, for example, until it is tested in some way. Or you may find yourself reacting strongly to a situation and not knowing why. After thinking about it and discussing it with your supervisor, you discover that the situation tapped an unresolved issue in your life. Such differences can create just the climate for your own reaction patterns to be become problematic. You will need to label those patterns and responses so that you can make sense of the issues you are experiencing. This student did and came to an important realization: *"I would have to say that the biggest and most important thing I have learned in all areas is that I am (only) one person. I am not helping myself or my patients by trying to please everyone. I learned that by spreading myself too thin, I can only hurt the patients and myself and really not accomplish anything."*

Think About It

Do You Have Problematic Patterns?

We encourage you to think about these and other issues/patterns that have arisen in your placement and to discuss them within supervision or in your seminar class. They are potentially problematic to your internship.

- You have dealt with the same issues as your clients, and you are not as "over them" as you thought.
- You have a hard time accepting criticism.
- You say "yes" when you want to say "no."
- You struggle to establish intimate relationships.
- You are extremely upset by confrontation.
- You have a need to smooth over conflict.

THINKING ABOUT MATCHES AND MISMATCHES

If you have not yet done so, now is a good time to think about what you have learned about yourself and the areas of matches and mismatches in the internship. Perhaps you feel as this intern did: "*I am in a constant struggle between my own emotions. When I am given something to do, I feel useful and productive. When I am left alone to find something to do, I feel useless and out of place, and unproductive.*" This intern's inability to self-direct through the tasks and involve himself in tasks that have meaning is creating an existential problem for him.

In another situation, an intern worked at a family planning clinic regularly fielding telephone calls asking for information on abortion. Over time, she became impatient with the phone calls, the clients, and even the agency. She said that many of the questions she answered were "stupid" and wondered why the people were so poorly informed. She chafed at the bureaucratic procedures employed by the agency. Upon further exploration, it appeared that a values issue was at work. The intern was firmly opposed to abortion for herself and had been open about that belief with her campus instructor and the agency. However, she insisted that she believed with equal fervor that everyone should make their own choices and that she was not opposed to abortion in general. However, as the semester progressed, she grew more and more uncomfortable with the fact that abortions were performed at her placement site; it seems that she had much stronger antiabortion beliefs than she thought. It took her some time to admit this, even to herself, because she believed she should be tolerant. Once she did so, she was able to direct her anxiety away from the agency and its clientele and view it as a poor match, given what she had learned about herself. The site arranged for her to work in their community education division, and the internship was ultimately successful.

The fit of a student to an internship site and site supervisor is quite a challenging task. In the workplace, fit has been noted as the reason why the hiring process can be less than perfect and can even fail. Considered a catch-all word that defies definition, the fit of a person to a position, just as the fit of an intern to a placement site, involves personal characteristics, professional behavior, history, temperament, and appearance (Barden, 2007).

THINKING ABOUT YOUR LIFE CONTEXT

You have been doing the juggling act for a while now, and more balls may have been thrown your way than you can manage. As you are learning, life does get in the way of your internship—illness, a lost job, an accident. You may find your energy level waning or you may be bending under the strain of multiple demands. You may even feel as this intern did: "*I have no time to be myself, or to be by myself, never mind spending time with friends or family.*" Spending that time is very important. Friends and family are central to your support system and your sense of well-being. Just as you cannot really know how emotionally demanding the internship will be, you cannot know how well your support system will respond when it comes to internship issues. It is time to take stock of your support network as well and decide what changes, if any, are needed.

WHAT HAPPENED TO MY INTERNSHIP?

When I had read about the stages of an internship, I had wanted to believe that I was prepared for the tough stages and so it would be easier to handle. Somehow the internship process challenges students differently depending on the site, the environment and the way the student engages the experience. In the end, I was both humbled and awestruck by how well my professor and other professionals were able to capture the critical components of internships for students.

P. ARNDT (SUMMER 2003, P.5)

For some interns, the problems they encounter are troubling aspects of an otherwise positive experience. Other interns reach a point where the emotional tone of the internship changes for them and not for the better. For example, you are realizing that you are not able to work collegially with co-workers or supervisors or within the organization's guidelines; and you find little compatibility of values or philosophy with the site. Instead of enthusiasm and nervous excitement, you feel anger, resentment, confusion, frustration, and even panic.

In Their Own Words   Interns Describe Their Problems

Some of the problems that our interns have encountered are listed here, written in the language of their student journals. Perhaps some of them look familiar to you. Perhaps you can add some of your own.

- The pace here is totally insane (or incredibly slow).
- My supervisor is too vague (or awfully blunt).
- My supervisor never seems to have time for me.
- My supervisor is always looking over my shoulder—he doesn't trust me.
- If I don't have any questions, my supervisor ends the supervision hour. That's not right. She's supposed to ask the questions.
- I don't like the way my supervisor treats the clients (or staff).
- The people here are cliquish—I don't belong.
- They are so cynical. BORING.
- They are trying to get me involved in their problems. I'm just an intern here. I don't need to know what's going on. It's not my problem.
- I'm just an intern. Why does everyone expect so much of me?

continued

> **In Their Own Words** *(continued)*
>
> **And of special relevance for helping professionals:**
>
> - I really can't accomplish much with these people—too much damage has been done, and I can't perform miracles.
> - I've never experienced what they have—they are writing me off. I can't work with them, and they know it.
> - A client lied to me. Or manipulated me. I thought they trusted me.
> - A client had a relapse. Great. Now what am I supposed to do?
> - There are some clients I just don't like. I can't find common ground with them.
> - They just won't respond to me; not all of them but some of them.
> - I have a good surface relationship, but I can't move beyond that to really challenge them and explore some issues.

Encountering the Unexpected

If the previous description and examples resonate with you, you are now coping with an enormous amount of the unexpected—not what you planned but what you are living with during your internship. For example, one of our interns had a long and successful career in a very emotionally demanding field in the helping professions before she returned to campus to complete her degree. She was not prepared for the emotional toll of the internship on her full and demanding life. *"I really needed a break. I dropped the ball last week because I was just about cooked to a crisp. This ole gal was running on fumes."*

What you also may find unexpected is the end of the hopeful feelings of earlier days in placement and the onset of more negative feelings. There is the unexpected drop in your enthusiasm and the subsequent drop in your productivity (negative feelings do interfere with learning!). Concerns at this time center on many of the same areas as in the Anticipation stage, except that there is a shift in concerns from *What if?* to *What's wrong?*: "*What's wrong with my internship? What's wrong with my clients? What's wrong with the organization? What's wrong with me?!*"

Realizing the Changes

The changes brought about by this shift in concerns are often subtle at first, but they persist and begin to become pervasive. This period is one of distinct character, a dip that leaves students, instructors, and site supervisors alike somewhat perplexed and overwhelmed at times. You may be complaining to friends or family without even realizing it or hesitating before commenting on your internship to friends or co-workers. You may notice that you are having trouble getting to your placement on time, not looking forward to going, or muttering to yourself under your breath about the situation. You no longer have problems to deal with in the internship; the *internship* has become the problem.

> ### In Their Own Words   Voices of Disillusionment
>
> I just want to SCREAM! This week was so much work. I feel like I just sunk down with a ship and need to find a way out. It wasn't so difficult for me at the beginning, but now, I'm finally seeing the "bad" side of things. . . . After reading this chapter, though, I have been able to think and focus and realize that this is one hurdle that I have to cross and I can look back and say that I survived.
>
> My life is crazy at this point. Everything I do is either for internship or work. On weekends, I work 20–30 hours while trying to keep up with housework and family. The roles/responsibilities that have been taking a huge hit is my role as daughter, sister, cousin, niece, granddaughter, and friend. I have an extremely connected support system, and my family is worried that my work may over consume my life. The biggest hit, though, is not having time for myself. Stress has started to settle in because of the sense of being overwhelmed. I have been taking one day at a time and hoping that I will figure out a balance. It's an adjustment, and eventually I will find something that works through trial and error.
>
> I have been through a wide range of emotions when it comes to mentally preparing myself to go to my internship in the morning. I really did not want to be there. I feel as though I have no idea what I am doing there, and I do not like being exhausted at the end of every day. I have been feeling totally wiped out by everything that has been expected of me in my internship, work, personal life, and my health. In the past couple of weeks, I have felt very reluctant to get out of bed and get myself to go in to the site, which could be due to the fact that I feel as though I do not really belong there—that it's not the right setting for me.
>
> I am going through the Disillusionment stage. There has been a loss of focus for three weeks in regards to my not doing what I went there to do. I should have discussed the issue sooner with my supervisor, but I kept telling myself that next week will be busier. I am feeling frustrated going in each week, and still there was no real work for me—and it is written right there in the contract that I will be doing it. I have decided that I need to face the issue with my supervisor and make my frustration known so I can feel like a member of the team again.
>
> Disillusionment seems to have hit me very hard. I feel very inadequate—lost among all the professionals. I am part of a team, but not. It is very frustrating always relying on others to tell me what to do. I am a grandmother! I'm not used to that. I do not feel like I have any control over my life anymore.

Managing the Feelings

One unexpected consequence of this period is the tendency to direct your feelings at those around you, especially those who are connected to the internship experience—your campus and site supervisors, co-workers, the clientele, or even your peers.

Sometimes, the seminar class itself becomes the focus of your feelings. The feelings can also be directed inward, and this can be a time when an intern's self-image can take a bit of a beating. For example, when clients do not improve as you expect, refuse to continue services, or react to you very negatively, it can call into question whether or not you were adequately prepared for the work, whether you are cut out to do the work, or whether you should be doing the work at this time in your education (Skovholt & Ronnestad, 1995). In addition, many interns in the helping professions are used to their friends telling them that they are helpful and very easy to talk with, and the interns are often filled with a sincere desire to help. However, there is a difference between wanting to help and actually being helpful and between being an easy person for friends to talk with and being effective with challenging students, residents, clients, or customers. When the inevitable difficulties and criticism come, these interns may become resentful, perfunctory in their job performance, or critical of the supervisor (Blake & Peterman, 1985).

Engaging the Seminar Class and the Supervisors

Earlier in the chapter, we talked about interns disclosing feelings of disappointment and frustration with their peers and campus instructor or supervisor and learning to deal with the feelings and situation through a supportive community in the seminar class. Not all interns have seminar classes for such support, but if you do, it is important to remember the confidential nature of the discussion when peers are talking about problems in the field. Respect is the operative word, and respect is evidenced in a number of ways. For example, it is very important to protect the identities when discussing details of the situation in the seminar class. Our rule of thumb is to advise our interns to assume that a family relative of the individual(s) involved in the problem is in the class and to protect all identities so that even family members wouldn't recognize their relatives from the intern's discussion.

The campus supervisor or instructor is another valuable resource and one that interns sometimes overlook. Perhaps this is because of their faculty or staff roles on campus and the interns feel uncomfortable approaching them or because the interns are just getting to know them in the class, which often meets only weekly. However, some interns have realized that, those factors aside, the campus supervisor has an intuitive sense of what is happening during this stage of the field experience and how to help the intern empower her or himself and move beyond the situation. Sometimes, it's a matter of taking the time to talk about the situation in a quiet space and in reflective ways. Sometimes, it's a matter of learning about resources on campus that would be helpful to you at this time. Sometimes, it's a technical matter, such as revisiting the placement agreement or the learning contract to enhance your awareness of why the situation is so troubling. Regardless, your campus supervisor is there for you when the going gets tough. Your success is your campus supervisor's work, and that is a priority.

It probably comes as no surprise to you that it is a lot easier to discuss feelings and issues with peers and the campus supervisor in the seminar class than it is to raise the issues with a site supervisor. In fact, it is the site supervisor that interns find most intimidating when it comes to dealing with these feelings in particular. The reasons vary considerably. When you think about it, your site supervisor is responsible for working with you to create and develop your field experience, and right now, that experience is disappointing. Not that you are blaming the supervisor, but would it not be very helpful if she or he changed things for you so that you would be more satisfied with your placement? Of course, the site supervisor may not know that you are disillusioned at this point; and even if the supervisor did know, the changes that are needed are not necessarily something the supervisor can control. Nor is it necessarily the supervisor's responsibility to "fix" your field placement.

The reasons interns find it so difficult to discuss the Disillusionment stage with the site supervisors vary considerably. Sometimes, the issue is one of not wanting to disappoint the site supervisor. Other times, it is a fear of the supervisor ending the placement. In reality, site supervisors are often a wellspring of resources and support for their students and welcome the opportunity to work with students in resolving such challenges.

However, if you are not feeling comfortable in discussing this stage with your site supervisor and it becomes apparent that you really need to, what do you do? Did you and your peers develop scenarios in seminar class that would be helpful? Has a peer experienced a similar reluctance? If so, how was it resolved? You may want to try what some of our students have done. Provide the supervisor with information about the Developmental Stages of an Internship, and during supervision time, bring the Disillusionment stage information to your supervisor's attention and mention that you think you are in that stage. Of course, you'll be asked why you think that, so you may want to prepare mentally for that moment because that tends to be the moment students most dread. Once that moment is behind you, you are ready to confront the issue and resolve the feelings. But that comes in the next chapter. For now, it's important to understand the bases for the feelings and your reactions to them.

Understanding the Crisis

Earlier in this chapter, we discussed the integration of theory and practice. Garvey and Vorsteg (1992) have suggested that this process can be quite difficult for some interns. It can be a time in placement when interns temporarily reject their previously held beliefs about the theories they work with and experience a *crisis in confidence* about the worthiness of the work they are doing. It is not unusual for interns to become disoriented about their learning and appear to go into a funk, experiencing intense frustration, blaming their clients, co-workers, or supervisors, and questioning their commitment to the work.

There is no predictable time when these changes in feelings occur. Often, though, these changes are noticed during the phase of learning when the workload increases and there are greater demands for skills. The sense of disillusionment can occur earlier in placement, such as when anticipation concerns are highest, or later in placement, such as when concerns about competence emerge. It can even occur during the final stage of your experience.

Think About It

A Sense of Loss

In general, we have observed four sources for the feelings of disappointment and frustration that reflect the Disillusionment stage, and all have to do with a sense of loss.

Loss of Focus   You are not doing what you went there to do because the focus of your placement has changed either intentionally or through neglect. Interns who lose the focus of their placements can feel it almost immediately; it is a feeling of being sidetracked or otherwise ignored or discounted.

Loss of Accomplishment   You are not doing what you went there to do, not because the focus changed but, rather, because you either are not able to demonstrate the skills or competencies needed to accomplish the internship or the design of your internship does not allow you to have the experience you wanted. You will probably feel frustrated and angry—feelings that you may direct at yourself or at the personnel or clients at your site.

Loss of Meaning   Essentially, you are unhappy with what you are doing in placement. Either the work does not matter to you, personally or professionally, or the internship has not been designed to provide you with a worthwhile field experience. Regardless, you may find yourself going through the motions of an internship but without the spirit that comes from being personally invested in meaningful learning and work.

Loss of Purpose   You feel as if your internship is not of value to the organization or to your supervisor. You do have assigned tasks, but they are not interwoven in such a way that you know what you are doing from day to day and are able to go about that business on your own under supervision. And that will not change in the future. Essentially, you do not feel grounded in the work because you do not have a clearly defined role or purpose as an intern. If you answer the question *What is your role as an intern?* by saying you don't really know, then you are experiencing a loss of purpose in your placement.

If you are in the middle of this sort of slump or if you hit one later on, try to remember that it is perfectly normal; in our experience, more interns go through it than do not. Many authors feel that some level of dissonance is necessary for growth. Perry (1970) argues that shifts in ways of thinking about the world are caused by conflict and dissonance. Similarly, Langer (1969) identifies conflict as playing an important role in progress and development and contends that for change to happen, you need to feel that something is wrong. He calls this a state of disequilibrium and says that it is necessary to raise the energy and emotions you need to make changes and put your internship back on track. Furthermore, many authors and researchers note that a period of disillusionment is a normal part of any internship (Garvey & Vorsteg, 1992; Lamb, Baker, Jennings, & Yarris, 1982; Suelzle & Borzak, 1981; Sweitzer & King, 1994).

Think About It

Are You Preparing for the Wrong Career?

It is worth noting that interns at this point in the placement may question their choice of career and their suitability for it. Doing so can be quite unnerving, especially if you are approaching the end of your academic studies. Questioning your choice of career may make sense, though, and not only because affirmation of the choice is very important. It also makes sense that the choice can withstand the scrutiny of reflection at any time in your career, especially at the beginning, just before you commit yourself to your first position. You may in fact conclude that the specific career you decided on is not for you after all; that does happen. Finding out now, and not in the middle of your first position, is a windfall for all concerned. Furthermore, if you have come this far, you have certainly gained useful knowledge and skills to use in life. However, this is no small problem you are facing, and it will need to be dealt with along with the others.

SUMMARY

This time in your internship is full of opportunities. You have to keep pushing yourself and find ways to keep your internship fresh and challenging. Problems, on the other hand, may seem more like a curse than an opportunity at this point in the internship. However, you will handle your problems better and learn more from them if you take a more constructive view of them.

Whether you are experiencing isolated problems or a more pervasive disappointment in your internship, facing and resolving the issues are necessary. You might want to ignore the problems, but chances are that it will make things worse. Making sense of the feelings and concerns of this stage is critical to growing through it. And growing through it is critical to succeeding in your internship.

Growth lies in the constructive activity of putting things back on track; resolving the conflict that's preventing that from happening is a prerequisite for progress to the next stage, for learning to occur, and for normal and healthy adjustment to take place. Remember, the new state of equilibrium, i.e., the feeling that all is well, does not cause progress to occur; rather, it is the disequilibrium, i.e., dealing with the problems, that is the source of growth and development (Langer, 1969).

You may recall that in Chapter 3, we described the Disillusionment stage as a *crisis of growth* and a time of risk and opportunity. If you do not try to face and cope with the problems that arise, you can end up feeling stuck; i.e., in a crisis of growth. Those feelings of resignation and resentment will not go away; instead, you will suffer bravely (or not so bravely) through them, just as you would a course that turned out to be a disappointment. On the other hand, you have an opportunity to learn more about your work and yourself. More importantly, you have an opportunity to feel empowered. If you can get through this time, you will feel a new kind of confidence.

This is not the false bravado or naiveté that tells you there won't be any more problems, but the confidence that you can resolve them and learn about yourself and others in the process.

The first step in moving through the challenges that face you is to take stock and have a clear inventory as well as awareness of the obstacles. This intern summarized her style of taking on challenges like this: "*When I am faced with an overwhelming demand, I tend to initially stress myself out with thoughts like, 'How am I gonna do this?' Then I am able to self talk and make a plan. I typically react by overreacting and as soon as 'the crisis' is over, I manage and become strong willed.*" Earlier, we asked you to consider how you were progressing in several areas. Perhaps some issues surfaced then. Perhaps others have crystallized as you read this section. The second step, of course, is developing a plan, and that is the subject of the next chapter.

For Contemplation

PERSONAL REFLECTION:
SELECT THOSE INQUIRIES THAT ARE MOST MEANINGFUL TO YOU.

1. Reread your answers to the Anxieties and Anticipations exercise at the end of Chapter 5. Have your concerns changed over these first weeks? If so, how?

2. Go back and reread your daily journal entries. What would you say are the major things you have learned so far about your clients, about working with them, about the agency, and about yourself?

3. When it is time to go to your internship, how do you usually feel? Be honest. Has that changed over the course of the semester? If so, what do you make of that change?

4. Take a look at the goals in your learning contract. List them in your journal, and indicate whether each has been met, not met, or partially met. In the latter cases, are you disappointed? How might you move beyond these goals to the next logical challenge?

5. What things have happened that you didn't expect? Include positive and negative things and think about the three areas (knowledge, skills, and self) as well as your responsibilities and your supervisor's.

6. In general, would you say the tasks you have been given so far have overchallenged you, underchallenged you, or have been about right? Explain.

7. What have you learned about your supervisor's style, and how good a match it is to your needs?

8. Have you and your supervisor disagreed yet? If so, how did that happen? How did it make you feel? How was it resolved?

9. Have supervisors or others given you verbal feedback on your performance? How did you feel about what they said?

10. In what ways have your co-workers met or not met your expectations?

11. Do you have any concerns about your relationships with your peers—other interns at your site and/or students in your seminar class?

12. How has the agency itself compared to your expectations? Does the pace approximate what you thought it would be? What are the unwritten rules, or values, of the agency? How do you feel about them?

13. Now that you have been at your placement for a while, does the support system you described in an earlier entry seem adequate? If not, in what ways is it failing you?

14. In light of your accomplishments in the field so far, are there some new goals you want to set for yourself? Be specific.

ESPECIALLY FOR THOSE IN DIRECT SERVICE

15. Are there clients you are beginning to feel close to? Are there some you have a hard time connecting with? What are your ideas about why in both cases?

16. Take a moment to think about the kind of relationships you have with clients. In what ways have you been challenged by the clients? Have they tested your boundaries? If so, how? What other boundary-related issues have surfaced up to this point? Discuss your answers with classmates, and try to help each other clarify the problems and perhaps some solutions.

SEMINAR SPRINGBOARDS

What problems have you become aware of at your internship? Make a list of them, and try to put them in order of importance. How does your list compare to those of your classmates in seminar class? What similarities are evident? What differences do you notice?

For Further Exploration

Albert, G. (Ed.). (1994). *Service learning reader: Reflection and perspective on service.* Raleigh, NC: National Society for Experiential Education.

Collection of inspiring and informative readings compiled by the Center for Service Learning at the University of Vermont used as a textbook in field study programs.

Benner, P. (1984). *From novice to expert.* Menlo Park, CA: Addison-Wesley.

Seminal book in clinical practice in nursing looks at five levels of competencies and the demonstration of excellence in actual practice.

Borzak, L. (Ed.). (1981). *Social work field instruction: The undergraduate experience.* Beverly Hills, CA: Sage Publications.

A compilation of articles, including students' perceptions of the field experience.

Dreyfus, S. E., & Dreyfus, H. L. (1980). *A five-stage model of the mental activities involved in directed skill acquisition* (Unpublished Report F49620-79-C-0063). Air Force Office of Scientific Research (AFSC), University of California, Berkeley.

Brief, informative paper describes a model of skills acquisition based on a study of chess players and airline pilots.

Garvin, D. S. (1991). Barriers and gateways to learning. In C. R. Christenson, D. A. Garvin, & A. Sweet (Eds.), *Education for judgment.* Boston: Harvard Business School Press (pp. 3–13).

> Power of dialogue discussed as part of the process of learning, self-discovery, and self-management.

Ronnestadt, M. H., & Skovholt, T. M. (1991). A model of professional development and stagnation of therapists and counselors. *Journal of the Norwegian Psychological Association, 28,* 555–567.

> A model explaining periods of growth and nongrowth in the counselor's professional development.

Rosenthal, R., & Jackson, L. (1968). *Pygmalion in the classroom.* New York: Holt, Rinehart and Winston.

> A classic text, illustrating the power of expectations on performance.

References

American Psychological Association. (2000). DSM-IV-TR. Washington, DC: Author.

Arnst, P. (Summer, 2003). They shred students, don't they? *The Human Service Educator*, p.5.

Barden, D. (June, 2007). A fitting end. *The Chronicle of Higher Education.* 53 (43) p. C2.

Battistoni, R. (2006). Civic engagement: A broad perspective. In K. Kecskes (Ed.), *Engaging departments moving faculty culture from private to public, individual to collective focus for the common good.* Boston: Anker Publishing (pp. 11–26).

Benner, P. (1984). *From novice to expert.* Menlo Park, CA: Addison-Wesley.

Blake, B., & Peterman, P. J. (1985). *Social work field instruction: The undergraduate experience.* New York: University Press of America.

Brill, N., & Levine, J. (2004). *Working with people: The helping process* (8th ed.). New York: Longman.

Burnham, C. (1981). Being there: A student perspective on field study. In L. Borzak (Ed.), *Field study: A source book for experiential learning.* Beverly Hills, CA: Sage Publications (pp. 65–72).

Deutsch, C. J. (1984). Self-reported sources of stress among psychotherapists. *Professional Psychology: Research and Practice, 15*(6), 833–845.

Dreyfus, S. E., & Dreyfus, H. L. (1980). *A five-stage model of the mental activities involved in directed skill acquisition* (Unpublished Report F49620-79-C-0063). Air Force Office of Scientific Research (AFSC), University of California, Berkeley.

Dubi, M., Webber, J., & Mascaari, J. B. (November 2005). Extreme conditions test counselors. *Counseling Today.* American Counseling Association (pp. 1, 13, 27–28).

Faber, B. A. (1983). Psychotherapists' perceptions of stressful patient behavior. *Professional Psychology: Research and Practice, 14*(5), 697–705.

Garvey, D. C., & Vorsteg, A. C. (1992). From theory to practice for college student interns: A stage theory approach. *The Journal of Experiential Education, 15*(2), 40–43.

Georges, R. A., & Jones, M. O. (1980). *People studying people.* Berkeley: University of California Press.

Gordon, G. R., & McBride, R. B. (2005). *Criminal justice internships: Theory into practice* (5th ed.). Cincinnati, OH: Anderson Publishing.

Howard, J. (Summer, 2001). *Service-Learning Course Design Workbook* (pp. 40, 42). OCSL Press, University of Michigan: Michigan Journal of Community Service Learning.

Kegan, R. (1982). *The evolving self: Problem and process in human development.* Cambridge, MA: Harvard University Press.

King, M. A., & Uzan, S. L. (1990). *The field experience: An integrative model.* Workshop presented at the annual conference of the National Organization for Human Service Education, Boston.

Lamb, D. H., Baker, J. M., Jennings, M. L., & Yarris, E. (1982). Passages of an internship in professional psychology. *Professional Psychology, 13*(5), 661–669.

Langer, J. (1969). Disequilibrium as a source of development. In P. Mussen, J. Langer, & M. Covington (Eds.), *Trends and issues in developmental psychology.* New York: Holt, Rinehart and Winston (pp. 23–37).

Miller, M.C. (August, 2007). Rethinking posttraumatic stress disorder. *Harvard University Mental Health Letter, 24*(2), 1–4.

Newmann, F. (1989). Reflective civic participation. Paper prepared at the National Center for Effective Secondary Schools and supported by the USDE, OERI, and the Wisconsin Center for Education Research, School of Education, University of Wisconsin-Madison.

Perry, W. G. (1970). *Forms of intellectual and ethical development.* New York: Holt, Rinehart and Winston.

Pfifferling, J-H., & Gilley, K. (April, 2000). Overcoming compassion fatigue. *Family Practice Management, 7,* (4), 39–44.

Porter, L. (1982). Giving and receiving feedback: It will never be easy but it can be better. *Reading book for human relations training* (7th ed.). Arlington, VA: National Training Laboratories.

Royse, D., Dhooper, S. S., & Rompf, E. L. (2007). *Field instruction: A guide for social work students* (5th ed.). New York: Longman.

Schon, D. A. (1983). *The reflective practitioner: How professionals think in action.* New York: Basic Books.

Skovholt, T. M., & Ronnestad, M. H. (1995). *The evolving professional self: Stages and themes in therapist and counselor development.* New York: John Wiley & Sons.

Sparr, L. F., Gordon, G. H., Hickham, D. H., & Girard, D. E. (1988). The doctor-patient relationship during medical internship: The evolution of dissatisfaction. *Social Science Medicine, 26*(11), 1095–1101.

Stanton, T. K. (1981). Discovering the ecology of human organizations. In L. Borzak (Ed.), *Field study: A source book for experiential learning.* Beverly Hills, CA: Sage Publications (pp. 208–225).

Suelzle, M., & Borzak, L. (1981). Stages of fieldwork. In L. Borzak (Ed.), *Field study: A source book for experiential learning.* Beverly Hills, CA: Sage Publications (pp. 136–150).

Sullivan, W. (2005). *Work and integrity: The crisis and promise of professionalism in America.* The Carnegie Foundation for the Advancement of Teaching. San Francisco: Jossey-Bass.

Sweitzer, H. F., & King, M. A. (1994). Stages of an internship: An organizing framework. *Human Service Education, 14*(1), 25–38.

Trippany, R. L., Kress, V. E. W., & Wilcoxon, S. A. (Winter 2004). Preventing vicarious trauma: What counselors should know when working with trauma survivors. *Journal of Counseling & Development, 82,* 32–37.

Wagner, J. (1981). Field study as a state of mind. In L. Borzak (Ed.), *Field study: A source book for experiential learning.* Beverly Hills, CA: Sage Publications (pp. 18–49).

Watzlawick, P., Weaklund, J. H., & Fisch, F. (1974). *Change: Principles of problem formation and problem resolution.* New York: Norton.

Wilson, S. J. (1981). *Field instruction: Techniques for supervisors.* New York: Free Press.

Yuen, H. K. (1990). Fieldwork students under stress. *American Journal of Occupational Therapy, 44*(1), 80–81.

CHAPTER 11

Breaking Through Barriers: The Confrontation Stage

If students don't have a chance to address their fears by having the chance to prove themselves, then at graduation, they will walk away with the same fears instead of having overcome them.

I had a breakthrough happen in what I really want to accomplish in the internship. . . . I can now focus . . . to find some solutions to some long-standing problems. Things are moving again and without so many struggles.

STUDENT REFLECTIONS

Afeter working your way through the exercises in the last chapter, you have a clearer and more concise understanding of the issues, both large and small, that you are facing in your internship. If you think about it, though, you will realize that you have already managed to prevent a number of difficulties from turning into problems. You have been encouraged to focus on those that are still unresolved, without losing sight of your previous victories—however small they may seem to you now—as you prepare to confront those that await you. This is the Confrontation stage—an opportunity for you to resolve some troubling issues, gain confidence in yourself, and continue to grow. It represents another level of "taking charge" of your internship and shaping it into what you want it to be, a process you began by taking an active role in your learning contract. In this case, taking charge means you will need to deal with that which most intimidates you or leaves you feeling ineffective in your internship.

THE TASKS YOU FACE

I chose to confront this situation and figure out just what was happening to me that made me feel unsatisfied. . . . Once I confronted these anxieties, though, everything worked out. I am beginning to confront this situation by making something good of this.

I recognized, sensed, and felt the great struggle exploring change can initiate within others as well (as me).

STUDENT REFLECTIONS

The Confrontation stage can be the most challenging time in your placement. The process of resolving the issues that continue to challenge you tends to take center stage. As you struggle to achieve a sense of independence, confidence, and effectiveness in your work, you are faced with the reality that such qualities are born not just of your skills and accomplishments but of your ability to overcome obstacles. Breaking through the barriers and taking charge of your internship is, for now, as important as the other work in your internship because it has become the most important and meaningful aspect of your internship.

In Their Own Words | **Voices of the Confrontation Stage**

The whole thing has made me realize how beneficial this crisis has been for me down the road.

I chose to confront this situation and figure out just what was happening to me that made me feel unsatisfied. . . .

Once I confronted these anxieties, though, everything worked out. I am beginning to confront this situation by making something good of this.

Our experiences have informed us of the trials and tribulations of interns trying to manage these barriers and bring their internships to a place where they are thriving in the experience, rather than just surviving the experience. To guide you, we have developed a model for approaching the problems you choose to take on. It will combine easily with any other models and approaches you may know. Regardless of the one you choose, though, a model itself is not going to be enough.

LEADING WITH YOUR HEART

Change can be very difficult. Perhaps you have had this experience either in or outside of your internship. You identify a problem and think you know clearly what is involved. You really want to do something about it. But every time you try to think clearly, you become confused and feel drained until you finally start to think about something else just to get away from those feelings. Or you may find that you begin making a little progress, only to drown in the frustration and anger that you feel about the problem and your perceived inability to do something about it. This sort of emotional paralysis can be extremely frustrating.

You may feel like you don't know what to do, but knowing is not the problem. And that is the limitation of even the best models. They are largely cognitive; they involve your *thoughts* more than your *feelings*. We have found that no model is going to work without three important ingredients—belief, will, and effort—which, interestingly, are driven not by your thoughts but rather by your heart. The lessons from the heart inform us that the affective domain must be an equal partner with the cognitive domain if you are to be successful in your confrontations.

The Power of Belief

If you do not believe that a situation can change, most likely it will not. Let's rephrase that. If you wait for a situation to change on its own, it might. However, the point of this chapter is to help you be an active agent of change, as this intern was: *"I have learned that after you have worked to put all the pieces into place . . . what you really want to accomplish finds its way into existence. I just knew it (it would happen)."*

If you do not believe that you can change a situation, then you probably cannot. Your perceived ineffectiveness will became a self-fulfilling prophecy. Gerald and Marianne Corey have written a marvelous book on self-change and effectiveness titled *I Never Knew I Had a Choice* (Corey & Corey, 2006). Perhaps *I Never Believed I Had a Choice* more accurately reflects the reader's experience. You need to *believe* that things can be different and that you can create the change that needs to take place. *"I had an incident where I didn't agree with the director's decision about a student. . . . I stood up for what I believed in and what I believed should happen."* This intern believed in the potential of the client. It was the intern's commitment to his own beliefs—which were coming from an intuitive feeling about what this client was capable of—that resulted in the client being allowed to take the upcoming GED— and the client passed the examination!

Of course, you can work on feeling confident, and that will help you succeed, or you can work on solving the problem, and that will help you feel more confident. For some people in some situations, it is the belief that is the foundation for the action. For other people, the converse is true. You know best what works for you. However, if you begin to try to bring about change and feel your confidence falter, you may want to step back and examine some of the sources of your lack of confidence (a review of Chapter 4 may help you here).

The Power of Will

Will is the intention and determination to make change occur. Intention and determination are powerful motivators, and the frustration and anger you may feel are normal parts of the emotional experience of trying to change something. You need to allow yourself to feel those feelings and not run from them or hide them. As with most feelings, that is how you begin to work through them and allow determination to come to the fore. As one intern recalled: *"I . . . just pondered . . . and just wondered whether I was actually helping the client or hindering the client. I was determined to know so I stayed with my decision of saying "no" and having the client stick it out. She did and I am happy that I did not give in. It would have been so much easier for me to say yes. But it would not have helped either the client or me."*

Another barrier to this determination is the resentment some interns feel about having to work on creating the change. You may feel that this problem is not of your making, and you may be right. So, if you didn't "do it," then why should it be you who has to fix it? Accepting responsibility can feel like you are accepting blame for the situation. But you are not accepting blame; you are empowering yourself, as is evident in this intern's comment: *"I need to speak up if I am being given too much work. Otherwise, I will burn out. Once I start saying 'hold off, I have too much to do,' I know it will affect my internship (in a negative way). But I have to do it."* The overriding goal is for you to empower yourself so that you feel able to take charge of your life and your situation when you want to, as this student learned to do.

A Case in Point

Belief, Will, and Effort

Belief, will, and effort, of course, are often interrelated. As you read the following account of a minority student's efforts to bring to the forefront an incident—and maybe a pattern—of questionable practices (from the intern's perspective) in one criminal justice agency, you can see how interconnected these three domains are and how the heart drives the commitment to one's beliefs, will, and efforts.

> *"I did an intake with a defendant who was in on a default warrant out of this court from nine years earlier, which he thought was taken care of because he had been on probation in this same court five years earlier and the issue of the warrant never came up. The warrant was not noted on his record and doesn't even come up in the record system. He was arrested on the default warrant after a minor automobile accident over the weekend and held for court after the police ran a check on him. This made no sense, so I decided to make sure it got resolved. Each department I went to said the same thing—let the other department handle it. Did someone forget that the court made this major error and discovered it nine years later? Now everyone wants to pass the problem onto another department, and no one wants to take care of it. He had been a model probationer and never tried to run from anything. I wanted to find out how such a thing could happen and not be noticed until nine years later. I heard about these kinds of things in my community and now I was seeing it happen to someone from my community. While I was working in one of the departments with the clerk and checking on this man's file like I was asked to do by this clerk, the director of another department came in and asked me why I was involved. I was without words. This was my case. I was checking out his file at the clerk's request. Then I was told by this director to leave the file and leave that office and let that office take care of it. I was bothered. It seemed as if no one cared that this man and his family were suffering because of the mistake this court made. The good thing is that they all knew that someone—me—was watching and that these matters were not going to be brushed under the rug as long as I worked there."*

This intern's efforts to bring about change actually did make a difference, although the internship ended before the intern was able to see those results.

As long as you wait for others to act differently, you are not in charge. *"I learned that I have to take responsibility for my internship . . . to take the bull by the horns, as you said. . . . I feel that it's an important step in the process. . . . I did not ask for an agency in crisis for an internship, but whether it is or not, I have got to get my internship on track."*

The Power of Effort

Finally, you have to do something; effort refers to what you actually *do*. It also refers to persistence and perseverance. Some of your difficulties will be relatively easy to analyze and resolve; others take more time. *"What my heart really wanted to do (in this internship) is now going to happen. . . . I am so excited about this great boost (to morale). I feel so motivated now and thrilled to be able to share from my heart something I know people will enjoy and benefit from. It was a lot of work but worth the effort."* Another intern said it more succinctly: *"At this point in my internship, it is reassuring to see that persistence does pay off."*

By now, some of you may be feeling impatient. *"Enough,"* you may be thinking. *"I am ready. I am willing. Let's go!"* It will not take long to read the model that follows, but it takes focus and practice to use it well. We hope you are willing to make the effort and that you find the effort worthwhile. But if you are struggling with it and you are sure you understand it, you may want to come back to this section and consider these issues again.

A METAMODEL FOR BREAKING THROUGH BARRIERS: EIGHT STEPS TO CREATING CHANGE[1]

The mere formulation of a problem is often more important than the solution.

ALBERT EINSTEIN

I knew that my internship might be a little more difficult to manage than most others because I was employed at my internship site. I also knew that it would be difficult to work full time, complete all my internship and seminar requirements, baby-sit part time, and to maintain all of my relationships and my new marriage. I knew these things, but I didn't know how I would get through them.

STUDENT REFLECTION

We hope this model will be both simple to understand and helpful to use. Before we explain the steps, though, a few introductory comments are in order. First, the model is designed to slow you down a bit. Yes, you read that correctly. If you have a problem that is really troubling, you are probably focused on what to do about it and what

---

[1] This model is derived from a model for interdisciplinary integration and problem solving (see Sweitzer, 1989) that in turn has its roots in the Behavior-Person-Environment (BPE) principle advanced by Kurt Lewin (1954).

action to take. Anything, you figure, has got to be better than the ways things are. Or you choose a strategy because it worked in another situation or because a friend used it successfully. Thinking intelligently about a problem involves some careful analysis, some goal setting, and only then deciding on and implementing a plan of action.

Focus on THEORY

Problem-Solving Models

The literature is replete with techniques for solving various kinds of problems and with generic problem-solving models (see, for example, McClam & Woodside, 1994). The generic framework is hardly a new idea; in fact, it has its roots in the ancient teachings of Buddha, who wrote about the Noble Truths, including suffering, the origins of suffering, and the path leading to cessation of suffering (Watzlawick, Weaklund, & Fisch, 1974). Most models include steps that answer the following five basic questions:

1. What is the problem?
2. What do I know about the problem so far?
3. What is my goal?
4. What are my alternatives for reaching that goal?
5. What is my plan of action?

This model is also designed to broaden the way you think about your problem. We want you to move beyond the initial stage, where you hold one individual responsible, and look at the perspectives and contributions of many other people. We also want you to consider the wider context in which the problem is occurring, including the systems that are involved, such as work groups, the office as a whole, and other agencies with whom you work. This model asks you to think and feel your way through a process, not simply to plug in answers to questions and in turn get a score that instructs you what to do next. It also involves a process of engaging others, namely your system of support and resources. For example, you may want to work on the problem with other interns at your site, discuss the issue(s) in seminar class or over the Internet with your support teams, work with your campus and site supervisors to find a resolution, or do all of the above.

A word to the wise . . . Collaboration with others does not mean dependence on them. When faced with a problem on new turf such as the internship, you may find yourself slipping into dependent habits. When engaging your support systems for help, do so with a list of what you already have done to move the problem along.

We suggest that you first read through the model in its entirety and follow the example we included. Next, return to the first step, and using an issue that is problematic to your internship, work your way through all eight steps. To help you through the steps, we have provided guiding questions and identified potential pitfalls. After you complete the tasks of each step, you will find guiding questions for reflection before moving on. We designed the reflection components so you can write about them in your journal if you wish. You will also be advised to make a reality check before moving on to ensure that you benefit from the support and perspectives of your support and resource systems.

One Step at a Time

In the middle of difficulty lies opportunity.

ALBERT EINSTEIN

STEP 1: SAY IT OUT LOUD

Quick—in one sentence—what is the problem? Don't think too much about it; just say what comes to you (in the next step, you will name the problem more carefully). Then write down what you said. You can learn something important about how you are thinking from the way you first state the problem. Consider an intern who is struggling with someone else at the internship, perhaps a co-worker, supervisor, or client. Here are some ways the intern might initially describe it:

- I can't get anywhere with her.
- She is not interested in changing.
- This co-worker needs a different unit to work in.
- No one told me how to handle this.
- Their suggestions aren't working with this co-worker.

Do you see how each of these statements locates the problem in a different place? Where did you initially locate your problem? Where are some other places it might be located?

Reality Check!   Does it make sense to involve anyone from your support system in the process at this early point? Your campus supervisor or instructor? Your site supervisor? Peers? Co-workers? Someone else?

STEP 2: NAME THE PROBLEM

Now it is time to let go of the feelings, the thoughts, and the impulses that you've been having and think clearly about the behaviors that are causing problems for you. Don't think about the people; think about the specific behaviors. Behaviors are actions that anyone can see if they are able to observe. You also need to think about the thoughts and feelings you are having as a result of these behaviors.

In the next few sentences, see if you can describe exactly how this problem affects you. Use the following questions to guide your thinking. Once you have worked on this group of questions, you will have a concise, clear statement of the problem in paragraph form to add to your worksheet.

- In what specific way does it create unpleasantness?
- Does it embarrass you?
- Does it make you angry?
- Does it hamper your ability to work?

Focus on PRACTICE

Using the Three-Column Method for Step 2

Here is one way to work on Step 2, using a process that will be somewhat familiar to you if you have been using the "three-column" method in your journal entries:

- Divide a piece of paper into three columns. In one, list the behaviors.
- In the next column, list your thoughts that result from those behaviors.
- Finally, list the feelings that result from each behavior.
- Now look at what you have written and try to state the problem in one sentence. It can be a long sentence, but try to use just one sentence.
- Think about when and in response to what this problem usually occurs and see if you can express that in a sentence. If it is not a recurring problem, this sentence will not be important.

Here is an example of what such a paragraph could look like:

> *When she runs away from the program, it angers me because I think about all the work I have done with her, and she just blows it every time. It's just like when I used to work with kids on the streets. They would do the same thing, not listen to me. It makes me feel helpless, sort of powerless, and useless in my work.*

A word to the wise . . . Keep in mind your initial notion of the problem may only be a surface issue. In our experience, many of the problems that interns choose to confront in this stage are issues that have come up again and again, although they may appear in different forms because the specific problems they create can vary. For example, an intern who has issues with acceptance may choose to eat lunch alone and thus not be included in the energy and discussions that go on in organizations where working lunches are part of the normal routine. Problems can develop if others interpret the intern's choice to eat alone as an indicator of a lack of investment or interest in the issues discussed at lunchtime. Spending time sorting out the real problem now will save time and energy later.

Reality Check! You need to ask yourself again: Does it make sense to involve anyone this early in the process? Your campus supervisor or instructor? Your site supervisor? Peers? Co-workers? Someone else?

STEP 3: EXPAND YOUR THINKING

This step is not unlike the reflective observation step of Kolb's learning cycle (Kolb, 1984). You are asked to think about the problem by examining its components from different perspectives and along different dimensions. Although taking perspective and examining it are insightful experiences, they can also be frustrating for reasons we have already discussed. But someone has to do the investigative legwork, and in this case, the only one who can is you. Begin with some scrap paper. As you work your way through the following four guiding questions, try to keep in mind the role of politics and other subtle dynamics. Be patient and persistent with yourself; some of these steps may not come easily to you, but they can all be mastered with a little perseverance.

- **Who are the players involved in this issue?**   Identify as many people as possible who are related to the problem. Including them does not mean that they are causing or contributing to the problem, only that they are part of the scene. Don't forget to include yourself.

- **How does each of them see the issue?**   Work your way through each person on your list and try to put yourself in that person's shoes. Try to see the situation from their points of view by imagining what it is like to have their positions on the staff, to have been there as long as they have, to work with an intern, or to have you in their work life.

Think About It

Where You Sit Is How You Stand

Some of our students have found it helpful to imagine all the individuals involved with the problem forming a circle, which in turn forms the hub of a wheel. Each person on the wheel sees the issue differently, depending on where they are located. Their perspective, in turn, influences their stand on the issue at hand. Try to imagine how each of these individuals views and feels about the problem. From your perspective, how might each of them be contributing to the problem? How about from their perspectives?

- **What are the major systems involved in the situation?**   Each of these individuals is part of and influenced by a number of systems. (You may want to review Chapter 7 if the notion of systems is not yet clear to you.)

  - Your first task is to make a list of all the systems that are somehow connected to the problem. You may find it easier to approach this task by brainstorming. Begin with those directly connected to the situation and move to those with indirect connections. Or you might find it easier to be more organized and work your way from the inside out or the outside in. The inside group of systems includes the staff (and any subgroups of the staff you can identify), your work group, the clients, and anyone else who connects directly with the agency.

  - Now think about the agencies that work with yours, either collaboratively, competitively, or in adversarial ways.

  - Next, identify the sources of funding and other resources.

  - Don't forget your systems of support and resources both on campus and related to the placement site. This includes your campus supervisor or instructor, professors, administrators, and, if you have one, the seminar class or web-based support team. There may be more; if so, keep going!

- Now go back to your list and eliminate any systems that, on further consideration, seem to have very little to do with the problem.

- Now think about how each system—as opposed to the individuals in it—contributes to the problem. To do this, you need to focus on the rules and roles that drive individual behavior (again, you may want to review this section of Chapter 7).

- **What about you?** Yes, you! In all likelihood, you are contributing to the problem in some way. Once again, we urge you to return to an earlier part of the book—this time to Chapter 4 and the exercises you did. Are some of your personal issues and values being touched by this situation? Your initial statement of the problem (Step 1) may be a clue here. Does your response to this situation remind you of any other situation or group of situations? Is there a pattern or a set of irrational beliefs at work?

- **How did it get so complex?** Now go back over all the ideas and insights you have had during this portion of the problem analysis. Write a paragraph discussing them. Notice how you have begun to create a new picture of the problem that is more complex and takes into consideration a number of perspectives on the issue.

Reality Check! At this point, you have done a considerable amount of deliberate thinking about this problem. If you have not as yet engaged those in your support system, it is time to do so. Ask yourself again, who makes sense to involve at this point? Your campus supervisor or instructor are valuable resources, as is your site supervisor. If you are not involving them and they *should* know about this problem, you need to ask yourself why you are not involving them and work with someone to resolve this new issue.

STEP 4: CONSIDER THE CAUSES
Now go back over your notes and summary from the last step and try to come up with a list of the potential major causes of the problem. Notice we did not say the major cause; there is always more than one possible cause. Be sure to consider all the individuals and systems you examined before concluding your list.

Reality Check! It's time to find out how you are doing seeing possible causes for this problem. It's best to check this out with others before moving on.

The Eight Steps to Creating Change

1. Say It Out Loud
2. Name the Problem
3. Expand Your Thinking
4. Consider the Causes
5. Focus Your Attention
6. Determine Your Goals
7. Identify the Strategies
8. Create the Change

STEP 5: FOCUS YOUR ATTENTION

You have identified a number of dimensions of the problem. You will need to make changes to and alter the dynamics discussed in the cause statement, but where do you begin? Practice the exercise below. After doing it, go back and add a new paragraph to the statement you are developing. This paragraph needs to answer two questions:

- What is blocking me from resolving this problem?
- What is it that I need to change?

To help you with Step 5, try dividing a piece of paper in half and labeling the columns "What I can change" and "What I cannot change." We are asking you to think about the issues over which you do and do not have influence. Let's look at these lists more carefully, this time in terms of what you can and cannot do about the issues. When you do have control over the issue, the responsibility is to do something about it; that is, do what you can do about it. When you have no control over an issue, your responsibility is to accept it; that is, resign yourself to accepting this as something that you must live with.

The possible pitfall in this commonly used exercise is the tendency to confuse what you *can* do something about with what you *cannot* do anything about. As you look over the lists, check for those issues that you continually try to do something about but actually have no control over. Being caught in this danger zone is like spinning the wheels of your car on ice: You are going nowhere fast and using up a lot of energy trying to get there! Keep in mind that there are also those issues that you actually have some control over and can do something about. That is where your energy must be focused to move this problem forward.

Reality Check! If you find issues on your list of things you do have control over but privately doubt you will do anything about, it is important that you consult with someone who you know will help you sort out this issue before moving on.

STEP 6: DETERMINE YOUR GOALS

For each cause of the problem that is on the "changeable" side of your list, you now need to develop goals for change. State as specifically and behaviorally as you can what you want to be different about the situation. There is no room for fuzzy thinking here. Just wanting the issue to go away may be your most salient emotion, but now is the time for clear and concrete statements. Be sure that you have covered all the individuals and systems that you believe have a substantial role in causing the problem.

One possible pitfall here is to confuse a goal with an action step. For example, suppose that one major cause you identified is that your duties are radically different from what you expected and were told. Changing placements is not a goal; it is an action step. The goal would be to bring your daily actions more in line with your stated learning goals. One reason this is important is that there are usually multiple ways of achieving any goal. If one approach doesn't work, take some time to be disappointed, but then think of another. In the example just given, if you made switching placements your action step and then were told you could not make the change, you might feel the problem is now unsolvable. If you keep your real goal in mind, though, you are more likely to come up with another approach.

Reality Check! Goals are a big deal. Before moving on, it is best to review them with someone who you know is able to have a clear perspective on the problem and provide you with useful feedback.

STEP 7: IDENTIFY THE STRATEGIES

Your task here is to develop a list of possible concrete actions you could take to meet each of the goals you identified previously. In developing the action steps for each goal, be sure the interventions reflect your knowledge about the nature, cause, and context of the problem. The principle of effectiveness is operating here; that is, each action must be effective for each goal. The picture you have created of the problem in this step is one that is action packed and future focused. When you write your paragraph about this step, let yourself be guided by the question: *What must I do to make my goals happen?*

You may remember that we called this model a metamodel because it allows you to incorporate many other approaches and strategies. You have probably studied many approaches to resolving interpersonal problems and conflicts, making constructive changes in yourself, and making change at the system or community level. But if you have not or if you want to do some more exploration before deciding on a strategy, we have included some resources at the end of the chapter.

Reality Check! Strategizing is an art and a science unto itself. You may want to involve at this point someone you know will be a faithful sounding board for you—who can stay with you and the process until the goals are clear and realistic.

STEP 8: CREATE THE CHANGE

This is the step you have been waiting for! Ever since the issue became a barrier to your internship, you have focused on this step. It is the time when change happens. If you've done your work with determination and effort, with genuineness and a willingness to be reflective and introspective in your thinking, and if you asked yourself the tough questions that needed to be asked, then you are ready to move on. It is time to do the achievable and implement your plan with commitment and perseverance. Yes, you've given it your best, and your best is good enough. The potential pitfall of this step is to fail to do what you can do about the issue, resigning yourself to the status quo when you have the capacity to make change occur.

The guiding question for this step is: *How will I know when things are different?* Let this question guide you as you implement your plan and move through the barrier that has prevented you from moving ahead in your internship. Let it also guide you as you write the last paragraph of your problem statement. Congratulations. You did it!

Reality Check! This is no time to forget those who have supported you throughout the process. Check back in with each of them and let them know how you worked things out. And, of course, thank them. What's much appreciated is to tell them how they made a difference in your decisions and the outcome of what once was a challenging problem for you.

Not Being Taken Seriously: The Case of the Rebuffed Intern

Some of you may be comfortable with this model and can work with it right away. But for others, it may be too abstract to be useful. So, we are going to work through an example for you. We will not detail all aspects of the steps, but we will show you how the eight steps work in this particular situation.

We are using the case of Kim, an intern at an adoption agency. Because of the nature of the problem, this case could apply to a wide variety of contexts for an internship. So, as you read about Kim's situation, substitute the focus of your internship in place of that of the adoption agency.

Kim is conducting an advanced internship at an adoption agency and is very excited about this opportunity. He thinks this is the kind of work he would like to do for a career, and he really likes the agency and his supervisor. He was told, and it was written into his learning contract, that he would eventually have his own cases and would write case notes and reports.

A few weeks into the internship, things are going smoothly. Kim has been working alongside his supervisor, and the supervisor thinks things are going well. The supervisor wants to give Kim two cases, one of which is a new case and one of which is a postadoption situation where the family only needs routine follow-up visits. In both cases, the supervisor is going to monitor Kim very closely. However, about that time, a new executive director comes to the agency.

The new director decides that it is not appropriate for an intern to have any cases, since the intern is not fully qualified, and directs the supervisor to take the cases away from Kim. The director understands that having cases was in Kim's learning contract, and he is apologetic to Kim and to the college supervisor; but the director is firm about the change. The supervisor feels badly about doing it but explains that it is the director's decision. Kim is very disappointed and considers leaving the placement.

Step 1: Say It Out Loud

Once able to talk about the problem, Kim said, "My internship is ruined! This new director is too rigid and doesn't know me. He doesn't trust me!" Just saying the words brought to the surface feelings that had been bubbling below for quite a while.

Reality Check! I need to contact my campus supervisor right away. Or maybe the placement coordinator on campus. They won't let the director get away with this. After all, an agreement is an agreement.

Step 2: Name the Problem

I know I can handle these responsibilities, and it doesn't seem fair that I am not at least going to get a chance to prove myself. I am afraid I will lose a major learning opportunity as a result of this change in my internship. I am bored because I don't have anything to

continued

Not Being Taken Seriously: The Case of the Rebuffed Intern *(continued)*

do that is worthwhile and makes me think; and I am frustrated because there is important work going on around me and I am not included in it. I'm beginning to think they don't trust me. I spend a lot of time trying to think of what I could have done differently.

Reality Check! I don't know. I left a message for my campus supervisor about this, and he's not called back, and it's been a whole hour. I realize he's not on campus today, but you'd think he'd call in for messages every hour!

Step 3: Expand Your Thinking

My Supervisor . . . is probably embarrassed. But there is a new director, and my supervisor has to be careful not to be seen as disruptive or disrespectful.

The Director . . . is under pressure as the new boss. I remember that there have been some concerns about the agency and its performance, and a couple of adoptions last year did not go well and the children had to be removed.

The Clients . . . would probably benefit from consistency in case work, and I am not going to be here for their whole time with the agency.

My Issues . . . I have to admit that I have always resented not being taken seriously. It goes back to being the youngest in my family and having all my brothers make fun of me and not listen to my opinions. Maybe I am overreacting in thinking that these people don't take me seriously.

Reality Check! Okay, it's time to get focused and do something about my situation. My campus supervisor did return the call, although it was two hours after I called him. I was a bit taken back when he observed that I seemed quite agitated and that this behavior seemed so unlike me. Maybe he was right in saying that usually I dig in to solve problems, and this time, I seemed to be stewing over the situation instead of grabbing hold of the problem and trying to solve it. Maybe he's right. I don't know. I think I'll send the interns in the seminar class an e-mail and tell them that I'm dealing with a nightmare of a situation and could use some of their ideas. That's what Jo did last week, and she solved the mess she was dealing with.

Step 4: Consider the Causes

1. The director has legitimate ethical concerns about confidentiality and privacy.
2. The agency is worried about its performance record.
3. I have been unwilling to compromise my position.

Reality Check! Okay, I'm not in this alone. That feels good. My peers had some good ideas and dished out some blatant reality for me to face. My campus supervisor surprised me as well. He asked me what I had done up to this point to move the problem along, and I told him about e-mailing the seminar class and working with a small support group some of us in the class set up over the Internet. He liked that.

continued

Not Being Taken Seriously: The Case of the Rebuffed Intern *(continued)*

It seemed like he was more willing to work with me when he saw that I was working on the problem.

Step 5: Focus Your Attention

I cannot change much about this situation, except my reactions to it. But maybe I can still have some valuable learning anyway.

Reality Check! Okay. I'm onto this one. There is no magic wand being used by the campus supervisor or placement coordinator to solve my problem. This is life, and life happens. I guess this is what is meant by "learning lessons in life." I can't change the director's decision, only she can. I don't want to throw away this internship because there is so much more about it that makes it quite valuable to my career. So, if I want this internship, cases or no cases, I guess I've got to find some other value in staying at this site.

Step 6: Determine Your Goals

I can think of three primary things that have to happen in order to make this internship work for me. I must:

1. Stop being so upset about this situation
2. Identify some other responsibilities that can help me meet my learning goals
3. Determine if this change in my learning contract meets with the approval of the academic program, my campus instructor, and the licensing board's requirements for direct service before graduation.

Reality Check! I better talk with the site supervisor. My peers were right in suggesting that if I work with my site supervisor. Together, we'll find other ways to keep me at the site and get me involved in work that has value for my future career.

Step 7: Identify the Strategies

- Develop some alternative suggestions for responsibilities before meeting with my site supervisor to discuss the future of my internship.
- Review my learning goals with my site supervisor and determine if there are other options for working directly with clients; e.g., at a sister agency.
- Ask my site supervisor for suggestions as now I have few of my own!
- Identify some new learning goals as a result of this situation, such as understanding some of the legal and ethical issues of internships.
- Clarify with my campus supervisor whether this change in the learning contract is acceptable to the standards of the academic program and to the licensing requirements for adoption work.
- Consider moving the internship to a sister agency for part of the time.

continued

Not Being Taken Seriously: The Case of the Rebuffed Intern *(continued)*

Reality Check! I can't believe how much my campus supervisor knows about this internship stuff. I should have known. He's been doing it for some time. I bet I'm not the first student who wanted to throw away the internship just because of a change in circumstances. And, boy, does he know how to brainstorm real good ideas! Guess the guy really knows how to get me to rescue my own placement from an early demise. I got some great ideas from a couple of interns in the seminar class too. I'm going to have a great internship after all.

Step 8: Create the Change

I believe that I can bring about the changes that are needed. But I must persevere until I see changes that satisfy the need for a challenging and meaningful internship for me.

1. I will e-mail the professors in my academic major and the placement coordinator about this matter immediately to determine if my academic program allows me to be in an internship that does not allow students to be involved in direct service work.
2. I will meet with the site supervisor and discuss the ideas I've developed, thanks to my peers and my campus supervisor. I just have to be sure that it is possible to take on some of these other responsibilities; they are challenging, and they are meaningful to me and to the site.
3. I will explore with the site supervisor the possibility of conducting part of the internship at a "sister" or contracting agency.

Reality Check! I am a bit surprised at what I can do to move this situation forward. It really is not all that difficult—but it sure felt that way before I started working on it as a problem to be solved, not a rejection by the site. I just have to accept that until I hear back from my professor or the placement coordinator. I don't know if I can stay at this site. If I can't, then I will have to be placed at a new site—maybe a sister site to this one. If I can stay, then I have to work closely with the site supervisor to create a new learning contract that ensures that I am satisfied with the new responsibilities. It's going to be okay. Makes me wonder why I have been stuck for so long.

SUMMARY

This chapter has prepared you to confront the issues that challenge you and has given you some tools by which to do it. However, sometimes it is also helpful to pay attention to how others approach and resolve problems. Look around you. You probably won't have to look far until you find someone—a friend or a peer—who thinks very differently than you about problems and their solutions. What are those differences? How might others have dealt with the issue(s) you are facing? Can you learn anything from their approaches?

We close this chapter with this student's thoughts:

I never thought the crisis I experienced and the change I brought about could be used as a selling point on my cover letter. . . . The whole thing has made me realize how beneficial this crisis has been for me down the road.

STUDENT REFLECTION

For Contemplation

PERSONAL REFLECTION:
SELECT THOSE INQUIRIES THAT ARE MOST MEANINGFUL TO YOU.

1. Go back to the problem you used to work your way through the model. Go through it again, only this time for each step and ask yourself two questions: "On what do I base my thinking? And what generalizations am I making?"

2. Learning to solve problems takes intentional commitment on your part to be an active player in finding a solution. Being active includes consulting with peers as well as supervisors. One mantra we encourage our students to use is "Bring potential solutions to the table." By this, we mean before approaching a supervisor or peer with a problem, think about at least two different ways of resolving it.

SEMINAR SPRINGBOARDS

In seminar class or in your support teams, select a problem you or a peer is having or make up a problem you anticipate occurring. Choose the most challenging of the problems for this exercise. This time, when you work your way through the 8-step model, have peers select their parts in a role play of the problem solving.

For Further Exploration

ON PROBLEM SOLVING

McClam, T., & Woodside, M. (1994). *Problem solving in the helping professions.* Belmont, CA: Brooks/Cole.

Clearly written and concise with a problem-solving model and lots of cases to study.

ON INTERPERSONAL EFFECTIVENESS

Corey, G., & Corey, M. (2006). *I never knew I had a choice* (8th ed.). Belmont, CA: Brooks/Cole.

Excellent book with a good chapter on relationships.

Johnson, D. W. (2006). *Reaching out: Interpersonal effectiveness and self actualization* (9th ed.). Upper Saddle River, NJ: Prentice Hall.

Clear and incisive with lots of examples and exercises—a classic.

ON MAKING INTRAPERSONAL CHANGE

Brill, N., & Levine, J. (2004). *Working with people: The helping process* (8th ed.). New York: Longman.

Self-understanding and self-modification are discussed throughout this book.

Ellis, A., & Harper, R. A. (1975). *A new guide to rational living.* Upper Saddle River, NJ: Prentice Hall.

Ellis's ABC model is explained with helpful self-help guides to work through your own situations.

Gibbs, L., & Grambrill, E. (1996). *Critical thinking for social workers: A workbook.* Thousand Oaks, CA: Pine Forge Press.

Provides a comprehensive skills-based approach to develop reasoning skills for effective decision making.

Grobman, L. M. (Ed.) (2002). *The field placement survival guide.* Harrisburg, PA: White Hat Communications.

Offers social workers an edited volume of writings that include skills and advice in confronting challenges.

Sweitzer, H. F. (1993). Using psychosocial and cognitive behavioral theories to promote self understanding: A beginning framework. *Journal of Counseling and Human Service Professions,* 7(1), 8–18.

Combines Erikson's work with personal reaction patterns and discusses implications for human service work.

Weinstein, G. (1981). Self science education. In J. Fried (Ed.), *New directions for student services: Education for student development.* San Francisco: Jossey-Bass (pp. 73–78).

An easy-to-follow model for interrupting reaction patterns that may be troubling you.

ON SYSTEMS CHANGE

Garthwait, C. (2005). *The social work practicum: A guide and workbook for students.* (3rd ed.). Needham Heights, MA: Allyn & Bacon.

Offers a comprehensive approach across contexts of practice for skills development in bringing about change.

Homan, M. (2008). *Promoting community change: Making it happen in the real world.* (5th ed.). Belmont, CA: Brooks/Cole.

Excellent, well-written text on community organizing with interesting thoughts on the change process.

Kotter, J. P., & Cohen, D. S. (2002). *The heart of change: Real-life stories of how people change their organizations.* Cambridge: Harvard Business School Press.

Offers an approach that affirms the importance of teamwork, visions, and communication in bringing about short-term wins that create change.

Shepard, H. A. (1975/November). Rules of thumb for change agents. *Organization Development Practitioner*, 1–5.

Offers aphorisms to think about when you are the change agent.

Watzlawick, P., Weaklund, J. H., & Fisch, R. (1974). *Change: Principles of problem formation and problem resolution.* New York: Norton.

Somewhat theoretical but very interesting. A groundbreaking book on why change can be so stubborn and what to do about it.

References

Corey, G., & Corey, M. (2006). *I never knew I had a choice* (8th ed.). Belmont, CA: Brooks/Cole.

Lacoursiere, R. (1980). *The life cycle of groups: Group developmental stage theory.* New York: Human Sciences Press.

Lewin, K. (1954). Behavior and development as a function of the total situation. In L. Charmichael (Ed.), *Manual of child psychology* (2nd ed.). New York: John Wiley & Sons.

McClam, T., & Woodside, M. (1994). *Problem solving in the helping professions.* Belmont, CA: Brooks/Cole.

Sweitzer, H. F. (1989). The BPE framework: A tool for analysis and interdisciplinary integration in human service education. *Human Service Education*, 9(1), 11–19.

Watzlawick, P., Weaklund, J. H., & Fisch, R. (1974). *Change: Principles of problem formation and problem resolution.* New York: Norton.

GOING
THE DISTANCE

*You have come quite a distance in your intern-
ship. You have dealt with the concerns of beginning, you have launched and nurtured a number
of relationships, and you have learned how to
identify and cope with problems. The next several
weeks should be a time of great productivity and
enjoyment, although they will undoubtedly have
their challenges. Chapter 12 examines in more
detail the joys and challenges of the Competence
stage. Along with this exhilaration, reality is ever
present as you begin to take notice of the complexity of the demands of the work and the workplace,
especially in the areas of legal, ethical, and professional matters. Chapter 13 provides you with
an overview of the types of ethical and legal concerns that you begin to encounter now that you
are grounded firmly in the work. Another aspect of
reality that creeps into consciousness at this time
is thoughts of ending the field placement. This is
yet another critical time in the internship. Ending
well is not easy, but it can be done and in many
different ways. Chapter 14 will help you meet the
concerns and challenges of the final stage—that
of Culmination.*

CHAPTER 12

Riding High:
The Competence Stage

The price of greatness is responsibility.
WINSTON CHURCHILL

There comes a time in nearly every internship when interns tell us that they are enjoying their internships in a whole new way. The initial anxieties have subsided, problems big and small are being resolved, and considerable confidence has been gained. There is a sense of feeling grounded—finally! If this sounds like what you are experiencing, then you have entered the next stage in internship: the Competence stage. Like any stage, this one has its concerns and its challenges, but for many interns and supervisors alike, it is the most exciting and rewarding time in the field experience. With confidence and with pride, the intern demonstrates each day in placement the knowledge, the skills, the values, and the attitudes expected of the field site, along with the interpersonal skills, motivation, and cognitive abilities needed to attain success (Klemp, 1977).

THE TASKS AT HAND

Students arrive at this stage at different times in their internships. Some of your peers may not be here yet, while others may have been here for some time. What is important is not how fast you and your peers reach this juncture but rather having the confidence in yourself that you will get here in due time. In this chapter, we review some of the pleasures you can expect from this stage and prepare you to face the challenges that are an integral part of it. Chief among those challenges

249

are learning coping strategies that allow you to thrive in this stage and sharing concerns openly, as appropriate, so that you can continue to move forward in your development and work. Such response strategies will allow you to deal effectively with the tasks at hand while enjoying the sense of competence that earmarks this stage.

In Their Own Words   **Voices of Competence**

Things that didn't seem to come together have finally come together.

I feel that all my years of schooling have been for this exact purpose. It's like nothing that I have ever experienced before.

This by far has been THE best experience I have ever had (in a workplace setting). For the first time in a very long time, I feel on top of what I am doing. I know my work, I am respected by my supervisor and co-workers, I get great pleasure from seeing the difference I make in the community, and I just can't get enough of the work. I finally know what I am going to do "when I grow up!"

I am confident now. I take a lot of pride in my work, and I think more independently. I am doing important work—busy work. My supervisor is great. I have developed important relationships with her and the workers, and I feel a part of the team. I have learned so much about myself too. This I didn't expect when I started out. This internship is everything I hoped for, and it happened.

TAKING CHARGE OF THE JOURNEY

Students have described this stage as the *emotional moment* of the internship. It is this stage—the Competence stage—that defines the internship for the students. Finally, they *know* what they are doing and feel proud of it! They tend to work independently at this point and make decisions as needed—under supervision, of course. They are also developing a greater appreciation for how their placement site contributes to the community and society as a whole.

Marked by "good" feelings, this is the stage of high accomplishment when the interns feel confident about their knowledge, their skills, and their sense of competence. Importantly, the learning contract is being fulfilled during this stage. The interns feel in charge of their work, and they feel in charge of their internships. They are invested in their work—work which they expect to be personally meaningful and valued by the organization. It is during this stage that quality takes on added importance—quality of the work, quality of the supervision, quality of the workplace behaviors, such as ethical regard by co-workers, and the quality of their seminar class and assignments. Why? Because during this stage, being competent is what it's about, and the need for that sense of competence extends to not wanting to be involved in what they consider to be a waste of their time when they could be attending to their work from which they derive the sense of competence they are enjoying.

As you join others in reaching this stage, keep in mind that interns have different styles of expressing their emotions, and while you may experience this stage as a subtle, yet profound, shift in feelings, others may experience it as a positively giddy swing in emotions. Keep in mind as well that no one gets to this point all by themselves. Along the way, there have been people who have been there for you, who have believed in you, who have given you the benefit of the doubt even when you might not have earned it. Perhaps it is a professor or staff on campus, perhaps it is your campus or site supervisors, or perhaps it is your seminar peers who made all the difference in being where you are today and feeling as you do. At this point in the internship, it is important to take the time to reflect on what you've been given by others and to think of ways to let them know how they made a difference in your internship.

The Concept of Competence

We use the concept of *competence* to refer to the intern's overall productivity and achievement, including interpersonal and intellectual skills (Chickering, 1969). The intern's concerns during this stage focus on developing a sense of competence and taking charge of the internship to ensure a quality experience. Striving for competence often means being all that you can be in your role as intern and doing your very best. You may recall the discussion of competence in Chapter 4 and Erikson's (1963) position that a sense of competence includes feeling like a capable person even when you stumble or fail. Most likely, you have experienced times during your internship when you were disappointed by your performance in some way. According to Erikson, if you have a strong sense of competence, moments like these will not shake the basic confidence you have in your ability to succeed and move ahead.

A Time of Transformation and Empowerment

You may realize by now that you have grown not only in terms of your knowledge and skills but in personal ways as well. The transformation that is occurring may seem slow and subtle, so much so that you didn't notice until someone commented on it. Or it may seem sudden and dramatic, leaving you to wonder what is happening. Either way, you are changing. You have successfully faced new situations, including ones that seemed overwhelming at the time; you have accepted greater and greater responsibilities, developed your competencies, and enhanced your faith in yourself; and you have learned to develop new ways of thinking about situations and the people with whom you work, including yourself. None of these changes have come easily, but they add up to a transformation that lets you see more, do more with what you see, and be more in charge of your internship. In short, you are empowering yourself. Each day, the changes are evident in your work and in your approach to it.

Your emotional landscape is changing as well. Many interns feel an enhanced sense of mastery as their skills become sharper, and their growing sense of autonomy means less need for constant supervision and direction. There is also a tremendous sense of confidence associated with this stage. The anxiety, awkwardness, and trepidation of earlier stages have gradually given way to a period of calm and inner strength. Interns often tell us that things have settled down at the site. However, life at the internship site is still hurtling along at the same pace. It is not the world around you that has changed and become more peaceful; rather, it is you who have developed an inner sense of calm from which you will derive strength as you move through this intensely productive period in your internship.

ENJOYING THE RIDE

This is the time in field placement when your experience most closely matches your ideals and the images you carried of what an internship is all about. You have greater responsibilities each week (and may even be wanting more!) as you strive to reach your goals, and most of your energy these days is devoted to fine-tuning your skills so that they become second nature and can be generalized to everyday situations.

Chances are that you are also seeing your internship through a set of new and different lenses. Having the knowledge to feel competent in doing the work is critical during the Competence stage. It's not as simple as what you know and what you don't know. Discovering knowledge you didn't know you had and discovering gaps in knowledge you thought you had can be overwhelming at times. One intern described it this way: *"I went to a training session yesterday . . . that made me think about my approach with all of my clients and how far I have to go. I found myself thinking about this new knowledge and how learning this little bit of knowledge was just the tip of the iceberg. It was all the stuff I didn't even know I didn't know!"*

This is often a time when interns are able to differentiate more clearly between what is important to their work and what is not (Dreyfus & Dreyfus, 1980). Subtle dynamics that escaped you before are probably visible now. Remember when you were concerned about acceptance issues? *"Now I look back on all those feelings of inadequacy and laugh!"* was one intern's reaction to that period in her placement. Even though you spent quite a bit of time trying to find your way around the organization while you were concerned about these issues, the organization's internal politics and stance in the community were virtually invisible to you because your primary concerns were focused elsewhere. Now, however, you are more savvy and in tune with the norms and rhythms of the organization and you feel office politics when they are affecting your work. Why? Because your primary concerns now are on the quality of the work and your competence in doing it. Office politics can affect both of these dimensions. So, too, can a difference in philosophies about the role your site should be playing in the community and society.

Redefining Supervisory Relationships

When we ask our interns to reflect on the changes that have occurred in their field placements, they immediately identify changes in relationships with their supervisors. They smile, for example, as they tell us that they no longer need to check in with their supervisors as they go about their work. *"At the beginning of my internship, I was constantly asking: 'What do I do next? What do I do now?' Now I am not asking what to do. I am just doing it."* We are often struck by the quality of the relationships our students develop with their supervisors. They not only are supportive and encouraging connections but often become redefined into mutually satisfying relationships that last long past graduation day.

AN EMERGING SENSE OF EQUALITY

Working with supervisors and co-workers becomes easier during this stage. Communication becomes more comfortable and open, issues can be approached without concern about rejection or conflict, genuine teamwork can develop, and supervision can be a source of insight and feedback for personal as well as professional growth. Often, a change in the openness of the supervisory relationship becomes evident at this time as well. Most of you would not have considered challenging your supervisors on theoretical grounds earlier during the placement. However, a combination of confidence and a sense of emerging equality may make a big difference in your willingness to engage the supervisors in academic discussions as well as in disagreements (Lamb, Baker, Kennings, & Yarris, 1982).

SHIFTING PERSPECTIVES

There has also been a shift in your primary focus. Chances are you are talking less about yourself during supervision and in seminar class, and more about the projects, the needs of clients or the community, and the contributions of your site to society's solutions for its problems. This is just one more shift that tells you important changes are taking place (Tryon, 1996). One explanation for such shifts is that your supervisory needs are changing (Hersey & Blanchard, 2008; Tryon, 1996). Your former dependence on your supervisor for direction and support has given way to working more autonomously and developing more equal relationships with your supervisors.

THINKING ABOUT A MENTOR

Some site supervisors develop more than a supervisory relationship with their interns. They develop what is traditionally called a *mentoring relationship*. The mentoring relationship differs from supervision because it is interpersonal in nature and is one in which the supervisor takes an interest in the intern's professional development and career advancement (P. Collins, 1993). It can be a very empowering relationship, leaving both the mentor and the mentee or protégé better professionals for it. The mentor not only oversees the development of knowledge and skills but provides "valuable insight into an organization's environment and culture, as well as psychosocial support for the student" (Stromei, 2000, p. 55). It may interest you to know that the mentoring relationship is often considered the most important aspect of graduate education (Mozes-Zirkes, 1993). At the undergraduate level, we have found that the mentoring relationship, when it develops, is a most instructive and influential factor in the quality of the internship experience.

The best mentoring relationships occur spontaneously between supervisors and interns. The effective mentor makes sure that the intern becomes part of the organization very quickly and is given highly visible tasks; this type of mentor ensures that the intern is introduced to the profession through such resources as networking, luncheons, and conferences. Although many supervisors do make sure that all these bases are covered for the intern, the mentoring supervisor personally invests time resources in coaching the intern for success (Tentoni, 1995); it is a *personal* investment in making the protégé successful.

Of course, this relationship, like others, takes time to develop (P. Collins, 1993). It moves from a relationship of positive role modeling, when much of the learning occurs through interactions, observations, and comparisons with the supervisor, through a time when you have grown to really like each other, and then to a point where you might be right now: valuing the relationship and recognizing how mutually rewarding it is for you and your supervisor. When this happens, the intern tends to become increasingly self-assured, competent, and autonomous, no longer needing the mentor in the same ways for guidance or support, and eventually becoming more independent of the supervisory relationship. In the best of mentoring relationships, a more equal relationship emerges over time. That becomes evident when the supervisor accepts you as a colleague, and you accept the supervisor on equal footing (P. Collins, 1993).

Becoming the New Me

A second area of significant change is a most visible one: a transformation in your sense of yourself from a student to an aspiring civic professional. You think differently, you act differently, and you feel differently. You might be asking yourself just how these changes happened.

You may not be aware of this, but how you think about yourself is continually redefined by your experience and by the process of reflection and meaning making. Have you noticed that you no longer refer to yourself as "just an intern" or "the intern"? Rather, you recognize yourself more and more as a member of the profession. One intern reported, "*I am fitting in around the office. Everyone is helpful and giving me tips on how to do things. They even ask me what I would do in certain situations. I feel they respect what I have to say.*" Another offered, "*I realize that people really do depend on me here. . . . my advice is important.*" You have a greater sense of self-awareness and self-respect. You are most likely learning that you can trust your knowledge and skills, and you are developing a good sense of your strengths and limitations. Your work might suddenly be taking on new meaning because you realize that you are making a difference in the lives of others. Your professional identity is beginning to emerge. We find that there are two hallmarks of this shift in identity that are discussed with frequency by students: a commitment to quality work and a commitment to personal integrity.

COMMITMENT TO QUALITY WORK

Not so long ago, in the throes of the transition from learning the work to doing the work, "good enough" may have been all you expected of yourself in your endeavors; just getting the work done seemed a lofty goal! That was especially so if you were dealing with a multitude of responsibilities in addition to your internship or you were experiencing the confusion and discouragement associated with the Disillusionment stage. Now, though, the "good enough" standard that prevailed earlier in the internship gives way to a standard of excellence as you enter the Competence stage. You become more demanding of yourself, regardless of the amount of responsibilities you have. This expectation of excellence extends not only to all aspects of your internship but the seminar class as well as the full range of issues impacting your personal, professional, and civic development.

COMMITMENT TO PERSONAL INTEGRITY

The second hallmark of a changing identity is a more conscious commitment to a sense of personal integrity. Changes here are reflected in the flexibility and adaptability of your thinking, what you value, and how your beliefs and values affect your behavior (Chickering, 1969). Although you may feel that you have matured substantially since you began your internship (much to your relief and, in some cases, to that of your loved ones and family!), changes in your basic level of integrity tend to be gradual and do not affect the values that are important in your life. Developing integrity is a lifelong process that begins during the college years (Chickering, 1969). This is not a goal of your internship but rather a gift that emerges from the maturing process.

We do not think an internship necessarily changes your core values; however, it certainly does push you to clarify them professionally and personally. When this happens and you are accomplishing a great deal, you may feel so proud of your work that you want to keep on doing it—and doing it better.

Focus on PRACTICE

Choose Your Companion: Excellence or Perfection?

At the beginning of the internship in the first seminar class, we ask our students, "Who here is a perfectionist?" Some interns acknowledge it immediately, with moans, grunts, or laughs. Other interns acknowledge it after thinking about it, with resignation. We take that moment to share these observations with them: Many students have learned from experience not to confuse the need to do work well (excellence) with the need to do work perfectly (perfection). Although both may guarantee success, one guarantees headaches as well—for all involved. When not giving yourself permission to make a mistake while learning, you deny yourself opportunities to learn how to recover from mistakes and how to solve problems. If perfection is your only or preferred way of doing things, you are bound to be chronically exhausted from the pressure you are putting on yourself. Our suggestion is that for the duration of the internship, if at all possible, give up the need to be perfect and replace it with a need to be excellent.

As you grow in integrity, your values become genuinely reflected in how you act. In other words, your behaviors truly represent your values. For this harmony or congruence to occur, it means letting go of your literal beliefs in the external rules you live by and reconstructing a new value structure for living. This is no small feat, and in fact, you are already working on it. By doing so, you are "humanizing values," which is essential for ongoing maturity. Although necessary, letting go of one's literal beliefs can be quite perplexing when in the throes of a life transition, such as the internship is for many students.

Through experience and the process of reflection, your meaning making system is constantly transforming itself. Especially now, because anxiety and tension can run very high, students experience a shift in their meaning making structure. This new structure that emerges is one that represents the essence of you. It is not your parents', your school's, or your friends' ways of thinking but yours and yours alone. You own it (emotionally speaking); it is a part of you, and it reflects your values and beliefs. The closer you are to adopting this value structure openly and without a façade, the closer you are to living with a sense of personal integrity (Chickering, 1969). This is a lifelong task, and your journey has just begun.

BRACING FOR THE BUMPS

Although the Competence stage is recognized primarily for its good feelings and intrinsic rewards, it has its challenges like all other stages in the internship. For some interns, the highs of this stage allow the challenges to be put into perspective and managed rather easily; for others, the challenges seem to threaten the good feelings that characterize this time in the placement.

Recognizing the Bumps

This incident helps me put things into perspective . . . (and) makes me realize more and more that as successful as this internship is and will be, there will be several bumps in the road.

STUDENT REFLECTION

These challenges are generally different from the ones you faced in the past; those may have felt like walls that stopped you from moving forward, whereas these are more like bumps along the way. A word of caution is needed here though. These bumps in the road can throw you off course if you do not know what to expect or how to manage them at the time. We have identified three potential bumpy areas that you need to pay attention to:

- The tendency to focus on only the good feelings and avoid the challenges that are an inevitable part of your placement
- The tendency to not enjoy the success you are achieving
- The tendency to experience predictable transition issues during this highly productive time in your internship

Freezing the Moment

Many interns fall prey to the misconception that the days of difficulties or problems are behind them. Some say they expect more difficulties in their internships, but when the difficulties arise, they ignore or otherwise do not acknowledge them. Other interns just want to focus on the good feelings of having resolved issues. And still others fall into the trap of believing that real professionals (like really good interns) should be beyond having difficulties.

Well, you are not beyond them—no one is. Your skills, your relationships with clients, supervisors, peers, and co-workers, and your understanding of the field site are all evolving, and you could continue to experience new difficulties in all these areas. The good news is that you are better equipped than ever to meet these challenges; in fact, your ability to handle difficulties can be a source of ongoing pride. The difficulties you face at this time may actually feel different now and not be so overwhelming or anxiety provoking. However, they still need to be managed or the difficulties will become problems.

Feeling Success on the Ride

Some interns become concerned and challenged at this point in their internships when they realize that they appear successful to others, but they do not necessarily *feel* successful about their accomplishments. Such a situation can become a problem for you because your experience of the internship is based on what you feel, not on the perceptions of others. So, feeling good about all aspects of the internship is very important to your sense of success. Being able to feel your success, often described as a sense of fulfillment, is what makes the success genuine (King, 1988); anything less can compromise those feelings.

SOURCES OF FULFILLMENT

> *The aspects of my experience which contribute to my sense of fulfillment are the feelings of satisfaction I get when I know I have connected with a client and really made their day better in some way. . . . In my internship, it is rare to receive gratitude from a client, but I feel pleasure just knowing that I am giving of myself in a positive, caring way.*
>
> STUDENT REFLECTION

There are three sources of fulfillment that nurture the emotional experience of feeling one's success: *worthwhile work*, *responsible relationships*, and *self-defined success* (King, 1988). As you read about them, think of your own internship and the ways in which it provides you with what you need to *feel* the success you have earned.

Worthwhile Work   The first source of fulfillment in success is embedded in the "outer" or social dimension of success (Huber, 1971) and involves the work you do every day—in this case, the work of your internship. Whatever that work is, it must be considered *worthy* or *worthwhile work* by you and by your supervisor. The work must be productive, with responsible activities that are personally meaningful, and the work must allow you to accomplish clear goals. There are five aspects of the work that allow these feelings to develop:

- **Accomplishments:** The work itself needs to provide you with opportunities to develop your skills and apply concepts in real-life situations. For your part, you need to take an active role in creating your work and your achievements. What you accomplish must be *purposeful in nature, constructive, challenging,* and be *a source of pride* for you. Additionally, both your work and that of the agency need to reflect *socially responsible goals* to have credibility in the community. Feeling good about the public relevance of your work reflects socially responsible goals, whereas interning at a site that is a known abuser of the environment or has a reputation for questionable practices with employees or clients would certainly call into question—by you and by others—the quality of your accomplishments.

- **Acknowledgment:** Being *recognized* and *respected* by your supervisors takes on paramount importance to how the achievements are experienced. Such acknowledgment is especially significant when your contributions are above and beyond what is expected of an intern.

Focus on THEORY

Sources of Fulfillment in Success

- **Doing Worthwhile Work** (Social, outer dimension)
  - Accomplishments
  - Acknowledgment
  - Self-determination
  - Self-actualization
  - Intrinsic rewards
- **Developing Responsible Relationships** (Personal, inner dimension)
  - Supervisors
  - Peers
  - Co-workers
  - Staff
  - Community
- **Self-defined Success**
  - Conscious choices
  - Freely made
  - Self-determined goals
  - Engaged involvement
  - Source of enjoyment

King, 1988

- **Self-determination:** This aspect of the work has to do with the potential for autonomy, which is evident in the *freedom to create* and *carry out* the tasks.
- **Self-actualization:** Opportunities for *creative expression* and *personal growth* need to be built into the work. Otherwise, the work can feel stagnant.
- **Intrinsically Rewarding:** Above all, the work has to provide you personal pleasure, regardless of the demands or stresses inherent in the work.

In Their Own Words   Student Reflections on Worthwhile Work

Accomplishments: By allowing me to take charge and work independently with a group of clients yet having the support of a professional . . . I am proud of my internship now that I feel like I'm making a difference and doing some meaningful work.

Acknowledgment: It really made a difference when my supervisor and my co-workers complimented me and told me that I had done outstanding work. Up to that point, I thought that I had done a good job on the project, but I didn't really know. I needed to hear it from someone there. Up until then, I began to question myself and my abilities.

Self-determination: I coordinated the event and actually made it happen. . . . I was given the opportunity to develop something on my own. It felt great.

Self-actualization: I was getting worried because everything I was expected to do was already laid out for me. There was no room for "me" in the work. I was just carrying out someone's prescription for how to do the work. I had some great ideas, really creative ideas, for how to do it better, and I felt so frustrated because I didn't think my ideas were welcomed. It's not really my work if my ideas aren't reflected in it. I hated how I felt. I was going nowhere just doing what I was told to do.

Intrinsically Rewarding: It made me feel good to help someone, and it was nice to be appreciated for my efforts. I feel like I am making a difference at my internship. It feels good.

Responsible Relationships   The second source of fulfillment in success concerns the "inner," or personal, dimension of success (Huber, 1971) and is derived from the personal relationships that are integral to your experience of the internship. These relationships need to be *genuine, cooperative,* and *mutually satisfying* for all involved; in other words, they need to be *responsible relationships* (King, 1988).

Real commitment on your part and on the part of co-workers, peers, supervisors, and the community is needed to make responsible relationships happen. Your co-workers must be receptive to your coming on board, and when that happens, it is an important milestone. This intern speaks to its effects: *"This past week, I have been looked at as a member of the staff, not just by the residents but also by the staff. That was so important."* The supervisor needs to establish an effective supervisory relationship with you and include you in work groups/teams, staff meetings, and social functions (when appropriate), and you, in turn, need to be willing to engage both the supervisor and your co-workers in responsible supervisory and collegial connections. You also need to be willing to engage community groups in genuine, cooperative, and mutually satisfying ways.

What is particularly interesting is that the sense of inner success, which reflects your success in being a person, is believed to render all of your achievements worthy (Huber, 1971; King, 1988). Not only do you need to achieve the goals of your internship (sense of outer success), but you also need to pay attention to yourself as a person (sense of inner success) if you want to reap such psychological rewards as the internship being *"the best experience of my life"* and *"a most fantastic experience."* Otherwise, the worthwhile work is reduced to just work.

Defining Success for Yourself The third source of feeling the success you attain is being able to *define success for yourself.* Self-defined success essentially means you are doing what *you* truly want to do, not what someone else wants you to do. The choice must be yours and yours alone. It is a choice you made through engaged involvement in the process, free of the influences of other people. It is a choice that will be a source of enjoyment, regardless of the nature of the work (King, 1988).

When it comes to your internship, making *conscious choices* can be evident in your involvement in the selection of your field site or in the development of the goals of your learning contract. Preferably, you were actively and consciously involved in both decisions, the decisions were freely made, they reflect self-determined goals, and they are a source of enjoyment or pleasure. Now, that's a tall order and not necessarily possible in all academic programs. However, if the field instructor determined

---

Think About It

Are You Doing What YOU Want to Do in This Internship?
It's time for some soul-searching about this issue of conscious, active choices about your internship, and now might be a good time.

- Think about what you really wanted to achieve in this internship, not about what others think you should achieve or what others want for you in this experience.

- Think about the extent to which your placement site is a good match for your personal, professional, and civic goals, including how compatible it is with your views of its civic responsibilities.

- Think about the work of the internship and whether it is really what you wanted to do in that experience.

- Think about the extent to which the choice of internship was made freely by you. Just how involved were you at the time decisions were made? These are the thoughts of one student who did engage in such soul-searching. *"Oh, my . . . I'm not sure that I even want to be a counselor now! After all this work. And putting off starting a family."*

your site, and the site supervisor determined your learning contract, and you were not actively involved in an informed way in these processes, that does not mean all is lost! Even if you went into the internship not consciously aware of what you really wanted, those overseeing your actual placement may have been able to create a context that allows you to eventually *own* the decision: i.e., emotionally avow the decision as your own. That context could be one in which the placement supervisor assessed your interests and goals quite well and placed you with a supervisor who provided considerable support for personal growth and awareness while at the same time entrusting you with worthwhile work. If that was the case, then there is a good chance that you will eventually fully accept the choice of field site and the learning goals and, in turn, feel the success in your achievements.

So, what do you do if you are not feeling your success? Actually, there are several things you can do:

- Spend time determining whether or not the design of your internship allows you to experience the steps of processing and organizing experiences so that the learning cycle described in Chapter 1 takes place (Kolb, 1984). How does the design of your internship measure up to that framework?
- Spend some time examining the relationships with your supervisors and others who are important to your field experience. Are they disappointing relationships in any way? If so, why do you think that is? If not, what makes them so good?
- Review the sources that nurture the feelings of success using the eight Steps Model. Can you identify the areas in your internship that you would like to see strengthened? If so, start strengthening it!

Transition Issues and New Realities

There are several issues that tend to surface as you enter this period of highly productive work. They all seem to be part of a natural transition to a more realistic way of living with the internship once the crises in growth are over and the climb has ended. If these issues are not recognized and understood for what they are, they can compromise your feelings of competence and the joys that come with this stage in the internship. A discussion follows of these three issues that you encounter at this point in your journey: the leveling-off effect; the need for a more balanced life; and the inevitable "crunch."

LEVELING OFF

This effect is experienced as a subtle change in the pace and intensity of your ascent on the ladder of responsibilities. Students often liken it to that point in an air flight when the plane is no longer ascending and begins to level off. The timing of this change in the internship is determined either by the calendar or by saturation in workload. When this happens, you will experience the cruising effect: The intensity of increasing responsibilities decreases, and you no longer feel the constant push to greater heights. Eventually, you will stop taking on new responsibilities and settle into a rhythm of working that is more realistic and predictable in nature. Here's how one intern described his experience: *"This was the first week that I can think of that nothing new happened at all. I need to learn that sometimes in this line of work, there will be days and weeks where nothing big will happen. So, having this happen was a good thing because it's teaching me things that I need to learn, but I don't want to learn."*

For many students, leveling-off means that the goals of the internship are being reached, and it is time to relax somewhat and enjoy the ride. For some, though, it is difficult to enjoy the ride because of the mixed feelings that come with reaching the final plateau of a journey. For a brief moment, you might wonder whether this is all there is to the internship and speculate that there are few surprises left in the work and little left to learn. Be assured that any emotional letdown you experience in reaching the apex of your journey is only momentary and will pass. You will adjust to these changes and get on with the challenges and personal rewards that come with being a productive member of the organization.

RECLAIMING A BALANCED LIFE

You are probably sensing by now that your total immersion into the role of being an intern is beginning to change (Lamb et al., 1982). If you are like many interns, you may suddenly realize that you had a life before the internship—a social life, a family life, a private life—and you find yourself needing to reclaim that life. One intern was stunned with this realization: *"I realized I was not giving enough attention to my relationships with my friends and family. It had been nearly three months since we had spoken. The incident made me stop and think about how important my relationships are and how, even though I am busy, I must not neglect them because these are the people I count on for support."*

Perhaps that awareness will come to you on your way home from the field site, when you are no longer so preoccupied and are thinking about aspects of your life that were left behind in all the excitement of the internship. Or perhaps it will happen during a morning break as your mind wanders to more personal needs—for relationships on-site or in one's personal life, for family time, for time alone. *"I can really understand the need now to slow down and pace myself and try to create a more livable, useful pace for other people in my life, including at the site,"* noted one intern. Another student's experience was a nagging feeling: *"I can feel a tug going on between family demands, work, and my academic requirements. . . . The holidays can also complicate my situation to the point where I am beginning to feel both pressure and stress."*

Interns usually find a need to get on with living at this point in their internship. In our experience, these changes signal a healthy shift of energies toward a more *balanced way of living* with your professional life. Although your internship still remains very important, it may no longer be the driving energy of your life. As one intern noted: *"The kids notice when I am not there. Continuity is important to them. This incident reinforced my need to balance my job and internship both for me and especially for the students I see."*

FEELING THE TIME CRUNCH

Up to this point in your internship, you have had to struggle with the ups and downs of becoming part of an organization, learning skills, developing competencies, resolving differences between real and ideal expectations, and contending with changing perspectives and life's big questions. Just when you thought it was clear sailing ahead . . . *wham!* You've run into yet another wall in your journey and just about everyone hits this one.

This wall is different from previous ones. This one you know you can manage; you just don't know whether you can muster the energy you need to do it. You feel some anger about the amount of work you have and frustration over the amount of time you do not have. Something subtle is happening as well. Your perspective seems to be changing again. You have little tolerance for trivialities, and you find yourself feeling indifferent toward some of your assignments and responsibilities, especially if they are not directly related to the internship or do not measure up to the importance of the work you are doing in the field. This is *not* a crisis in growth you are having; this is a *crisis in management* of your time and the workload. The crisis takes on added importance because it is in fact a real threat to your schedule and to the pace of your internship.

You are encountering what we refer to endearingly as *the crunch*. It is an early warning signal to let you know that the end of the internship is fast approaching. This crisis in management is no different from the crunches you faced at the end of every semester (usually at the end of the midsemester slump, remember?) when papers, exams, and projects all become due at the same time. However, this time, much more is at stake academically, financially, and professionally. And, you also are carrying the greatest number of responsibilities you have had in your academic career.

RECOGNIZING SLIPS AND SLIDES

If any of the following questions or declivities best describe you at this time in your internship, know that you are not alone! Most of your peers are or will be going through the crunch about the time you do. It's important to be aware of the extent of its effects and how vortex-like these effects can be so that you can begin to manage it better.

Have you begun to. . .

- question the merit of each assignment?
- challenge requests for more work?
- struggle to avoid the wall of ineptitude?
- feel overwhelmed?
- feel stuck in time?
- think you have. . .

    too many deadlines?

        too many details?

           too few resources?

                too little social life?

                    too little family life?

                      too high standards?

                          too many responsibilities?

                            OOPS!

                              YOU ARE SLIDING!!!

MANAGING THE CRUNCH

The time has come to rethink your situation and regroup yourself, emotionally and physically, for the last mile of the journey. You know yourself best in times like this. What works for you? What allows you to manage time and workload crunches? Some people actually thrive when they are up against a wall of deadlines, while others crumble under the same circumstances. Perhaps one of the following suggestions will work for you as they have for many of our students.

Taking Time for You   A day will do; an afternoon or evening could do just as well. The important thing here is to stop. Just stop everything. Take a holiday from the assignments, the responsibilities, and the timelines you are facing. Of course, it is not just the internship that is overwhelming you. Many of you have responsibilities beyond your internship, and you may be overloaded for what you can effectively handle at this time. However, the internship is still the new family on the block, and its workload is most demanding right now. It is very important that you do something that you enjoy. Sing. Dance. Shop. Bike. Hike. Climb. Swim. Draw. Paint. Read. Run. Play. Ski. Work out. Create. Sew. Meditate. Sleep. Listen. Make music. Take in a movie. Go out to dinner. Do something, anything, whatever it takes to help you . . . *let go!*

Doing What Needs to Be Done   Things simply must get done. So, it is time to make schedule decisions and prioritize your tasks. You will need to decide what is doable in the time you have and what can get done only if you make significant changes in your schedule. This means prioritizing and scheduling as efficiently and effectively as you can.

Eight Stepping It   What is obvious is that something has to change to put you back on track and in charge of your internship. You already have had to deal with difficult challenges when you confronted the barriers in your internship that prevented you from moving ahead. This hurdle is far less complex and demanding and actually quite manageable by you simply, if for no other reason, because you have gotten this far in your

---

Focus on SKILLS

Organizing for Time Management

Here is an exercise that our students find helpful in accomplishing this task. First, list all the tasks that you have to complete. Then, next to them, indicate the time frame within which they can be finished, not necessarily must be finished. For example, if you have a final culminating statement to write for your seminar class, it obviously cannot be completed before the internship ends. However, you can work on much of it before the end of your placement. Then, when the internship is over, there is relatively little content to add. The time frame for this assignment actually begins on the date you start this exercise and ends on the due date. By listing tasks, assignments, and responsibilities in a similar manner, you will have a more informed framework for prioritizing tasks and assigning calendar dates to work on them.

internship. So, if you are struggling with this challenge, roll up your sleeves and get to work. Using the 8 Steps Model discussed in Chapter 11 (*Eight Steps to Creating Change*) will help you clarify the biggest hurdles you face during this time-and-work-load crunch and bring this crisis in management to an end.

PREPARING FOR THE PROFESSION

As the headiness of the Competence stage continues to define your daily feelings about the internship, you may find yourself thinking more and more about the profession for which you are preparing. Writing your résumé, seeking out job opportunities, and deciding where you want to live after graduation perhaps come to mind immediately. And those are certainly important ways to begin your journey to your professional home.

Leaving Your Footprint

We find that interns are preparing themselves in other ways as well, ways that reflect their growing awareness of the importance their profession will have in their lives. Two of those ways are a need to "give back" to their internship site and a need to feel a continued part of the organization and profession after the internship ends. There certainly are ways to ensure that you are not forgotten at your site. Whether you are in public service settings, corporations, civic organizations, research laboratories, or nonprofit agencies, you may be feeling a need to make contributions to the field site in meaningful ways. As one intern put it: "*I think, no, I know that the (police) station needs to have the back room cleaned out. I want to do that for the chief. He has been very supportive, and I want to give back in some way . . . I guess it means a lot to me to know that I somehow am still going to be part of the department in some way after I leave.*"

These contributions are not always part of the learning contract; they tend to be given from the heart. They are contributions that benefit the agency but have not been done up to now because of the workload demands, fiscal constraints, and task priorities. The contribution may take the shape of creating an internship manual for the site, developing databases, creating a resource library, or reorganizing workspaces. A popular contribution for our students has been reorganizing and downsizing data files that need to be recatalogued into new data systems. For the most part, the contributions tend to be on the "wish lists" of site supervisors or others in the organization. If you have been thinking about making such a contribution, now is a good time to think about how to do it and when. You might want to use the predictable downtime during the last few weeks of the internship.

Moving Beyond the Textbooks

A second area of awareness that typically develops in students during the Competence stage has to do with an appreciation for what is beyond textbook knowledge. By now, you realize that remembering facts, working with efficiency and speed, approaching problems in logical ways, and effectively using the environment to meet the needs of the work are effective ways of going about your work. In fact, they may be exactly what is expected of you in your future profession. But you also realize by now that it takes more than these skills to handle challenging situations and resolve complex problems in the workplace. There are two invaluable resources that interns look to when they realize that what they need to thrive in their future professions is not all found in textbooks.

The first is wisdom. Wisdom transcends textbook knowledge. It is wisdom that makes a supervisor empowering and a remarkable leader. The differences between these two modalities reflect the differences between what can be instilled and developed in you (*intelligence/skills*) and what you must learn to develop through experience and commitment to a more visionary way of dealing with the work and the issues (*wisdom*).

Think About It

Your Supervisor as a Source of Wisdom
If your field site or your supervisors encourage and affirm a tolerance for ambiguity, an ability to intuitively understand and accurately interpret situations, and an ability to identify and frame a problem accurately, then it is wisdom as well as intelligence that is being valued by them. If you are interning at a site where empathy, concern, insight, and efficient coping skills are valued, again wisdom as well as intelligence are being valued (Hanna & Ottens, 1995). Ideally, your supervisor possesses both the qualities of intelligence and wisdom and has modeled the importance of them for you.

The second is a combination of the habits of reflectivity (Schon, 1995) and practical reasoning (Sullivan, 2005) as ways of being. When professionals go about their work in these ways, they make a practice of reflecting deliberatively as they consider circumstances and situations as well as the decisions. They are constantly integrating and reintegrating theory and practice. Developing these habits takes focus, practice, and conscious intent.

DEVELOPING AS A CIVIC PROFESSIONAL

As you continue to prepare for your arrival in your professional field and develop the knowledge, hone the skills, and assimilate into its culture, you will probably find yourself thinking more and more about your civic responsibilities as an adult and as a professional. The goal of civic development is to prepare you to participate fully, responsibly, and actively as a member of a community. Active participation refers to engaging in the responsibilities and being an owner of government instead of being a consumer of government, a passive recipient of the benefits of membership in a given community. Ownership reflects a level of civic capacity that suggests that the individual is able to manage in useful ways the natural tensions that exist between an individual's competitiveness, personal autonomy, and civic cooperation (McKensie, 1996).

The internship has provided you an opportunity to learn about the public relevance and social obligations of the profession for which you are preparing and how these social obligations are or are not carried out at your site. The internship has also provided you the opportunity to think about yourself in relation to the community and what that means in terms of your civic learning and preparation for civic effectiveness. Does that mean more is expected than voting in elections and obeying the laws? That is a question you will have to come to grips with yourself, for only you can determine what your place will be in your community. However, during this stage of the internship, these are the areas of awareness that tend to give interns reason to pause and reflect while performing competently and realizing the largeness of their work and the field to society.

The Public Relevance of the Work

The profession for which you are preparing is replete with values about how the work gets done and the intent of the work. For example, fairness could be a value of your profession, as could integrity, justice, beneficence, and so on. You know those values by now because you work with them each day at your site. You may be expected to uphold them in how you present yourself to co-workers and to the general public, to the community, and to clients, whether it is to provide services to military families, make the community greener, or build safer bridges. Some of you are in organizations that serve the public and are publicly funded to do so. Some of you are in the private sector, where generating wealth is requisite in order to exist and provide products or services. Others of you could be in private, not-for-profit agencies providing services to the public that are funded through a variety of sources of revenue. And yet others of you could be in foundations, where private monies are distributed for social benefits. Regardless, and whatever the profession, your work has a purpose in the community and society at large through a social type of contract for community betterment.

As we mentioned in Chapter 7, the purpose of your site in the community tells you a lot about the relevance of your work to the community. If you are in the helping professions, the mission of the organization—be it government, private, or not-for-profit—spells out that purpose and describes its relevance to the public good. If it is government administered, most likely the mission and its relevance are part of a statute that authorizes its existence and funding by tax dollars. If, on the other hand, you are in an organization that is private and not serving the public in such a way, the company probably has a statement of responsibilities to the public or the community, the relevance of which is beyond maximizing profits and generating wealth. If your internship is at such a site, you might be impressed with the organization's operational philosophy, although you still find room for philosophical and political debate about its commitment to the community.

Issues in Civic Development

You may already know what is expected of the professional workforce at your site in terms of providing service time to the community each year, whether sitting on boards

The Public Relevance of Your Work

As you reflect upon the public relevance of your work and that of the site, think about:

- The social obligations the site has to the communities it affects or serves
- The ways you see the professional workforce carrying those obligations to the communities
- If the site is not carrying out its social obligations, how do you know and how do you make sense of that?
- The ways in which you have been engaged by your supervisor or co-workers to carry out the organization's social obligations
- Your reactions to being included or not being included, if that is the case

of directors or advisory groups, picking up trash, or being part of river cleanups; what you might not be aware of, though, is the company's position on ecostability in the community and its commitment to a greener presence there. You also might not be aware of the company's policy about being part of the solution for social problems, even though it is not part of the problem. For example, it could be that the company has a commitment to employing a percentage of the immediate or neighboring community or it contributes financially and with in-kind donations to a local shelter for battered women (Sullivan, 2005).

Civic Participation

When you think about being prepared through your education for responsibilities to the democracy in which you live, what comes to mind? Do you think in terms of your knowledge of civic matters, such as public problems, the causes of social problems, the challenges individuals face on a daily basis, or government laws and institutions? Perhaps, instead, you think in terms of the civic skills you are developing, such as communication competencies, organizational analysis, or advocacy for change in the workplace. Or you may be thinking in terms of the civic values that you are seeing in practice, such as being a good neighbor, involving minority opinions, or an obligation to the common good (Battistoni, 2006). Whether it is the skills, values, or knowledge needed to develop your civic learning, they are being nurtured in you so that you can contribute in direct ways to your community.

So, you wonder, perhaps, if you have the basics in place to grow into a civically effective individual. Most likely, you have been laying the groundwork for that for some time now. Being politically informed is important to civic effectiveness, as is sensitivity to diversity issues, having leadership know-how and awareness of social responsibilities, taking academic coursework in basic civic knowledge, and developing personal competencies to work with others in the community. How you learned them—through studying, employment, volunteer work, or this internship–is not important. What is important is that you are aware of how your decisions and judgments can make a difference, now and in the future; it is that ability to make good choices that is the essence of civic effectiveness (McKensie, 1996; Howard, 2001).

SUMMARY

The rewards of this long journey in experiential learning are finally realized when you enter the Competence stage. Although hardly without its concerns and challenges, this stage is one in which you can indulge yourself and enjoy the feelings of finally reaching your goals. The transformations in development that you have experienced and the sense of empowerment that you have developed are evident not only to you in how you feel and go about your work but to those around you as well.

As you continue to develop competencies and a sense of professional identity, you begin to realize that you also are developing awareness of the behaviors of professionals in general and of your co-workers in particular. This awareness tends to take the form of curiosity about politics and staff actions and their effects on the quality of the work that, up to this point, did not seem to carry the importance they do now. Your status as an emerging professional has a consequence of knowing your profession's social obligations and your responsibility to civic competence. In addition, you have the responsibility of an awareness of the ethical and legal issues common to the workplace. And, that is the focus of the next chapter.

For Contemplation

PERSONAL REFLECTION:
SELECT THOSE INQUIRIES THAT ARE MOST MEANINGFUL TO YOU.

1. Take a moment to think about just how fulfilled you are with your internship. What aspects of your experience contribute to this sense of fulfillment? Which aspects tend to interfere with it?

2. Take time to think about who made it possible for you to be at this point in your internship and feel the highs of being competent in your work and accepted as staff. How do you thank them? If you struggle with either of these questions, you might want to think about whether it is hard for you to engage others when you need to, to be modest or humble, or to thank someone else for your achievements.

3. On a piece of paper, jot down all the factors you can think of (including the ones mentioned in this chapter) that are important to your internship (positive and negative). Then, next to each one, indicate the amounts of "too much," "too little," or "just right." Chances are that what's just right leaves you with the greatest feelings of satisfaction.

SEMINAR SPRINGBOARDS

As you read over the following questions, think about the ways in which your perspectives about professionalism have changed during the internship. Share them with your peers and discuss your differences. What do you think contributes to these differences?

- What is your operating definition of a professional? What behaviors and values do you associate with being professional?
- What professional behaviors are the norm for your field site? What behaviors contradict the norms? What values are reflected in both instances?
- Think about someone you met in the course of your internship who meets your definition of professional. What is it about this individual that you find particularly admirable?
- How do you see yourself measuring up to your definition of professional? (Be as specific as you can be.)
- In what ways are you already becoming a civic professional?

For Further Exploration

Baird, B. N. (2007). *The internship, practicum, and field placement handbook.* Upper Saddle River, NJ: Prentice Hall.

Comprehensive and covers a wide range of topics relevant to the clinical/counseling dimension of human service work.

Jackson, R. (1997). Alive in the world: The transformative power of experience. *NSEE Quarterly,* 22(3), 1, 24–26.

Explores what it is about experiential education that "gives experience the power to transform" an individual's thinking.

References

Battistoni, R. (2006). Civic engagement: A broad perspective. In K. Kecskes (Ed.), *Engaging departments moving faculty culture from private to public, individual to collective focus for the common good.* Boston: Anker Publishing (pp.11–26).

Chickering, A. W. (1969). *Education and identity.* San Francisco: Jossey-Bass.

Collins, P. (1993). The interpersonal vicissitudes of mentorship: An exploratory study of the field supervisor-student relationship. *Clinical Supervisor,* 11(1), 121–136.

Dreyfus, S. E., & Dreyfus, H. L. (1980). *A five stage model of the mental activities involved in directed skill acquisition* (Unpublished Report F49620–79-C-0063). Air Force Office of Scientific Research (AFSC), University of California, Berkeley.

Erikson, E. H. (1963). *Childhood and society.* New York: Norton.

Hanna, F. J., & Ottens, A. J. (1995). The role of wisdom in psychotherapy. *Journal of Psychotherapy Integration,* 5, 195–219.

Hersey, P., & Blanchard, K. (2008). *Management of organizational behavior: Utilizing human resources* (9th ed.). Upper Saddle River, NJ: Prentice Hall.

Howard, J. (Summer, 2001). *Service-learning course design workbook.* OCSL Press, University of Michigan: Michigan Journal of Community Service Learning, pp. 40, 42.

Huber, R. M. (1971). *The American idea of success.* New York: McGraw-Hill.

King, M. A. (1988). *Toward an understanding of the phenomenology of fulfillment in success.* Unpublished doctoral dissertation. University of Massachusetts, Amherst.

Klemp, G. (1977). Three factors of success. In D.W. Vermilye (Ed.), *Relating work and education.* San Francisco: Jossey-Bass (pp. 102–109).

Kolb, D. A. (1984). *Experiential learning: Experience as the source of learning and development.* Upper Saddle River, NJ: Prentice Hall.

Lamb, D. H., Baker, J. M., Jennings, M. L., & Yarris, E. (1982). Passages of an internship in professional psychology. *Professional Psychology, 13*(5), 661–669.

McKensie, R. J. (1996). Experiential education and civic learning. *The National Society of Experiential Education Quarterly, 22*(2), 1, 20–23.

Mozes-Zirkes, S. (July 1993). Mentoring integral to science, practice. *APA Monitor,* 34.

Schon, D. A. (1995). *The reflective practitioner: How professionals think in action* (2nd ed.). Aldershott, UK: Ashgate Publishing.

Stromei, L. K. (2000). Increasing retention and success through mentoring. *New Directions for Community Colleges, 11*(2), 55–62.

Sullivan, W. (2005). *Work and integrity: The crisis and promise of professionalism in America.* The Carnegie Foundation for the Advancement of Teaching. San Francisco: Jossey-Bass.

Tentoni, S. C. (1995). The mentoring of counseling students: A concept in search of a paradigm. *Counselor Education and Supervision, 35*(1), 32–41.

Tryon, G. S. (1996). Supervisee development during the practicum year. *Counselor Education and Supervision, 35*(4), 287–294.

Considering the Issues: Professional, Ethical, and Legal

I have to make sure that I am not in the wrong place at the wrong time. I have to be careful . . . because I don't want to be in a bad situation. I try to stay out of situations where I could be forced to make a bad decision.

STUDENT REFLECTION

As you go about working independently now, making decisions and testing the limits of your own competence, you will encounter situations about which you will feel uncomfortable. These are situations that will give you cause to pause and think. You already may have found yourself feeling this way and were unsure about what conclusions to draw, what decision to make, or what action to take. Welcome to the world of professional work and its many challenging issues. An *issue* refers to a point that is in question or in dispute, thus internship issues pertain to some aspect of an internship that can become a matter of debate among others, usually supervisors, co-workers, clients, or members of the community. Some of these issues have ethical dimensions; some have legal dimensions; some have professional dimensions; and some have a combination of all three. This chapter will provide you with a way of thinking about these issues and identify tools and resources to help you deal with these matters in the course of your work in the field.

We have one reminder before we move forward with this discussion. Although many of you using the text are interns, not all are. Many students are conducting practica or are in other field-based learning experiences. The academic level of your field experiences varies considerably, from two-year undergraduate programs through doctoral programs. Although this text is intended primarily for the helping professions, there are a wide variety of professional roles to consider: human service professionals; social work; mental health, school guidance, school adjustment, vocational, rehabilitation, and pastoral counselors; and psychologists. In addition, many of you are in public service as well, such as criminal justice placements like probation, parole, corrections, and law enforcement; environmental agencies; immigration agencies; and positions in all levels of government. Consider, too, that your academic majors vary considerably, including English, communications, nursing, business administration, social justice, political science, and public administration, to name a few. Consequently, as you move along in the chapter, keep in mind that at times, there will be content of significant relevance to what you are doing because of the similarities across all these disciplines; at other times, you will need the resourcefulness and guidance of your instructor or supervisors to access information relevant to your specific field.

There has been a growing need since the late 1980s for helping professionals in particular to know about their rights and responsibilities in the helping process and for others to be aware of those specific to their professions. Although a detailed discussion of these issues is beyond the scope of this chapter and text, a useful way of organizing the most common issues is to think of them in terms of the distinct aspects of the work you do on-site (Chiaferi & Griffin, 1997). We use this method, with some adaptation, and incorporate issues across the spectrum of the helping professions (American Psychological Association, 2003; Baird, 2007; Chiaferi & Griffin, 1997; D. Collins, Thomlison, & Grinnell, 1992; G. Corey et al., 2006; Goldstein, 1990; Gordon, McBride, & Hage, 2005; Martin, 1991; Schultz, 1992; Wilson, 1981).

INTERNSHIP ISSUES

We have grouped issues pertaining to internships into four aspects of the work: general *internship role issues* across disciplines, *practice* issues, *integrity* issues, and *intervention* issues. The first three categories apply to all internships; the remaining category is particularly relevant to the helping and service professions. Keep in mind that your role, first and foremost, is that of an intern. However, it is easy to lose sight of that aspect of the field experience as you assimilate into the professional field of your internship. In each of these four categories, we identify situations that interns may face. Some of the issues actually overlap categories due to the nature of the work. We have listed them in detail to underscore the vastness of responsibilities and potential issues. As you read over the examples, think in terms of your own internship and what kinds of situations could arise. You are expected neither to remember them nor know them, athough in time and through practice, they will become quite familiar to you. At the very least, you will know where to go to learn more about them. The issues are listed by categories in Table 13.1. The list should look familiar because you were introduced to it in Chapter 2, where we discussed essential knowledge for the internship. Now that you have been an intern for a while, some of the items on the list might have greater meaning for you in your role.

As we mentioned, the role, practice, and integrity issues apply to interns across all disciplines. However, if you are in the helping professions and have the responsibility (role) of being a care provider to others, then accountability is demanded in all four areas of issues. The potential for liability in all these areas is very real. Interns may be held to the same standards and ethical documents as employees. In addition, there are a number of issues that are specific to experiential education and internships in particular. For example, as an intern, you have the multiple roles of being a student and a recipient of services by virtue of your need for supervision. Keep in mind that much of what you are reading about in this chapter is applicable to all internships. As you consider the examples that are given, think about how the situations could apply to you and your work.

TABLE 13.1
Internship Issues Across Disciplines

Intern Role Issues

- Right to quality supervision
- Responsibility to confront situations in which educational instruction is of poor scholarship and nonobjective
- Disclosure of risk factors to all potentially affected parties (to site supervisor about intern; to intern about site supervisor)
- Behavior consistent with community standards and expectations
- Awareness of risk status of agency
- Active involvement in the placement process and consideration of more than one placement site
- A clearly articulated learning contract that identifies mutual rights, responsibilities, and expectations
- A service contract with the agency that defines the limitations of the intern's role
- Liability insurance
- The prior knowledge clause
- Assurance of work and field site safety
- Assumption of risk as limited to ordinary risk
- Employer-employee-independent contractor relationship
- Compensation: stipend, scholarship, taxable/tax-free

- Deportment
- Negligence
- Malpractice
- Implication of federal funds and related statutory and regulatory requirements
- Use of college work-study funds for interns
- Grievance processes
- Informed consent in accepting the internship
- Respecting the prerogative and obligations of the institution
- Responsibility to confront unethical/illegal behaviors
- Public representation of self and work
- Disclosure of status as intern
- Boundary awareness
- Boundary management
- Personal disclosures
- Criminal activities
- Political influences/corruption
- Subversion of service system
- Office politics

Internship Role Issues

These issues deal specifically with your role as an intern and the issues of academic integrity, competence, and supervision. You were first introduced to these issues in Chapter 2 when you were a fledgling intern; now that you have been in the role for some time, review these issues with consideration of how well they are being addressed in your internship.

- *Academic integrity* issues include a quality field site, responsible contracts, and a seminar class that ensures a "safe place" for reflective discussions (Rothman, 2000).

- *Competence* issues include knowing your limitations and finding a balance between challenging work and a realization you have exceeded your level of competency. It is important that you know the limits of your skills and seek help as needed (Gordon, McBride, & Hage, 2005; Taylor, 1999, p. 99).

- *Supervision* issues include the assignment of an appropriate supervisor who knows how to supervise interns in particular and can appropriately deal with such complex issues as client abandonment, the dynamic of attraction in the supervisory relationship, and quality evaluations of the intern's competencies.

An example of a dilemma in the internship category is your becoming aware of unethical and possibly illegal practices at a certain field site that you want for your internship and where you have been offered a paid field experience by the agency's director. The site is out of state and is not one of your campus's regularly used sites. Your family is relocating to that state because of financial reasons, and having a paid internship in that area of the state would be of great help financially and prevent greater hardship for your family. You doubt that the field placement coordinator for your academic program is aware of the improprieties at the agency. You are questioning whether to inform the campus about what you know or be silent and help your family. Both choices make sense to you, and that makes it the dilemma that it is.

Practice Issues

This set of issues has to do with how you engage your profession and include such issues as educational preparation, diversity awareness, and dressing for the role. An example of a dilemma that falls into this category is your having information about a teacher in a residential school who is demonstrating insensitivity to the culture of a student's family, is misrepresenting qualifications to work with such families, and is now being promoted to supervisory teacher of your work group at the site. It happens that you and the student share a similar cultural identity. It also happens that you and this teacher have had difficulties working together in the past. You are wondering whether you should disclose your concerns to your site or campus supervisors, talk directly with the teacher, or say nothing.

Integrity Issues

These issues concern the way you approach your work on a daily basis and include such issues as confidentiality, disclosure, and recordkeeping standards. An example of a dilemma in this category is your being out to dinner and overhearing the conversation of a technician who works at your placement site who does not know you but whom you recognize. The conversation is about a customer, and the technician

---

TABLE 13.2
Internship Issues Across Professions

Practice Issues

- Competence in doing the work
- Frequency and focus of supervision
- Consultation
- Education
- Diversity awareness
- Grievance issues
- Limitations in the scope of practice
- Credentialing/license standards and requirements
- Advertising for services
- Dressing for the role
- Relationships with supervisors and staff
- Managing the risks of physical danger and legal liabilities

Integrity Issues

- Dual/multiple status relationships
- Obtaining information
- Disclosure of information
- Recordkeeping
- Informed consent
- Privileged information
- Right to privacy
- Confidentiality
- Upholding the values of benevolence, autonomy, nonmaleficence, justice, fidelity, veracity

- Exceptions to confidentiality, including abuse/neglect cases
- Dangerous-client cases (self and others)
- Third-party payer requests
- Responses to court orders
- Release of information to clients
- Duty to warn and protect
- The integrity of clients
- Attraction/intimate relationships (emotional, physical, sexual)

Intervention Issues (Direct Service)

- Clinical issues (transference and countertransference)
- Limitations on scope of responsibilities
- Client's right to self-determination
- Management of referrals
- Size and nature of caseload
- Termination
- Working with special populations
- Abandonment by therapist
- Obtaining information
- Release of information
- Sharing information with colleagues
- Emergency response during nonworking hours
- Differences in legal and ethical practices
- Individual vs. group vs. marriage and family interventions

---

identifies enough data so that you recognize the customer from a meeting earlier that week. You know from your orientation period that the organization has a policy that company personnel are not supposed to talk about customers outside of the office, or at the very least, they must not disclose any identifying information about the customer's profile. The person with whom the technician is speaking is your child's teacher. Neither of them saw you, and you are not sure what to do. You realize that something inappropriate has occurred, but you are hoping to be hired at the end of your internship. You are hesitant to pursue the issue for fear of making the wrong move.

Intervention Issues

The final set of internship issues is specific to the helping and service professions and has to do with working directly with clients. They include managing clinical issues, making referrals, and overseeing a caseload. An example of a dilemma that would fall into this category is finding out at an Alcoholics Anonymous meeting, which you attend for personal reasons, that your supervisor's patient is planning to leave the country in the next couple of weeks. Of particular concern is the fact that a friend told you that she overheard the patient threatening to hurt a former girlfriend. You had been told that what goes on at Alcoholics Anonymous meetings is confidential, and you know how strongly confidentiality is valued in your future profession. You wonder how to uphold your responsibilities to all the parties involved (i.e., the patient, the friend, AA and your profession), many of which seem to be in conflict, and how to determine which responsibilities take priority over others.

PROFESSIONAL ISSUES: A WORLD OF RESPONSIBILITIES AND RELATIONSHIPS

Once your concerns focus on developing competence, it is quite common to pay attention to aspects of your professional deportment as well as, and in relation to, that of your supervisors and co-workers. You are discovering another layer of what it means to be a professional.

Focus on THEORY

What Is a Professional?

Professional is a loosely defined term that originally referred to "the honorific occupations of medicine, the bar and the clergy" (Sullivan, 2005, p. 35). Since then, of course, the term has expanded to refer to matters that pertain to many other occupations, all of which have three common features: specialized training and codified knowledge acquired through formal learning and apprenticeship; public recognition that the practitioners have certain autonomy to regulate their profession's standards of practice; and, very importantly, a commitment by the individual to provide service to the public that surpasses their personal economic gain (Sullivan, 2005, p. 36). As noted previously, an *issue* refers to a point that is in question or in dispute. So, when we talk about professional *issues*, we are referring to some aspect of how one goes about doing one's professional work that has become a matter of debate among others.

Questioning the Professional Conduct of Others

You may not have even been aware of when you began noticing how others manage themselves in their professional roles, but at some point, it started to take on meaning for you. For example, you may find yourself paying a lot of attention to the ways staff members go about their work, deal with clients, or conduct themselves with colleagues. Perhaps you are beginning to look at others not just as professionals with roles but as professionals with moral, ethical, legal, and professional responsibilities to the profession, to clients, to the organization, to the community, and to you. You may even be tuning into the subtleties of their overall conduct or their specific behaviors and becoming aware of possible improprieties in how they go about their work.

The improprieties that you observe did not begin when you first noticed them. In all likelihood, it is you who have changed: You can now see what has been there all along. The behaviors and attitudes were just not within your sphere of awareness. There are a number of possible reasons for this change. First, staff members tend not to disclose questionable or surreptitious aspects of themselves so readily with interns or new employees but rather tend to act as they are expected to act in their roles (Kanter, 1977). Consequently, they tend to shield questionable behaviors and attitudes from interns until they get to know them better. Another explanation is that you have been so busy with your increasing responsibilities that you have not had time to notice these behaviors or even think that there was something to notice. A third possibility is that you held stereotypes that needed to change before you were willing or able to see situations for what they were.

Think About It

Why Never in a Million Years?
There are a number of situations that lend themselves, conditionally or not, to questionable behaviors. Some situations are so obvious that mentioning them seems absurd. However, they do need to be mentioned because aspiring professionals, like seasoned professionals, are people too, and they have personal frailties that compromise their ethical standards at times. We both have known and worked with individuals who committed improprieties that neither we nor they ever would have expected. As you peruse the following list (Royse, Dhooper, & Rompf, 2007, in part), to which improprieties, if any, could you or someone you know be vulnerable?

- Being sexually intimate with supervisors and clients
- Being dishonest or fraudulent in your actions
- Libelous or slanderous actions against clients
- Threatening or assaultive behaviors against clients or co-workers
- Misrepresenting one's status or qualifications
- Abandoning a client in need of services
- Failing to warn or protect appropriate parties of a violent client
- Failing to use reasonable precautions with self-injurious clients

A word to the wise . . .   Regardless of the reason, you are bumping up against issues now that did not concern you in the past. And they can become complicated enough to be potential pitfalls for any aspiring or seasoned professional. When the issues are of an ethical or legal nature (we discuss the legal aspects in another section), the situation needs to be managed with reason and sensitivity. What is at stake at the least is someone's feelings and at most someone's career, integrity, and family. For now, it is important to realize that when the stakes are that high, they are high for you, too, because your opportunities for employment and the future of your career could be compromised if your concerns are not justified and handled professionally.

Questioning Your Own Professional Conduct

I recall not long ago having an ethical issue. I was attracted to another worker in the program. I knew it was not an ethical thing to do and that it was against work policy to date someone within staff. I must admit it was hard because we were attracted to each other. We both made the decision to date even though we knew we were not supposed to. Personally, I would not put myself in that predicament again.

STUDENT REFLECTION

When it comes to your own behavior, you have a number of dimensions to consider. Two especially important ones are illustrated in the example that follows: the dimensions of direct service and the dimension of civic professionalism. Let us suppose you are a frontline staff member working with female victims of domestic violence. As is sometimes the case, your position and the agency are fully funded by foundation money, and you and your co-workers are working in donated space at a corporate training facility that is well furnished and comfortable, supported with in-kind donations from the company. The person who was instrumental in making the center and the foundation money happen is a good friend of your family. In your capacity as a developing civic professional, you reflect on your membership in the Sustainability Club on campus and in your campus's chapter of the Responsible Endowments Coalition, with which you have been quite involved. As you take in your surroundings at the field site, you do not see very many green initiatives built into either the in-kind donations or the overall maintenance of the facility. Then, when you spend some time exploring information about the corporation, the donated space, and the in-kind resources, you learn that the company has investments that you consider to be socially harmful: namely, investments in Sudan. Do you have a problem with any of this? If so, what do you do? This is not an integrity issue as you are used to thinking about them. However, you certainly are pausing to think about the situation you find yourself in, as it is in direct conflict with your personal values as a community member, and you are not feeling good about it.

In terms of your own behavior, it comes down to whether you are able to recognize issues as ethical ones when dealing with them. In this instance, the situation just became more complicated. A client of yours at the center bought a newspaper subscription for you for daily delivery to your office (what a convenience!); she also has been bringing pastries from her bakery for her weekly meetings with you (how tempting!).

Do you see a problem with these situations? Do you struggle for the "right answer" after realizing this is an ethical issue?

We find that the areas of professional *relationships* and professional *confidences* are particularly challenging for some interns. Perhaps they are for you also. We'll be exploring those areas in the next section.

Focus on THEORY

Codes and Values

If you believe that there is a problem with the examples of the gift of a daily delivered newspaper or weekly pastry deliveries, what would you say if you learned that even though you think it is "wrong," there is more than a fleeting possibility that you might accept the daily deliveries and the weekly pastries? Some of you smiled about that. Why? It seems that if guidelines about such situations are not clearly spelled out, practitioners rely on their personal value systems and interpretations of the profession's ethics documents, and therein lies a known problem: Among practitioners, there is a discrepancy between knowing "what's right" to do and actually doing it (Jennings, Sovereign, Bottorff, Mussell, & Vye, 2005). So, if you know what's right to do and do not do it in this instance, you probably would not be alone in your failure to act. It appears that if professionals believe there is an infraction of a clearly articulated professional code, they tend to do what they thought they should do, especially if legal precedent exists. But if professionals don't believe that to be the case, then they tend not to act consistently according to what they thought was the right thing to do (Bersoff & Koeppl, cited in Jennings et al., 2005; Bernard et al., and Smith et al., cited in Jennings et al., 2005). So, why, when you rely on your personal value system in such situations, might you not do "the right thing"? It is suggested that some professionals are not honest and do not have integrity (Smith et al., cited in Jennings et al., 2005). There is also speculation that the professional might not have the courage to act on what is "right" (Rest, cited in Jennings et al., 2005). The central issues here are personal values and personal character and whether there is compatibility between personal competence and the values of the profession.

RESPECTING RELATIONSHIPS

When an intern is engaged in more than one relationship (role) at the same time (or sequentially), with a client, the situation that develops is referred to as *dual* or *multiple relationships* (Corey, Corey, & Callanan, 2006). The nature of these roles can vary from being an acquaintance, friend, or business client, to being an intimate—emotionally, physically, or sexually (Dorland, 1974, cited in Malley & Reilly, 2001; Gordon, McBride, & Hage, 2005; Royse, Dhooper, & Rompf, 2007). Dual relationships can occur not only with clients but with professors, supervisors, and co-workers (Malley & Reilly, 2001). All such relationships are fraught with complexity and ethical issues. If you are in one of them, it is best to seek consultation immediately. Another relationship that can be difficult to manage is the *collegial* one, which refers to your relationships with the co-workers, peers, and supervisors involved in your internship. These relationships need to be managed so that your professional obligations are upheld. You may want to take a moment to think about what you *could* do and what you *would* do if your commitment to the profession was being compromised by how you were handling one of these relationships (Rothman, 2000). Again, if you are in a potentially compromising relationship, it is best to seek consultation immediately.

SAFEGUARDING CONFIDENCES

The other area that tends to challenge interns is that of *confidences*. This issue is rather pervasive and involves *clients' rights* (to privacy, privileged communication, and confidentiality) and *information disclosure*. These matters are ones that you are obliged to honor in your role as intern, so it is important that you understand them. Although confidentiality and disclosure are professional issues, they can also be ethical and even legal issues, depending on the laws that govern the work of the profession.

Privacy, Privileged Communication, and Confidentiality   The clients' rights that you deal with on a daily basis probably include privacy, privileged communication, and confidentiality. *Privacy* refers to the constitutional rights of your clients to decide when, where, how, and what information about them is disclosed to others by you (Corey, Corey, & Callanan, 2006; Eisenstat, personal communication, 2007). *Privileged communication* is a legal concept, and the right to such communication belongs exclusively to the client. This concept protects your client from the forced disclosure of information in legal proceedings (Corey, Corey, & Callanan, 2006). You need to check with your supervisor to determine whether clients who see you in a helping capacity lose their right to privileged communications because of your status as an intern. If so, the client needs to know that, and your supervisor needs to guide you in informing the client. *Confidentiality* is a legal, professional, and ethical matter that protects the client in a therapeutic relationship from having information disclosed by you without explicit authorization (Corey, Corey, & Callanan, 2006). Your status as an intern also has direct implications for whether the clients you work with are protected by confidentiality statutes. Again, you need to discuss this matter with your supervisor, and your clients will need to be informed.

Disclosure of Information   The issues of what information can be disclosed and to whom can become complicated when you consider disclosing information to your colleagues (while on-site or socially) or to your peers (in the seminar class). In both instances, you have a responsibility to the privacy rights of the client and, at the very least, to be familiar with and follow the policies of the agency, disguise all identifying data, and give much thought to the question "*Why do I have a need to tell this information to someone?*" The issue of disclosure in the world of the helping professional also applies to cyberspace and its information highway. Be it e-mail, chat rooms, bulletin boards, blogs, or even text and instant messages, interns need to have a working command of this aspect of their workplace. Also, remember that when you use e-mail, a chat room, or a bulletin board to discuss clients or other sensitive issues, those means of communication are not always secure.

The electronic obtaining and releasing of information, along with maintaining that information for the agency, can be difficult for the intern to negotiate appropriately.

Whether you are working with databases or releasing information, it is important that you make informed decisions and do so carefully. The agency has its policies to safeguard the maintenance of information as well as forms to ensure that all criteria are met before information is released by you. If you have not yet reviewed them, this is a good time to do so.

Focus on PRACTICE

HIPAA and the Privacy Rule

If you are interning in the health care industry, including mental health service organizations or in companies that provide support or contract services to health care providers, then you might well be affected professionally by the U.S. Department of Health and Human Services Privacy Rule (August 2002). This rule, which is administered by the Office for Civil Rights in the Department of Health and Human Services, is a set of standards that address the use and disclosure of an individual's health information and also standards ensuring that individuals can understand and control how their health information is used. You will be required to become familiar with and trained in these standards and those select circumstances under which such information can be used and disclosed without an individual's permission. Despite all the limitations of the privacy rule, there are a substantial number of exceptions that you will need to know, such as those concerning the public interest and national priority purposes, including but not limited to public health matters; victims of abuse, neglect, or domestic violence; health oversight activities; judicial/administrative proceedings; law enforcement purposes; decedents; research; cadaveric organ, eye, or tissue donation; serious threats to health or safety; essential government functions; and workers' compensation. Be sure to ask your supervisor if your company or agency is affected by HIPAA's Privacy Rule (http://www.cms.hhs.gov/HIPAAGenInfo) and if so, in what way that will affect your work.

ETHICAL ISSUES: A WORLD OF PRINCIPLES AND DECISIONS

I try to confront ethical decisions by first asking myself what I think is right. If this gets me nowhere, I ask my coworkers, friends, relatives, professors, or anyone who I feel might have knowledge in that area or who has been in a similar situation . . . Then I choose what I think is the best solution.

STUDENT REFLECTION

You might be wondering how you have managed to survive for so long without knowing about ethical issues! This is exactly how our students feel after studying them in a semester-long course.

Chances are that your basic values have served you well. However, the list of potential issues can be overwhelming, even after several readings. Most likely, there are many that you have never heard about and many that seem remotely familiar. Reading through them is a good start. Now that you are becoming familiar with the language of ethics, even if you do not yet know everything it entails, it is a good time to go ahead and identify the dilemmas that already have surfaced in your internship and how satisfied you are with your handling of them.

Talking the Talk

In order to have a useful discussion about these issues, there must be a shared language for communicating and a common understanding of the problem. In the section that follows, we identify some of the terms that are frequently used in discussions of ethical matters and their working definitions. It might take you some time to take in the essence of the meanings, so do not be concerned if there seems to be too much to grasp the first time you read through this section.

Let start with the term *standards*. When used generically, *standards* refers to guidelines or codes that govern the behavior of members of a given profession. *Ethical* suggests that someone is acting in accordance with professional standards, codes, guidelines, or policies, and *unethical* suggests that they are not. *Legal* suggests that someone is acting in accordance with the law, and *illegal* suggests that they are not. Besides these terms, which are related directly to the responsibilities of a profession, the terms *values, moral,* and *ethics* are very important to any discussion of ethical, legal, or professional issues. *Values* refer to what is intrinsically good, useful, and desirable; *moral* refers to what is right or wrong conduct in its own right, based on broad mores such as religious principles; and *ethics* refers to the moral principles or rules of conduct of a particular profession (G. Corey, Corey, & Callanan, 2006; Pollock, 1998), as in helping professions ethics, business ethics, political science ethics, and so on.

Rules of the Trade

In addition to having a common language, it is important to have access to common resources. For one thing, although your understanding of the issues in your professional field has just begun, you are still responsible for acting in accordance with the values and standards of that profession. These values and standards are embodied in ethical documents variously referred to as *guidelines, standards, regulations, policies, principles,* and *codes.*

It is precisely these types of documents that are reviewed when conflicting or questionable situations arise. Unfortunately, they don't necessarily lend themselves to clear interpretations or resolutions of ethical or professional conflicts. However, they do provide guidelines for behavior and discussions. Technically speaking, there are differences between these documents, which are described in the box that follows. If you are not interested in that level of detail, skip the box and read on.

Focus on PRACTICE

Differences in the Documents

It would be helpful to your understanding of the discussion that follows if you have a copy of the ethics document that guides the professionals in your agency.

Begin by taking the time to notice the document title. If it is titled *guidelines*, it reflects recommendations from professional groups for acceptable behaviors for the profession. If the title reads *standards*, the statements reflect the rules of behavior for the profession that are drawn up by members of the profession itself and that often carry civil sanctions and set the parameters for ideal behaviors. If you have heard your supervisor or field instructor talking about accreditation teams or visits, they are referring to an assessment of the site or academic program based on designated standards of behavior for the profession.

If, on the other hand, the document is called *regulations*, it contains dictates typically from governmental authorities and often specifies sanctions for not complying with them. Undoubtedly, you were given a copy of your site's policy manual the first week in the field. A *policy* refers to the procedures or course of actions set forth by an organization to ensure expediency and prudence in getting the work done. All organizations have policy manuals that they give employees and interns to read as part of their orientation to the work and workplace. *Principles* are fundamental doctrines of the profession that are rooted in commonsense morality.

Finally, if the document is called a *code*, its statements reflect beliefs about what is right and correct professional conduct. Codes often include standards of practice along with statements that embody the values of a profession. Like the learning contract, codes tend to be living documents that continually evolve. They promote professional accountability and facilitate improved practice by protecting the professional from ignorance (e.g., from malpractice suits so long as the professional acts in accordance with acceptable standards), protecting the public from the profession (i.e., protects the consumer from harm), and protecting the profession from the government (i.e., the profession governs and regulates itself, protects itself from internal struggles, and establishes agreed-upon standards of care) (VanHoose & Kottler, 1978, cited in Bradley, Kottler, & Lehrman-Waterman, 2001).

If you have not done so already, ask your supervisor for a copy of the ethical code that guides your co-workers, for that is the document that should also guide you. We suggest that you take the time to read it. Some of you will be looking at a two-page document; others will be reading twenty pages of professional rules! Reading the shorter documents is a reasonable exercise. However, if you are bound by the American Counselor Association's (ACA, 2005) code or that of the American Psychological Association (APA, 2003), reading such lengthy documents is not a reasonable undertaking. What is expected of you, though, for the purposes of the field experience is to become familiar with the categories of codes and those categories relevant to

Focus on PRACTICE

Principles and Standards

There are three documents, listed below, that we suggest you take the time to examine. They will help clarify responsible ways to carry out the work of the internship or the profession and will also provide considerable insight into the next section, where we identify many of the issues that could potentially become problematic for you or any practitioner.

- *Eight Principles of Good Practice for All Experiential Learning Activities* (National Society for Experiential Learning, www.nsee.org)
- *Ethical Standards of Human Service Professionals* (National Organization for Human Services (NOHS) and the Council for Standards in Human Service Education (CSHSE) (www.nationalhumanservices.org; www.cshse.org)
- Two documents published in 2006 by The Council for the Advancement of Standards in Higher Education (CAS) relevant to experiential education: *The CAS Professional Standards for Internship Programs* and *The CAS Professional Standards for Service-Learning Programs* (www.cas.edu)

your specific field experience. Bring the document to seminar class. Unless you and all your peers are in the same specialty (e.g., the helping professions or mental health work), the documents you collectively bring to class will be varied in specialty area, length, and intent.

A word to the wise . . . It is important that you know the resources of your future profession. Your site supervisor can be very helpful in identifying them for you. For example, a very helpful resource is the book *Codes of Ethics for the Helping Professions*, which includes the full text of codes for fifteen professional specialties (Wadsworth Group, 2004). In addition to the codes, you will want to have the addresses (electronic and postal) of the professional organization that guides your work. Membership in such organizations can be expensive, but it is well worth the fee (most have a student category of membership with lower fees). The majority of national organizations can provide members with much information and have access to many resources. In addition to a variety of support services, such as casebooks and libraries of instructional videos, many also offer legal counseling and services when necessary.

Ethical Principles and Ethical Values

I believe that it is better to overthink than to be impulsive and regret a choice later on. I do know that the ethical decisions and choices I make in life are a reflection of my values and character. I know that not every ethical decision I make in life is going to be easy or the right choice. But I believe that I have good values, and that will help me with my decisions.

STUDENT REFLECTION

Regardless of how detailed the codes may be or how many times you read them, there may be no answers forthcoming to help you resolve the conflict you experience. When ethical codes fail to provide a direction toward a solution, then the *ethical principles* of the profession can be taken into consideration to guide your decisions. *Ethical principles*, as mentioned earlier, are fundamental doctrines of the profession rooted in commonsense morality.

If you are not in the helping professions, you may want to locate those principles that guide the work in your field and spend some time thinking about them. For those of you in the helping professions, we list here six ethical principles that are commonly accepted as reflecting the highest level of professional functioning (Corey, Corey, & Callanan, 2006). The principles are based on the works of Kitchener (1984) and Meara, Schmidt, and Day (1996) and probably look very familiar to you. You may not realize, though, that these principles have more than one purpose: They can guide your work as well as your decision making. When using these principles to guide your decisions, be as honest with yourself as possible, as this type of authenticity will be invaluable to the quality of the decisions and the personal insight you develop.

- *Autonomy* refers to the clients' freedom to control the direction of their lives by making decisions that reflect their wishes; this principle affirms the clients' right to self-determination.

- *Beneficence* refers to commitment to do "good," as demonstrated by carrying out work with competence and without prejudice. This principle affirms the clients' dignity and promotes the clients' welfare.

- *Justice* refers to treating others with fairness, regardless of gender, ethnicity, race, age, religious affiliation, sexual orientation, ability, religion, cultural background, or socio-economic background. This principle affirms the clients' right to equality in services.

Focus on THEORY

Principle Ethics and Virtue Ethics

Kitchener (1984) identified the first five of the six ethical principles as a way for practitioners to make responsible ethical decisions in their work when the ethical codes that are there to guide them fail them because of their broadness or narrowness. Meara, Schmidt, and Day (1996) differentiated Kitchener's principle ethics from virtue ethics, which focus on character traits and ideals (Meara et al., cited in Jennings, 2005, p. 33). Virtue ethics were considered more useful in providing a sense of an individual's moral life than principle ethics; instead, principle ethics were grounded in prescribed rules that were formal and obligatory in nature. Together, the principle and the virtue ethics provide a seminal framework to guide helping practitioners in their ethical decision making, which is rarely a simple matter (Jennings et al., 2005).

- *Nonmaleficence* refers to avoidance of doing harm; this principle affirms the clients' right to respect.

- *Fidelity* refers to having a trustworthy relationship of honest promises and honored commitments; this principle affirms the clients' right to informed consent before committing to interventions.

- *Veracity* refers to being truthful in dealings with clients; this principle affirms the clients' right to full disclosure.

How might you answer the question "What ethical values are relevant to your profession?" Would they be the values that you read about in your academic coursework preparing you for your profession? Presumably so. Would they be the values that you see in practice at the internship site—the ones you are encouraged to demonstrate? Again, presumably so. There is an operating assumption that the ethical values you see in practice are the ones that the professionals believe best inform them when it comes to making ethical decisions, which we know will draw upon their personal value systems at times.

Think About It

How Do Your Ethical Values Measure Up?

In recent years, eight ethical values used by master therapists in the helping professions were identified. These ethical values take on added importance when you think about how practitioners use codes of ethics and their personal value systems when making decisions about ethical issues and dilemmas (Jennings et al., 2005). How do the values cited here figure into the decision you made about the earlier example of the daily newspaper delivery and the weekly delivery of baked goods?

Building and Maintaining Interpersonal Attachments

- Relational Connection
- Autonomy
- Beneficence
- Nonmalificence

Building and Maintaining Expertise

- Competence
- Humility
- Professional Growth
- Openness to Complexity and Ambiguity

LEGAL ISSUES: A WORLD
OF LAWS AND INTERPRETATIONS[1]

An important part of making an ethical decision is knowing the laws that are relevant to and affect your work. Some of you are developing a familiarity with the law, especially if you are interning in legal settings or your work is closely directed by statutes and legal guidelines. Others of you may know little about this aspect of your work. In this section, we will give you a way of thinking about legal matters so you can make better sense of this aspect of your field experience.

For those of you interning in the criminal justice system, legal mandates govern much if not all of your work. For those of you working with dependent individuals, such as minors, elders, and those with special needs, the intent and extent of your work are largely affected by legal statutes, especially in the area of protection; i.e., abuse, neglect, and exploitation. If you are interning with a legislator, advocating for clients in class-action suits, or interning in hospitals, human resource departments, or in mediation services, you are working with laws. If you are interning in a government agency, your work is affected by the statutes or laws that govern the agency. For that matter, all internships are affected to some degree by legal issues.

A number of legal issues are particularly relevant to interns who work directly with clients, most of which have ethical dimensions as well (see Kiser, 2008). Such issues include, but are not limited to, liability and malpractice; confidentiality, privileged communication, and privacy; disclosure of information; end-of-life decisions; consultations with specialists; crisis intervention; suicide prevention; termination of interventions; intimacy with clients; duty to protect intended victims from violence; and informed consent (Birkenmaier & Berg-Weger, 2007; Kiser, 2008). The list does not stop here, but we will.

Your responsibility is to know the legal basis, if there is one, for your organization; the laws that affect and govern your work; and the ways in which you are bound by those laws in carrying out your responsibilities (Berg-Weger & Birkenmaier, 2007; Gordon, McBride, & Hage, 2005). The best way to learn about these matters is to bring your questions and concerns to your supervisors. The following sections will help to prepare you for those discussions.

Talking the Talk

As was the case with ethical matters, there is a terminology specific to legal matters in field work that you need to know. Again, we will take some liberty and use working definitions where possible so that you have a sense of the language and implications. We know that the information in this section is typically what interns most want to know about legal matters. However, the information is very technical, and it is not possible to describe it without a great deal of detail. So, on the one hand, we risk oversimplifying a complex body of information, and on the other hand, we risk boring or causing you

---

[1] Appreciation is extended to Professor Steven Eisenstat of Suffolk University School of Law, who specializes in civil tort law, for his assistance with the section of this chapter on legal issues.

undue concern. We will do our best to choose a middle ground. We advise you, though, throughout this discussion and throughout your internship, to bring all matters to your supervisors if you do not have a working understanding of them.

NEGLIGENCE

A tort is a civil wrong or injury done to another that is not based on an obligation under a contract. There are three types of torts: *intentional torts, negligence torts,* and *strict liability torts.* For the purposes of your internship, it is the negligence tort that is of most concern to you, your site supervisor, and your campus instructor.

For an act to be a negligence tort, there must be a legal duty, owed by one person to another, a breaking (*breach*) of that duty, and harm caused as a direct result of the action. For example, if you are interning at a home health care agency and you voluntarily assume responsibility for an elder in the community, you are then in what is referred to as a *special relationship* with that person. Your duty to your client would be considered breached if you fail to provide the standards of care of the home health care profession; you could do this either by failing to take certain required actions or, if you did act, doing so in a way that does not reflect the standards of care for the home health care profession. It makes sense to raise the issue of *breach of duty* with your supervisor so you can better understand how you could be at risk for such lawsuits (Eisenstat, personal correspondence, 2007). Negligence torts, then, can result when you fail to exercise a reasonable amount of *care* (standard of care) in a situation that causes harm to another person or to a thing. The basis for the negligence tort can involve doing something carelessly or failing to do something you are supposed to do.

There are two forms of negligence: ordinary negligence (i.e., failing to act as a reasonable person would) and aggravated negligence (i.e., reckless or willful behavior). It is the *ordinary negligence* tort that is the more likely concern in your internship. An example of ordinary negligence occurs when officers (e.g., police, probation, corrections, parole) fail to perform duties owed to clients or inmates or when they perform duties inadequately (Eisenstat, personal correspondence, 2007). For example, correctional officers have a duty to check regularly on the inmates under their care. If a correctional officer fails to do so and an inmate commits suicide, then the officer could be found negligent in terms of his or her supervisory responsibilities. So, it makes sense to raise the issue of negligence and the potential pitfalls you face with your supervisor so that you can better understand how you could be at risk for such lawsuits.

MALPRACTICE

The term *malpractice* is one you are sure to have heard and know enough about that you do not want it to be a part of your field experience! Malpractice, a form of ordinary negligence, refers to an act that you perform in your professional capacity for which you are being sued. This type of lawsuit charges professional misconduct or unreasonable lack of skill on your part that results in injury or loss to your client. For example, if you are an intern at a residential facility for emotionally disturbed adolescents and you fail to take the precautions that are ordinarily provided by other residential facilities/workers in the profession, and your actions or lack thereof result in one of the residents committing suicide, then you most likely would face a malpractice lawsuit (Eisenstat, personal correspondence, 2007). Again, it is important to bring up the issue of malpractice with your supervisors so you can better understand how you could be at risk for such lawsuits.

Although not reassuring, it is important to know that there are situations that tend to increase your liability for a malpractice lawsuit. (Being *liable* or having a *liability* means a breach of duty or obligation to another person.) For example, if you fail to use acceptable procedures, or you break a contract, or you use interventions for which you were not trained, or you fail to choose the most helpful interventions, your risk of liability increases significantly. And it does not stop there. If you fail to warn others about or protect others from potential danger, or you fail to secure informed consent appropriately, or you fail to disclose to your client the possible consequences of services and interventions, then your risk of liability could increase. Again, it is important to discuss potential malpractice situations with your supervisors.

Rules of the Trade

In addition to having an understanding of the terminology, it is important to have an understanding of the legal framework. Let's start at the beginning with laws. A useful way of thinking about laws is how they are classified. For example, laws can be classified according to how they come into existence. Laws in the United States derive from our constitutions (*constitutional law*, from state and federal constitutions), our legislatures and governmental agencies (*statutes* and *regulations*, respectively), and our *common law* (*case law* from prior decisions by trial courts or appeals courts).

Another way laws are classified is by the nature of their focus: criminal or civil. *Criminal law* refers to a group of laws that seeks to resolve disputes between the government and people. Criminal law seeks punitive measures such as imprisonment and fines to right a wrongdoing (Eisenstat, personal correspondence, 2007). Many interns work with criminal law on a daily basis; for example, those in criminal justice settings, legal offices, domestic violence agencies, and protective work with dependents.

On the other hand, *civil law*, of which tort law is an example, seeks to resolve disputes between people by enforcing a right or awarding payment or what is referred to as *damages*. Its primary intent is to repair rather than punish behavior, as is the intent of sanctions for criminal matters (Eisenstat, personal correspondence, 2007). An aspect of civil law that many interns work with is mental health law. This body of law regulates how the government takes care of or responds to people with mental health challenges. If you are interning in a mental health clinic, hospital, residential setting, or a community shelter, your work is affected by this body of law.

In some instances, students deal with both civil and criminal law. For example, interns at offices of the American Civil Liberties Union deal with constitutional rights, which involve both civil law as well as criminal law.

Another way of thinking about law and the helping professions is to separate the laws according to the aspects of the work (Garthwait, 2005). For example, there are laws that regulate the *services* or *actions* that you can give to a client. There are laws that regulate the *work of the human service agency*, such as working with youth, working with elders, and working with mentally ill individuals. And there are laws that regulate the *practice* of the profession, such as deportment, licensing, and supervision issues.

Relevant Legal Matters[2]

I just wished I had known exactly what the supervisor is legally supposed to do . . . Then I would have been able to point to something in writing. It was very hard knowing the supervisor was wrong but not having anything to point to and say "this is what you are supposed (sic) to be doing for me."

STUDENT REFLECTION

The legal matters of most relevance for interns come under the general categories of *standards of care* and *supervisory malpractice*. Again, these are complex areas of inquiry, and it is not possible for us to address them adequately in one chapter on ethics and laws. However, we hope to give you a way of thinking about them so that you can discuss these matters with supervisors and become better informed about these important areas of practice.

STANDARDS OF CARE

A matter that is both ethical and legal in nature and affects the work of the intern in the helping professions on a daily basis is that of *reasonable standard of care*. Interestingly enough, this area of potential legal matters is neither universally nor directly addressed in the ethical standards for the helping professions (Kiser, 2000). Kiser (p. 122) has described the components of a reasonable standard of care as including, but not limited to, knowledge of the clients and services being given; delivery of services and interventions based on sound theoretical principles; reliability and availability of services to clients; taking the initiative and acting on behalf of client and public safety; adherence to ethical standards of the profession in relation to client care; and systematic, accurate, thorough, and timely documentation of client care.

SUPERVISORY MALPRACTICE

The second legal matter is that of *supervisory malpractice*. In this instance, it is the behavior of the supervisor that comes under legal as well as ethical scrutiny. As you may be aware, your supervisor is liable for your work because when your supervisor agreed to supervise you, he or she accepted responsibility for all of your work, including your work with clients, and for your behavior (deportment) during the internship. If this responsibility sounds pretty serious, it is.

Quality of Supervision Failure to supervise the professional staff appropriately has been the cause of a growing number of malpractice suits (Sherry, cited in Falvey, 2002). This type of lawsuit concerns the quality of supervision given to the intern. The legal scrutiny a supervisor faces in such a lawsuit results from alleged negligence in carrying out supervisory responsibilities and from subsequent injury or damages.

---

[2] We based the discussion that follows on the work of a number of writers and experts in this field, including Birkenmaier & Berg-Weger (2007); Champion (1997); Corey, Corey, & Callanan (2006); Eisenstat (personal correspondence. 2007); Falvey (2002); Gordon, McBride, & Hage (2005); Garthwait (2005); Malley & Reilly (2001); Oran (1985); and Pollock (1998).

You, the intern, along with whoever may have been injured as a result of improper supervision, become the *plaintiffs* (i.e., the ones who bring the complaint), and your supervisor becomes the *defendant* in such a lawsuit for negligence. For example, you have been directed to conduct an in-home assessment to determine the removal of a child based on alleged neglect by the parents. In the process of conducting the interview, the mother becomes despondent, leaves the interview, goes into the bathroom, and slashes her wrists. If your supervisor did not prepare you adequately to respond to and manage the range of possible reactions to such an interview, your supervisor's risk for liability for failing to train you adequately increases substantially. Such preparation could include, but not be limited to, having you observe and/or conduct such an interview under the direct supervision of an experienced worker or talking with you about the potential for self-destructive reactions to such interviews. In this scenario, the family of the mother could also bring suit against the supervisor (Eisenstat, personal correspondence, 2007). We hope you will never have such memories as part of your internship.

Vicarious Liability   Another type of negligence liability on your supervisor's part is *vicarious liability*. Under vicarious liability, your supervisor could be held responsible for your negligence even if your supervisor did not act negligently. This area of law can become very complicated.

Ordinarily, there must be some form of salary to create an employer-employee relationship, as in cooperative education placements. Practica or service-learning placements, however, typically do not pay for the student's work; internships can be paid, but many are not. Ultimately, the question comes down to whether there is sufficient oversight of the intern—how to do the work, the hours of the work, salary, and so on—that one could argue the level of control that the company exercises over the intern is sufficient to create an employer-employee relationship. This is decided upon by the facts of the specific case.

In the case of paid internships where the employee acts negligently, it is the agency, not the supervisor (since it is the agency paying the salary), who is vicariously liable. Of course, if the supervisor also acts negligently, the agency can be vicariously liable for the supervisor's actions as well. In the case of unpaid internships, vicarious liability can still exist, but once again, it would exist between the "employer" and the intern/employee and not the supervisor and the intern/employee, unless the supervisor is also the employer. If there is a campus supervisor involved in the internship in addition to the organization's employee who oversees the intern's day-to-day work, the agency could argue it lacks sufficient control over the student because a second supervisor is involved and that it is the college or university that should be held vicariously liable for failing to adequately supervise the intern. This scenario of vicarious liability is not likely to arise in a typical internship. The more likely scenario would be the injured party suing the field site and/or the campus for their negligent supervision of the intern (Eisenstat, personal correspondence, 2007).

Obligations to the Intern   At this point in your understanding of liability, you may be wondering under what circumstances does your supervisor incur potential liability because of her or his failure to meet obligations to you? Four major sources of such liability have been identified (Harrar, et al., 1990, cited in Falvey, 2002). We think it's

---

Think About It

The Rights of Interns

These rights should look familiar to you. You were first introduced to them in Chapter 2 as your internship was beginning. You've been an intern for a while now. In what ways are these rights being respected in your placement? What rights do you believe are not being respected?

- The right to a field instructor who knows how to supervise, i.e., has been adequately trained and is skilled in the art of supervision
- The right to a supervisor who supervises consistently at regularly designated times
- The right to clear criteria when being evaluated
- The right to growth-oriented, technical, and theoretical learning that is consistent in its expectations

---

Munson, cited in Royse, Dhooper, & Rompf, 2007

important to note them so that you are more informed about what you can request of the supervisory relationship.

- If the supervisor is derelict in carrying out supervisory duties for planning your internship, the direction of your internship, or the outcome of your work, then your supervisor's liability can increase.
- If your supervisor gives you inappropriate advice about a treatment intervention that you use, and the intervention is to the detriment of the client, the supervisor's liability can increase.
- If your supervisor fails to listen attentively to your comments about a client and in turn fails to understand the needs of the client, the supervisor's liability can increase.
- If your supervisor assigns you tasks that the supervisor knows you are not trained adequately to perform, then the supervisor's risk of liability can increase.

All these conditions make good common sense, and our experience is that students know intuitively when they are being shortchanged or otherwise not being given quality supervision. However, seeing them in print can be most affirming for an intern. Similarly, it is helpful for practicum students to be aware of these rights as they go about their work in the field.

GRAPPLING WITH DILEMMAS

This is how I try to decide what to do: Think about it, talk about it, and try to look at it from every angle. Then I make a decision.

STUDENT REFLECTION

One of the most wrenching aspects of working in the helping professions is dealing with a dilemma that involves the welfare of another individual, family, or community.

A *dilemma*—be it ethical, legal, professional, or all three—refers to a struggle over alternative courses of action that might resolve a situation. To complicate things, the choices of courses of action tend to be correct in their own right, but they conflict with each other. A dilemma, then, is a situation in which you can find yourself facing more than one justified course of action; that is, two "right" ways of responding.

Our interns tell us that once they develop an understanding of the issues and become comfortable with the language, they begin to see the issues in their daily work. It just so happens that you tend to develop this awareness at the same time that you are moving into a collegial-like relationship with your supervisors. Consequently, you are much more apt to talk about incidents, behaviors, and concerns at this point in your field experience than you were a couple of months ago.

Recognizing Dilemmas

The next hurdle is to recognize a dilemma when you see one, which is no easy feat. Our experience tells us that recognizing dilemmas as such is quite challenging for both undergraduate and graduate students and for experienced professionals as well. Often, students cannot readily name ethical issues when they see them, and they do not necessarily see them in a given situation. Nor do they have a language to describe situations that might be ethical in nature or to discuss them. A course in applied ethics taught in a professional studies programs at the very least will help the student become aware of the ethical issues of the profession, learn the "language" of applied ethics so issues can be named as such and discussed, and begin to develop the reasoning and problem-solving skills necessary to succeed in the profession. Academic programs in human services education sometimes require a course in professional issues but more often will offer it as an elective or cover the content in other coursework (M. DiGiovanni, personal communication, 2007; CSHSE, 2005). A study of the effects of a one-semester, elective course in criminal justice ethics suggests that students' values do appear to shift in one measurable way: from having concerns about their personal gain to having concerns about issues of social justice and the welfare of others (Lord & Bjerregaard, 2003). Given that police work is a service profession, this particular shift in values is right on target for the students.

If you have not studied or discussed ethical issues academically, you may feel particularly underprepared for these challenges. We hope this discussion will help you to frame your understanding of the issues and learn what questions to ask.

There are three basic types of ethical dilemmas that you may encounter:

- Those dilemmas that result from your own decisions, behaviors, or attitudes (e.g., *considering whether to engage in a dual relationship with a client, such as knowingly working with a client whose sister you have dated*)

- Those dilemmas that result from another person's decisions, behaviors, or attitudes and directly affect you (e.g., *working with a client who, unbeknownst to you, is your cousin's intimate partner*)

- Those dilemmas that you pay attention to from a distance but that do not directly affect you (e.g., *observing or hearing about a co-worker dating a relative of his or her client*)

> ### Focus on SKILL
>
> #### Anticipating Ethical Issues
> One way of dealing with an ethical issue is to anticipate it. A useful and effective way of doing this is to rehearse in your mind the best possible response (how you would like to respond to it), the worst possible response (your worst nightmare in responding), and a more realistic response (found somewhere between the two extremes). Then discuss with peers or your supervisors the implications of the more realistic response.

Walking the Walk

You certainly will be exposed to ethical issues during your fieldwork. You may even experience an ethical dilemma. If that happens to you, remind yourself that you are no stranger to facing difficulties and that you have what it takes to work your way through yet another challenge. And like the challenges you faced in the past, an ethical issue can become a problem if you do not manage it effectively.

Regardless of the situation, you will need a way of thinking—a framework—for making ethical decisions. Thinking critically about ethical situations is important for making responsible decisions about them. Although most of you have heard this term constantly throughout your education, it is quite possible that you do not know the skills necessary to think critically about an issue, a situation, a decision, or an action. To think critically means to use standards of reasoning such as clarity, precision, accuracy, relevance, depth, and breadth. It involves evaluation techniques, weighing alternative perspectives, and genuine efforts to evaluate all views objectively. Examples include such skills as articulating ideas, asking significant questions, problem solving, and openness to contradictory ideas (Alverno College Productions, 1985). Why so much thinking? you are asking. Thinking thoughtfully and in demanding ways will help you make wise choices—choices that most likely will help your clients reach their goals (Gibes & Gambrel, 1996, p. 6).

Reasoned Steps to Resolving Dilemmas

> *I find that when I think about the whole situation (ethical decision), I realize that unethical decisions have a lot of repercussions . . . People lose their jobs, and I don't want to get into trouble.*
>
> STUDENT REFLECTION

There are many decision-making models to guide you in developing critical thinking skills. Some are specifically intended to deal with ethical or legal matters in the helping professions (see, for example, Corey & Corey, 2006; Corey, Corey, & Callanan, 2006; Gibbs & Gambrill, 1996; Kenyon, 1999; Neukrug, 2000; Rothman, 2000; Schram & Mandell, 2005; Tarvydas, 1997, cited in Tarvydas, Cottone, & Claus, 2003; Woodside & McClam, 2005). The model we offer is one that we use when teaching about ethical and legal issues. It is based in part on the Eight Steps to Creating Change in Chapter 11 and incorporates adaptations of other models as well (Close & Meier, 1995; G. Corey et al., 2006). The model offers opportunities to think critically in practical ways and to develop a reasoned response to and action plan for the presenting problem.

1. Name the Problem   Collect as much information as you can about the situation. Clarify the conflict. Is it moral? Professional? Ethical? Legal? Given that there are no right or wrong answers to the situation, anticipate ambiguity and challenge yourself to consider the problem from multiple perspectives.

2. Narrow the Focus   Once you have gathered as much information as is reasonable, list the issues you are confronting. Some are more important than others. Describe the critical issues and players; discard the unimportant ones.

3. Consult the Codes   Review the ethical documents of your profession, the policies and regulations of your agency, and related laws to determine whether possible solutions are suggested. Identify aspects of the documents that apply. How compatible are your personal values with those of the profession?

4. Consider the Laws   Chances are you are just becoming familiar with the laws—both civil and criminal—that are relevant to your work. Consult those laws. Once familiar with them, you can contact the legal counsel for your field site (your supervisor should be made aware of this first!) or a law librarian for guidance.

5. Consult with Colleagues   Consult with informed colleagues to discover other ways of considering the problem. Given the responsibility to make a reasoned decision, consulting with colleagues is one way to "act in good faith" and test your justifications. Choose your colleagues wisely.

6. Determine the Goals   Think through what change you hope to bring about in the attitudes, behaviors, or circumstances in question. Question your motives carefully and repeatedly. Is your client's voice heard in the goals you want? Talk with a colleague about whether there may be motives on your part of which you are not aware. Choose your colleague wisely.

7. Brainstorm the Strategies   Identify all possible courses of action, including the absurd. Some may prove useful, although unorthodox. Consider the client's perspective. Is the client's voice heard in your list of options? Discuss options with others. Choose your colleagues wisely.

8. Consider the Consequences   Think about the consequences of each strategy for all involved in the situation. Thoughtfully assess plans. Identify consequences from various perspectives, and question each of the consequences. Remember to include the client's perspective among those you consider.

9. Consult the Checklist   Use the following checklist to evaluate potential areas of ethical and legal misconduct. The questions are based in part on a model of ethical decision making that identifies six fundamental principles of moral behavior: autonomy (*self-determination*), beneficence (*in the best interest of the client*), nonmaleficence (*to do* *no harm*), justice (*fairness to all*), fidelity (*honest promises and honored commitments*), and veracity (*being truthful*). This model includes such qualities of ethical acts as universality, morality, and reasoned and principled behaviors (G. Corey et al., 2006; Kitchener, 1984; Meara, Schmidt, and Day, 1996; Pollock, 1998).

- Is the action in the best interest of the client? *Consider the six fundamental principles of moral behavior.*
- Does the action violate the rights of another person? *Consider constitutional rights as well as your duty to justice.*
- Does the action involve treating another person only as a means to achieve a self-serving end? *Consider the end-in-itself motive and the utilitarian perspective.*
- Is the action under consideration legal? Is it ethical? *Consider the laws and your legal duties; consider your civic and ethical duties and the components of an ethical act.*
- Does the action create more harm than good for those involved? *Consider the principles of nonmaleficence and beneficence.*
- Does the action violate existing policies, regulations, procedures, or professional standards? *Consider the duty to your professional role.*
- Does the action promote values in culturally affirming ways? *Consider the principles of nonmaleficence and beneficence and the duty to care.*

10. Decide Diligently   Consider carefully the information you have. The more obvious the dilemma, the clearer the course of action; the more nebulous the dilemma, the more difficult the choice. Although hindsight may teach you differently, the best decision under these circumstances is a well-reasoned decision—one with which you can live.

TABLE 13.3
Reasoned Steps to Resolving Dilemmas

1. Name the problem
2. Narrow the focus
3. Consult the codes
4. Consider the laws
5. Consult with colleagues
6. Determine the goals
7. Brainstorm the strategies
8. Consider the consequences
9. Consult the checklist
   - Principles of autonomy, beneficence, nonmaleficence, justice, fidelity, and veracity
   - Duties to care and civic responsibility
   - Responsibility to profession's laws, ethical documents, policies, procedures, and regulations
10. Decide diligently

MANAGING A PROFESSIONAL CRISIS

One of the potential pitfalls to a chapter such as this is the tendency, regardless of age, to worry unnecessarily about your vulnerability to the issues raised in the reading. The fact is that you would not be in the field if your academic program did not prepare you adequately for the experience and you would not be doing the work you are doing if your site supervisor and campus instructor did not believe you were academically and professionally ready for the experience. Even so, issues and situations will present themselves, and they can arise quite quickly.

Sometimes, a situation develops through no fault of your own, and you find yourself at the unenviable end of allegations, complaints, or legal charges. When that happens, it is a very stressful time for all involved, especially for you. You are well aware of what you have invested in your field experience and the importance of your grade and learning to your academic work and/or career plans. However, you may not be aware that such situations do arise and that *how* you handle the situation is very important to your supervisors' assessments of you in a crisis.

Knowing how to respond to such a crisis can make all the difference in you having a future in the profession. Managing this type of crisis is no different from managing other crises you have lived through or half expect to occur at some time. There are many approaches to managing crises, and it is not our intent to suggest a particular approach to you. You should know what works and what does not work for you. However, we do suggest that you think in terms of a four-pronged approach to managing a crisis.

Have Resources in Place

Regardless of the approach you use, it is very important that you know what resources are readily available to you and those that you need to develop. The resources should include, but not necessarily be limited to, legal, emotional, physical, academic, and

---

Case in Point

Being Tested on the First Day

On the first day of a practicum in a courthouse, one of our students was asked to assist in gathering information in interviews resulting from a drug sting. Those arrested were of the same ethnicity and spoke the same first language as did our student. The student was approached by those arrested and asked to help them evade court processing. Another student was asked on that same day for a date by a convicted offender within the first hour of her practicum at a day reporting center. Both students were shocked at how quickly the situations happened. However, neither of them experienced dilemmas, although they were surprised to learn how easy it was to find themselves in potentially compromising situations.

professional supports. Knowing beforehand what, who, and where the resources are and how to mobilize them is essential to a healthy and effective response. Otherwise, you can find yourself responding in ways that are not helpful to yourself or the situation.

For example, becoming so upset in times of professional crisis that you do not know which way to turn may be an understandable but not a very useful way of responding. Drinking alcohol, taking drugs, or engaging in other self-destructive behaviors is neither professional nor helpful to the situation. What is useful, though, is identifying your resources beforehand so you know who to call, where to go, and what you can expect from them. Your supervisor can be a wellspring of information for you about such a crisis in terms of who should be on your list and what information you should have beforehand. Also, having membership in the profession's national organization (which usually is at a substantially reduced rate for students) allows access to information and often legal advice. It may be helpful at this point to review the sections in Chapter 4 on support systems and identify supports that you need to develop for a professional crisis, which may be quite different from supports for a personal crisis.

Expect to Learn from the Crisis

Make a resolution to do the best you can under the circumstances and to learn from the crisis in your internship. It is very important to think through the value of such a resolution now, when you are not in a storm of emotions that makes it very difficult to appreciate the benefits. Not only does a resolution to learn provide you with an understanding of how you function in a crisis within a professional context, but it gives you insight into an aspect of the profession that you otherwise would only read about. Taking care of yourself legally, emotionally, physically, academically, and professionally is your responsibility in a crisis. It is also an essential factor for riding out the storm in ways that leave you the stronger for it.

Lay Out a Crisis Response Plan

In addition to knowing your resources and knowing yourself in a crisis, it is critical that you have a crisis plan of action: i.e., a plan that allows you to be most helpful to yourself and the situation even in times of high anxiety. An important piece of such a plan is a crisis team for a professional crisis—the people you can call on in an instant to give you the help you need, whether it is legal council or chicken soup, literally and figuratively. It is a team of first responders that you create for yourself. In putting together your critical

---

TABLE 13.4
HELP: Self-Care in a Professional Crisis

- **H**ave resources in place.
    - **E**xpect to learn from the crisis.
        - **L**ay out a crisis response plan.
            - **P**ractice self awareness.

support team, be sure to identify how to reach them when you need them (i.e., electronic, cell, telephone, and postal). Next, you need to think through what you must do to take care of yourself emotionally, physically, and academically throughout the storm so that you stay afloat when the waters get rough. Maybe a physical outlet is best for you, like running, biking, or workouts at the local gym. Perhaps it is mindfulness meditation or yoga that makes a difference in your ability to cope under pressure. Some find prayer or other meditative activities most helpful. Many find counseling to be comforting and affirming. Whatever it may be, you need to be aware of it, keep it foremost in your mind, and make it part of your agenda in a crisis.

Practice Self-Awareness

It is very important that you understand your reactions, strengths, and weaknesses in a crisis. How you go about solving crises, what works most effectively for you, and what is ineffective in such situations are all informative. Thinking these through before, as opposed to when in a crisis, is very important because your objectivity is not compromised by the pain of the situation. This may also be a good time to revisit Chapter 4, especially the parts that help you understand how you function under stress.

SUMMARY

You have done a lot of reading about a lot of issues in this chapter, most of which are probably new to you. At best, you have become familiar with some terminology and have become aware of areas of concern that you could face in both the field and the profession. This chapter is intended to give you a way of organizing your thinking about the wide variety of issues—professional, ethical, and legal—that are part of the profession and about ways to respond should a crisis develop in these aspects of the work. We encourage you to use this chapter as a resource throughout your field experience and to consider taking related academic coursework if you have not already done so.

As you continue to develop your sense of competency and professional identity and deal with your concerns about professional behavior, you are fast approaching the last mile of your journey. We discuss that final mile in the next chapter and the feelings interns experience when their good-byes are bittersweet.

For Contemplation

PERSONAL REFLECTION:
SELECT THOSE INQUIRIES THAT ARE MOST MEANINGFUL TO YOU.

1. In thinking about the personal system of ethics that you live by, keep in mind that yours might be similar to another person's in some ways, but you will always find areas of differences. Your system of ethics is unique, regardless of the shared ethics that your profession mandates. What is it about your system of ethics you like, and what would you like to change?

2. How do you go about confronting ethical decisions? What words best describe your style: Thoughtful? Impulsive? Expedient? Other ways of responding?

3. What values define your personal set of ethics? Your professional set of ethics? Are there areas of potential conflict between the two?

4. Are there legal aspects of your work in the field that are particularly challenging for you? Why do you think that is so?

5. Given a situation you've experienced, in what ways did your decision affect others? Would you make the same decision if you were on the other side of the issue? How proud are you of your decision? How proud would your family be?

6. What similarities and what differences are you noticing about the professional, ethical, or legal issues you are seeing in your internship work, and what you have seen in other internships or employment you've had?

SEMINAR SPRINGBOARDS

1. Think about an ethical issue. It does not have to be a dilemma or situation that you would not want to face in your field experience but one that you could face if the circumstances presented themselves. Share the issues with peers in class, and select one of particular interest to role-play.

2. Returning now to the issues you and your peers shared in the previous Springboard, what similarities and what differences became apparent as the issues were being discussed?

3. Think about how you might react to an allegation, complaint, or legal charge brought against you. Would it matter if you were not responsible for the resulting damages? Would it matter if you were responsible and denied being so?

For Further Exploration

Brill, N., & Levine J. (2004). *Working with people: The helping process* (8th ed.). New York: Longman.

> Thoughtful consideration of a variety of professional, moral, and ethical issues in the helping professions.

Corey, G., Corey, M., & Callanan, P. (2006). *Issues and ethics in the helping professions* (7th ed.). Belmont, CA: Brooks/Cole.

> Seminal text with a focus on the legal, ethical, or professional dimensions of the helping professions.

Cottone, R. R., & Tarvydas, V. M. (2003). *Ethical and professional issues in counseling* (2nd ed.). Upper Saddle River, NJ.

> Comprehensive and informative text focusing on professional, ethical, and legal issues in counseling.

Goldstein, M. B. (1990). Legal issues in combining service and learning. In J. C. Kendall & Associates (Eds.), *Combining service and learning* (Vol. 2, pp. 39–60). Raleigh, NC: National Society for Experiential Education.

> Offers an excellent guide to the legal issues relevant to service-learning.

Goldstein, M. B. (Undated). *Legal issues in experiential education.* Panel Resource Paper #3. National Society for Experiential Education.

Identifies key legal issues in experiential education for academic administrators

Kenyon, P. (1999). *What would you do? An ethical case workbook for human service professionals.* Belmont, CA: Brooks/Cole.

Comprehensive and useful workbook that uses actual field situations to inform student awareness of ethical issues and develop decision-making skills to respond in an ethical manner.

Loewenberg, F., Dolgoff, R., & Harrington, D. (2000). *Ethical decisions for social work practice.* Itasca, IL: F. E. Peacock.

Comprehensive treatment of the topic and resources for social work students, useful glossary of terms, and listing of Internet sources of information about ethics and values.

Rothman, J. C. (2000). *Stepping out into the field: A field work manual for social work students.* Boston: Allyn & Bacon.

A useful and comprehensive guide to varied professional issues for social work students.

Steinman, S. O., Richardson, N. F., & McEnroe, T. (1998). *The ethical decision-making manual for helping professionals.* Belmont, CA: Brooks/Cole.

Offers students a pragmatic and focused approach to awareness of ethical issues and responses and developing decision-making skills.

Wadsworth Group. (2004). *Codes of ethics for the helping professions.* Belmont, CA: Wadsworth.

A compilation of the full text of ethical documents for fifteen helping professions.

References

Alverno College Productions. (1985). *Critical thinking: The Alverno model.* Milwaukee, WI: Author.

American Counseling Association. (2005). *ACA Code of Ethics.* Alexandria, VA: Author.

American Psychological Association. (2003). *Ethical principles of psychologists and code of conduct.* Washington, DC: Author.

Baird, B. N. (2007). *The internship, practicum, and field placement handbook: A guide for the helping professions* (5th ed.). Upper Saddle River, NJ: Prentice Hall.

Birkenmaier & Berg-Weger. (2007). *The practicum companion for social work: Integrating class and field work* (2nd ed.). Boston: Allyn & Bacon.

Boylan, J. C., Malley, P. B., & Reilly, E. P. (2001). *Practicum & internship: Textbook and resource guide for counseling and psychotherapy* (3rd ed.). Philadelphia: Bruner-Routledge.

Bradley, L. J., Kottler, J. A., & Lehrman-Waterman, D. (2001). Ethical issues in supervision. In L. J. Bradley & N. Ladany (Eds.), *Counselor supervision: Principles, process, and practice* (3rd ed.). Philadelphia: Brunner-Routledge, pp. 342–360.

Champion, D. J. (1997). *The Roxbury dictionary of criminal justice.* Los Angeles: Roxbury Publishing Company.

Chiaferi, R., & Griffin, M. (1997). *Developing fieldwork skills.* Belmont, CA: Brooks/Cole.

Close, D., & Meier, N. (1995). *Morality in criminal justice.* Belmont, CA: Wadsworth.

Collins, D., Thomlison, B., & Grinnell, R. M. (1992). *The social work practicum: A student guide.* Itasca, IL: F. E. Peacock.

Corey, G., Corey, M., & Callanan, P. (2006). *Issues and ethics in the helping professions* (7th ed.). Belmont, CA: Brooks/Cole.

Corey, M. S., & Corey, G. (2006). *Becoming a helper* (5th ed.). Belmont, CA: Wadsworth.

Council for the Advancement of Standards in Higher Education (2006). *CAS Standards and Guidelines for Internship Programs.* Washington, DC: Author.

CSHSE. (2005). Council for Standards in Human Service Education Handbook/Guidelines. Unpublished manuscript.

Dougherty, A. M. (1995). *Consultation: Practice and perspective in school and community settings* (2nd ed.). Belmont, CA: Brooks/Cole.

Faiver, C., Eisengard, S., & Colonna, R. (2000). *The counselor intern's handbook* (2nd ed.). Belmont, CA: Wadsworth.

Falvey, J. with Bray, T. (2002). *Managing clinical supervision: Ethical practice and legal risk management.* Belmont, CA: Brooks/Cole.

Garthwait, C. (2005). *The social work practicum: A guide and workbook for students* (3rd ed.). Needham Heights, MA: Allyn & Bacon.

Gibbs, L., & Gambrill, E. (1996). *Critical thinking for social workers: A workbook.* Thousand Oaks, CA: Pine Forge Press.

Goldstein, M. B. (Ed.) (1990). Legal issues in combining service and learning. In J. C. Kendall & Associates (Eds.), *Combining service and learning* (Vol. 2, pp. 39–60). www.nsee.org.

Gordon, G., McBride, R. B., & Hage, H. (2005). *Criminal justice internships: Theory into practice* (5th ed.). Cincinnati, OH: Anderson Publishing.

HIPAA/Standards for Privacy of Individually Identifiable Health Information (August 2002). Washington, DC: U.S. Department of Health and Human Services.

Jennings, L., Sovereign, A., Bottorff, N., Mussell, M. P., & Vye, C. (2005). Nine Ethical Values of Master Therapists. *Journal of Mental Health Counseling, 27*(1), 32–47.

Kanter, R. M. (1977). *Men and women of the corporation.* New York: Basic Books.

Kenyon, P. (1999). *What would you do? An ethical case workbook for human service professionals.* Belmont, CA: Brooks/Cole.

Kiser, P.M. (2000). *Getting the most from your human service internship.* Belmont, CA: Brooks/Cole.

Kiser, P. M. (2008). *Getting the most from your human service internship* (2nd ed.). Belmont, CA: Brooks/Cole.

Kitchener, K. S. (1984). Intuition, critical evaluation and ethical principles: The foundation for ethical decisions in counseling psychology. *The Counseling Psychologist, 12*(3), 43–45.

Lord, V. B., & Bjerregaard, B. E. (2003). Ethics courses: Their impact on the values and ethical decisions of criminal justice students. *Journal of Criminal Justice Education*, 14(2), 191–211.

Malley, P., & Reilly, E. (2001). Ethical issues. In Boylan, J., Malley, P., & Reilly, E. (Eds.), *Practicum & internship: Textbook and resource guide for counseling and psychotherapy* (3rd ed.). Philadelphia: Brunner-Routledge, pp. 93–128.

Martin, M. L. (Ed.). (1991). *Employment setting as practicum site: A field instruction dilemma.* Dubuque, IA: Kendall/Hunt.

Meara, N. M., Schmidt, L. D., & Day, J. D. (1996). Principles and virtues: A foundation for ethical decisions, policies and character. *Counseling Psychologist*, 2(1), 4–77.

National Organization for Human Service Education (2000). Ethical standards of human service professionals. *Human Service Education*, 20(1), 61–68.

National Society for Experiential Education. (1998). Standards of practice: Eight principles of good practice for all experiential learning activities. Presented at the Annual Meeting, Norfolk, VA. Washington, DC: Author.

Neukrug, E. (2000). *Theory, practice and trends in human services: An overview of an emerging profession* (2nd ed.). Belmont, CA: Brooks/Cole.

Oran, D. (1985). *Law dictionary for nonlawyers* (2nd ed.). St. Paul, MN: West.

Pollock, J. M. (1998). *Ethics in crime and justice: Dilemmas and decisions* (3rd ed.). Belmont, CA: Wadsworth.

Rothman, J.C. (2002). *Stepping out into the field: A field work manual for social work students.* Boston: Allyn & Bacon.

Royse, D., Dhooper, S. S., & Rompf, E. L. (2007). *Field instruction: A guide for social work students* (5th ed.). New York: Longman.

Schram, B., & Mandell, B. R. (2005). *An introduction to human services: Policy and practice* (6th ed.). New York: Macmillan.

Schultz, M. (1992). Internships in sociology: Liability issues and risk management measures. *Teaching Sociology*, 20, 183–191.

Steinman, S. O., Richardson, N. F., & McEnroe, T. (1998). *The ethical decision-making manual for helping professionals.* Belmont, CA: Brooks/Cole.

Sullivan, W.M. (2005). *Work and integrity: The crisis and promise of professionalism in America* (2nd ed.). The Carnegie Foundation for the Advancement of Teaching. San Francisco: Jossey-Bass.

Tarvydas, V., Cottone,R., & Claus, R. (2003). Ethical decision-making processes. In Cottone, R., & Tarvydas,V. (Eds.). *Ethical and professional issues in counseling* (2nd ed.). Upper Saddle River, NJ: Prentice Hall.

Tarvydas, V., O'Rourke, B., & Malaski, C. (2003). Ethical climate. In Cottone, R., & Tarvydas, V. (Eds.), *Ethical and professional issues in counseling* (2nd. ed.). Upper Saddle River, NJ: Prentice Hall.

Taylor, D. (1999). *Jumpstarting your career: An internship guide for criminal justice.* Upper Saddle River, NJ: Prentice Hall.

Tentoni, S.C. (1995). The mentoring of counseling students: A concept in search of a paradigm. *Counselor Education and Supervision*, 35(1), 32–41.

Travers, P. (2002). *The counselor's helpdesk.* Belmont, CA: Brooks/Cole.

Tryon, G. S. (1996). Supervisee development during the practicum year. *Counselor Education and Supervision, 35*(4), 287–294.

Wadsworth Group. (2004). *Codes of ethics for the helping professions* (2nd ed.) Belmont, CA: Wadsworth.

Wilson, S. J. (1981). *Field instruction: Techniques for supervisors.* New York: Free Press.

Woodside, M., & McClam, T. (2005). *An introduction to human services* (5th ed.). Belmont, CA: Brooks/Cole.

CHAPTER 14

Traveling the Last Mile: The Culmination Stage

I've been taking in as much information as I can before "it's all over" and have been concerned with career goals. The past couple of weeks have just been filled with an overwhelming amount of anxieties and mixed emotions.

STUDENT REFLECTION

As incredible as it must seem to you at times, the end is in sight. For most of you, it is the end of something special, although for some of you, it may not have been the experience you hoped for. In either case, the beginning of your internship may seem like yesterday; it may seem like years ago; and you may go back and forth between those two extremes. Regardless, this is a time to look forward to the future and to be proud of what you have accomplished; it is also a time to reflect on the experience that you have had in the field. So, why do you not always feel clear about the experience and about ending it?

MAKING SENSE OF ENDINGS

We refer to this last stage as the Culmination stage because everything is reaching a crescendo. Another stage, another crisis, and another set of risks and opportunities awaits you. Like other critical junctures discussed in this book, this one is normal, as are the concerns and issues associated with it. Endings are a necessary part of anyone's development (Kegan, 1982). Normal does not mean easy though. Remember the distinction between difficulties and problems discussed in Chapter 11? You could think of endings as difficulties that can be solved, resolved, or aggravated into problems (Watzlawick, Weaklund, & Fisch, 1974).

A word to the wise . . . As your internship ends, you may feel more than ever the pull to do more and more. There may be a lot that can or needs to be finished, but becoming too involved in a frenzy of activity can also be a way to avoid facing your feelings about the ending of your field experience. Once again, we remind you—and it is especially important now—that you need time and energy for reflection. You will have to protect that time more jealously than ever, as the external pressures mount and the temptation to avoid your feelings grows stronger.

A Myriad of Feelings

In our experience, interns approaching the end of their experience report many different feelings. There is often pride and sometimes feelings of mastery. Others report relief and anticipation of freedom. A different kind of relief is expressed by interns who have struggled or had a disappointing experience. However, we also hear about sadness, anger, loss, and confusion, regardless of how satisfying the internship has been. In a survey of psychotherapy interns, Robert Gould (1978) found that many of them reported increased anxiety and depression as the end approached as well as decreased effectiveness. The interns in his study also reported feeling moodier. That has been our experience as well; you would not be unusual if you found yourself experiencing several—or even all—of these feelings, simultaneously or sequentially, with your emotional landscape shifting by the hour. What is going on here?

Changes and More Changes

For one thing, you have a lot on your plate right now. Your internship is ending, and you may or may not be ready. The calendar says the internship is almost over, but you may feel as though you have just gotten started, especially if it took some time to hit your stride (Suelzle & Borzak, 1981). In addition, you may or may not have completed the work you set out for yourself. Projects may not be finished; clients may not be quite where you would like them to be; customers may not be quite satisfied. On the other hand, some of you may feel as though you have been finished for a while.

The web of relationships that has been the social context of your internship is changing yet again. Many relationships are ending; others will be redefined as you leave the role of intern. Endings are part of most relationships and a necessary part of helping relationships (Brill & Levine, 2004). The more the relationships mean to you, though, the harder it will be to end them: "*Now I am really sad. I loved my internship and don't want to lose the relationships I've formed here. I'm trying to make plans with everyone so we can keep in touch.*"

The external context of your life is shifting as well. There may be new demands on your time and energy as you feel the pressure of endings and beginnings (Suelzle & Borzak, 1981). Your attention may be pulled toward papers and final projects for your internship seminar as well as to papers and exams in other classes. Your summer or holiday plans may need to be finalized. If you are graduating and have not yet found a job, your job search is occupying more and more time and attention. Of course, your internship is not the only thing that is ending. Your seminar class, the semester, the school year, even your college career may all be drawing to a close. That's a lot of endings, a lot of beginnings, and a lot of good-byes. Good-byes are never easy, and for some they can be very stressful. The prospect of beginning in a new school, a new town, a new job, or all of the above is daunting as well as exciting.

In Their Own Words   **Voices of Culmination**

The most important part of this internship was how it affected me personally . . . my thoughts, feelings, and beliefs. I went into this internship thinking I knew myself but learned very quickly that was not true. I am going to miss that kind of learning.

I am amazed at just how much my life can change over the course of just a few months. I am no way near the same person I was back at the beginning of my internship. I am proud of who I have become and where I am headed. I just feel sad it's all ending.

Now that my internship has come to an end, I feel excited to be moving on. I feel inspired and more prepared. I feel I have been challenged, and I have proven to myself that I am able to handle those challenges. It has been a beautiful ride and one of unexpected fulfillment.

I went from "that college student" to a professional. It took some time to adjust, and it wasn't glamorous at times, but I made it through. They took me under their wings; they didn't take me by the hand. They let me have responsibility as soon as I walked through the door. They treated me as a professional, and now I'm ready for what's ahead.

I look forward to the future. My internship experience will continue to help even after it's over. I will be able to draw from the confidence and skills I learned. I can honestly say that I am not nervous about the future. Rather, I have a sense of confidence about it and a feeling that whatever I end up doing, I will be able to do it well if I work hard enough at it.

SEIZING THE OPPORTUNITIES

I was given the opportunity to prove to myself that I could do it. This alone has allowed me to feel competent. I tested out my skills and got a professional feel about them. Now I have the key in my hand. I feel ready to move on. I am still not quite sure which doors this key will open, but I am sure that whatever I face, I will deal with as best I know how.

STUDENT REFLECTION

There are opportunities here for you to grow, to develop new insights, and to learn more about separation and moving on, but there are also risks. It is natural to want to avoid the conflicting feelings that can surface at this time, and that avoidance can take many forms, including lateness or absence, devaluing the experience (suddenly it doesn't seem all that great anymore), or putting on rose-colored glasses and forgetting all the struggles you had and may still be having (Gould, 1978). There is even a name for these behaviors: "premature disengagement" (Kiser, 2008). The risk in these behaviors is that you can turn a difficulty into a problem. Lateness and absence will not help anything and will create new problems for you. Becoming out of touch with your own emotions may make you less available to others. You also risk losing the opportunity to learn how to face the issues and feelings associated with endings. Finally, interns who do not face these issues often report a hollow feeling as they leave their placements, even though it may have been a good experience in many ways.

The rest of this chapter is designed to help you minimize the risks, meet the challenges, and maximize your opportunities. Making time to reflect on your experience, taking a new self-inventory, and facing the tasks of ending are all essential components of this effort.

Thinking About Endings

As you head into this phase of your internship, take some time to think about yourself. All these endings and beginnings are likely to tap a variety of emotional issues for you. Some of the issues you thought about in Chapter 4 may resurface or surface for the first time. You may also discover some new issues. Don't let them catch you by surprise.

Although this may be your first internship, don't be so sure the experience of culmination is all new for you. You have had endings, separations, and loss before, and these experiences can leave you with some unfinished emotional business. Your experience of ending the internship will be colored by those experiences as well as by your response patterns, your cultural norms, and even your family patterns.

Think back on the experiences of separation and loss in your life. Going away to camp or to college, having close siblings leave home, moving, divorce, ending an intimate relationship, and being fired from a job are examples of separation experiences you may have had. Those experiences are never easy, and for some people in some circumstances, they can be traumatic. Perhaps some of the hurts from those experiences have not healed. Perhaps you wish you had behaved differently or that others had.

Now think about how you say goodbye and how people have said goodbye to you. Some people just leave and don't say a word; others write long letters or schedule goodbye lunches or dinners. The way you say goodbye probably depends somewhat on the nature of the relationship that is ending, but perhaps you can discern some patterns in how you approach this task. Remember the discussion of dysfunctional patterns in Chapter 4? This is an area in which many people have those patterns, and they can surface in an internship. Here is one we have seen many times:

> *Saying goodbye to someone I don't care that much about is pretty easy. When it's someone I'm really invested in, though, I don't handle it well. Usually, what I do is get really busy. I keep promising myself to go to lunch, or dinner, or something with the person, but I never seem to make the time. Then, all of a sudden, there is no time, and I end up saying a hurried goodbye. I know I have hurt people's feelings that way.*

Scott Haas (1990), in a book about his internship, notices himself falling into a similar pattern as the end of his internship approaches:

> *I create obstacles to avoid thinking about the end. I distract myself: I make lists. I ruminate about minor inconveniences (like wondering for days whether the gas company will correct their bill). I develop new projects and interests. I go shopping, and then in the store can't remember why I ever wanted the thing I'm about to buy. When all else fails, I pretend there is no end. I'm just imagining it: it really isn't happening* (p. 171).

Considering Your Competence

Another area that may be touched by the Culmination stage is your feelings about the issues of competence. You may recall from Chapter 4 that this is one of the components of your psychosocial identity. Some interns report that they feel competent for the very first time during their internship, and that experience is very hard to leave behind. Others report feeling less competent at the end than in the beginning, even though that is almost certainly not true.

Remember the relationship of competence to success? If your sense of competence is tied to success, then you may be trying to get in one more success before you leave, pursuing perfection in a project or one more milestone with a client. Richard Schafer (1973) believes that many helping professionals strive for a "sense of goodness." Although achievement is part of this sense, another component is feeling like a good person, regardless of one's faults. Gould (1978) points out that this particular sense of goodness can be difficult for an intern to achieve. Many people entering the helping professions, he says, want and need to be liked and seek approval from their clients. Their sense of competence, and goodness, is tied to client success or to some other form of success at their site.

On the other hand, all that some of you can think about is how you did not succeed or the ways in which you did not succeed. If you had some substantial struggles or if you found out that the sort of work done at your internship is not for you after all, that is a hard lesson. But they are also good, important lessons and they do not make you incompetent.

The Tasks at Hand

According to Naomi Brill and Joanne Levine (2004), there are three tasks associated with what we are calling the Culmination stage: dealing with unfinished business, expressing feelings, and planning for the future. They are not necessarily to be completed in sequential order; in fact, you will surely see that they are interconnected and are usually dealt with simultaneously.

- **Identifying and Dealing with Unfinished Business**   These are issues with supervisors, with co-workers, with clients if you have them, and with yourself that have been present for some time but that often take on added urgency as the internship draws to a close (Shulman, 1983).

- **Identifying and Expressing Feelings**   You may find strong feelings about supervisors, peers, and others coming to the surface. It may or may not be appropriate to express these feelings directly to the person who has engendered them. At the very least, you will want to find a place where you can express them to someone else and say whatever you want to say before worrying about just what to say to the people involved.

- **Planning for the Future**   Finally, you need to make future plans for the work you have been doing, for the relationships you have developed, and for yourself. Such planning takes time and emotional space, so you will want to set aside that time when you are ready to think through your plans in each of these areas.

These three tasks can be applied to the work you are doing, the people with whom you are interacting, the placement site as a whole, and, of course, to yourself. In some cases, the tasks can be attended to informally, and each intern will do them differently. In other cases, however, we are going to suggest that you be formal and structured about it.

Think About It

The Importance of Rituals

Some of the tasks discussed in this section will be taken care of almost by themselves. They will happen naturally or someone else will initiate them. Others will not, though, and in those cases, you will need to be more proactive. In writing about ending an internship, some authors have discussed the need for rituals (Baird, 2007; Rogers, Collins, Barlow, & Grinnell, 2000). That may sound like a strange term to you, even calling up religious or pagan scenes. However, any formal way of marking an event or passage (such as a going-away luncheon) can be considered a ritual. In this case, rituals can help provide a sense of completion or closure. Rituals can add a sense of "specialness" to the ending; they can help you recall what was significant and important. They can also help ease the transition by connecting the past to the future. Finally, rituals can create a formal, structured opportunity to experience and express emotions that might otherwise be repressed (Baird, 2007).

FINISHING THE WORK

As you enter the last weeks of your internship, you need to think about what tasks remain and what you want to and can do about them. The nature of the tasks makes a difference here. If you have been given a series of small, concrete projects, such as a report to write or an event to plan, and they are complete or near completion, then ending the internship will be somewhat easier (Suelzle & Borzak, 1981). If you are part of a large project that will not end until after you have gone, then feeling some closure about your involvement can be harder. Perhaps there is a component of the project you can complete or a summary of your work that you can write. Finally, you may be involved in a complex project that must be completed before you leave. Perhaps, for example, you have done some research for your supervisor and have agreed to summarize your findings. There may be the temptation to read one more article or interview one more person; in some cases, there is always more research you can do. At a certain point, however, you have to stop generating and start summarizing. The work needs to be completed in such a way that your supervisor has all the necessary information to carry the work forward, whether it is project-based work, work with community groups, client or customer work, or research-based work. Additionally, it is important that you offer recommendations for how the work should be continued and that all "loose ends are tied up" before leaving the work for someone else.

A word to the wise . . . Planning for the future also means considering the nature of your involvement, if any, with the site and its work after the internship is over. Some interns are offered part-time work, relief work, or even full-time jobs. We often tell our students to be flattered, but be wary. Make sure that accepting is really your best option and not just a way to avoid bringing things to an end. In other cases, you may want to come back for a visit or to see and help with an event you have been working on. If you have patients, students, residents, or clients, returning for a visit to the site can be fine, depending on the nature of the work (this issue will be discussed more in a subsequent section). Seeing a project, event, or contract you have worked on come to fruition can be wonderful as well. Just take care not to promise more than you can deliver. If the agency is counting on you and you get caught up in your life and don't follow through, you leave a bad impression about you and possibly about your campus or program.

CLOSURE WITH SUPERVISORS

Many interns have one or more campus supervisors or instructors as well as more than one supervisor at the site. Such an arrangement calls for culminating tasks with each of these supervisors. How you pursue closure with the campus supervisor or instructor will depend on the nature of the relationship. Touching base by e-mail with a campus supervisor once or twice during the internship is a different relationship than meeting regularly with the supervisor and communicating frequently by electronic means. In some instances, you may want to ask the campus supervisor or supervisor to be one of your references as you begin your employment search. How you decide to bring closure on that relationship will certainly affect the supervisor's decision to be a reference for you.

For most interns, the site supervisor is the person who has most powerfully affected the experience. Your site supervisor is in a position of power, at least with regard to you, and is often in a position of leadership in the agency. Usually, the site supervisor is the person with whom you have worked most closely, shared the most with, and learned the most from. Now it is time to bring closure on that relationship or perhaps move it to a new phase of relationship. There are two dimensions to this closure process: the formal and the informal phases (Birkenmaier & Berg-Weger, 2007). In the formal phase of the process, there is a final evaluation with the site supervisor and perhaps an evaluation of that supervisor by the intern as well. In the informal—and equally important—phase, the supervisory relationship is processed and discussed, as are the feelings about the ending of that relationship.

The Final Evaluation

You have probably had evaluations at your internship before this point; at least we hope so. Perhaps you are less nervous about them than you once were; perhaps not. In any case, you have one more to go through, and this can be the most important one of all. Many internship programs and placement sites use a written final evaluation of some sort. Most of them have their own format, so we are not going to provide one here. However, if you would like to see some examples, we refer you to work by Wilson (1981), Stanton & Ali (1994), Baird (2007), Garthwaite (2005), and Birkenmaier & Berg-Weger (2007). As was the case with the initial or midsemester evaluation, we recommend that you become familiar with the form used at your field placement or reach an agreement about what it will be before the evaluation is actually completed. That way you will have a better idea what to expect and there will be far fewer disappointing surprises.

Regardless of what form you use, or whether you use one, it is important that you have a final evaluation conference with your supervisor. This conference can either precede or follow the written evaluation. Some of the time can be used to prepare for that evaluation or to go over it. In any case, be sure to schedule an adequate block of time—at least an hour (Baird, 2007; Faiver et al., 1994; Shulman, 1983). There may be lots of tempting reasons for both you and your supervisor to avoid this session. You may be nervous about the feedback; your supervisor may not be good at endings either; both of you may be pretty busy trying to wrap up projects or cases. This is one of those times when a formal, scheduled time—a ritual—will ensure that the task actually gets done.

Preparing for the Final Conference

This conference will undoubtedly be more productive if you spend some time preparing for it, and we suggest several steps for doing so. First, before you even consider what is on the form, take some time to review the internship experience with all its joys and frustrations. You have been reflecting on a regular basis, but it is easy to miss some important moments or broader themes; so, it is time to think reflectively once again and to think introspectively as well.

FRAMEWORKS FOR REFLECTING

Reread your journal and make some notes for yourself. If you had a midsemester evaluation, look it over as well. Stanton & Ali (1994) have suggested dividing your reflection into two sections: work performance and learning. As you will see, these two areas have some overlap, but they are worth considering separately. In discussing work performance, you and your supervisor should cover the areas where you seem to have

---

Think About It

Growth, Knowledge, and Interests

If you are stuck in reflecting on your work performance and on what you learned, here are some topics from Birkenmaier & Berg-Weger's work (2007) that you may find helpful:

- Areas of personal and professional growth
- Knowledge and skills you gained
- New or confirmed areas of interest
- Goals met, unmet, and partially met
- Your level of confidence
- Your ability to function as part of a team
- Your ability to make good use of supervision

---

been especially effective, the clients with whom you have worked most successfully, and the service areas where you have demonstrated the most skill (Faiver et al., 1994). Both of your perceptions of these issues are important. In discussing what you have learned, use your written goals and objectives as references, and consider how well you met each goal, the reasons why, and how what you have learned will be useful in your professional and personal life.

Another way to reflect on your experience is suggested by Baird (2007), who says that there are both positive and negative lessons to be learned about systems, co-workers, human problems, those to whom you provide services, and yourself. For example, you may have learned that some people are in the health care, education, or helping professions for all the wrong reasons. On the other hand, you may have learned that there are quiet, heroic efforts being made day in and day out by these same people. You may want to try this exercise for yourself.

PAYING ATTENTION TO AFFECT

It is also important to prepare for the affective dimension of this experience (Stanton & Ali, 1994). Being evaluated is always an emotional experience to some degree, and for some interns, the emotional stakes are very high. Baird (2007) has suggested completing a self-evaluation beforehand, perhaps using the agency form, in which you downplay your strengths and call more attention to your areas for improvement. As you do so, monitor how you feel and how satisfied you are with your reaction. We also suggest you spend some time thinking about your emotional reactions to praise and criticism, as you did earlier. Some interns have a difficult time listening to criticism; others chafe at praise. Still others find it hard to tell other people how they really feel about them. Anything you can do to help yourself participate honestly and openly in this process will pay dividends.

The Final Conference

During the evaluation itself, it is important to balance the positive and the critical (Baird, 2007; Shulman, 1983; Wilson, 1981). It is also important that the feedback be productive and constructive. It is normal to want to focus on the positive and bask in the glow of your accomplishments and your supervisor's praise, but constructive feedback or criticism is just as important. There are bound to be things you need to work on and perhaps areas where you didn't perform as well as you or your supervisor would like. Helping professionals are working with clients on just this issue. Just as their clients need to know where they still have work to do, so do they. Ignoring these areas is a little like saying, *"I'm perfect. I learned everything I need to know to be a professional in this field in just a few short weeks."* That sounds more like a quote from a late-night TV advertisement than the stance of a reflective, thoughtful professional. In some cases, you may need to be assertive about asking your supervisor to identify areas for growth.

It is also important, during the evaluation interview, to listen as carefully and critically as you can, clarify things if you need to, and ask questions (Birkenmaier & Berg-Weger, 2007). If you don't understand what your supervisor is saying or on what it is based, it will be hard to learn from the positive or the negative, and the process will not be productive or constructive.

You may not like some of what you hear or you may think it is unfair. It is important, though, to try and keep those two issues separate (Rogers et al., 2000). We suggest you listen quietly and respectfully, ask for clarifications and examples, and then go away and think about it for a while before responding. Take some time to vent your feelings and then consider whether some or all the criticism is valid. Try to reframe the criticisms into opportunities to learn and grow, and remember that this is a professional issue. It is neither a personal attack nor a comment on your overall worth and value (Munson, 1993). If after careful reflection you believe that comments have been made about you that are not true or not grounded in facts, you should discuss this with your campus instructor or supervisor.

Feedback for the Supervisor

The end of an internship may also be a time to offer some feedback to your site supervisors (Baird, 2007; Rogers et al., 2000; Faiver et al., 1994). Some supervisors will request this from you, and they may or may not give you advance notice. There are even forms available to offer written feedback (Rogers et al., 2000). You probably have mixed feelings about this prospect, and we don't blame you. On the one hand, this is an opportunity to tell your supervisor what went well for you in the relationship and where there may be areas for improvement; and it is an opportunity for both of you to learn more about yourselves. You also have an obligation to future interns (Rogers et al., 2000). On the other hand, it is important to think carefully before offering criticism, regardless of how constructively put, unless it is requested (Baird, 2007). After all, your supervisors have and may continue to have power over you in the form of grades, letters of recommendation, and word of mouth in the community. Your campus instructor or supervisor may be able to help you decide what and whether to share with your site supervisor. If you decide to give some feedback, we encourage you to practice or role-play and to remember the principles of effective feedback (see page 36).

Ending the Supervisory Relationship

The informal part of the closure process with your site supervisor is at the same time easier and more difficult. For better or worse, there is no form to fill out, no big meeting, and no high-stakes evaluation; just the world of feelings. For most interns, the relationship with the site supervisor is among the most significant of the internship. For many, the relationship has been close and positive, both intellectually and emotionally. For many others, the relationship has not been close or satisfying, but still one in which a lot was learned. Now that the relationship is ending, the ending can engender a variety of feelings for both of you. If your relationship with your supervisor has been mostly an intellectual, dispassionate one, it may make saying good-bye easier (Haas, 1990). Some supervisors will initiate a conversation with you about these feelings; some will avoid it; and some will engage but need to be prompted by you. If possible, it is important to try to give voice to your feelings.

Think About It

Getting Letters of Recommendation

If you have had a good experience in the field, you will probably want a letter from your site supervisor and maybe from a co-worker, either for your next internship or for future employment. It's a good idea to ask for it now, while you and the internship are still fresh in the supervisor's and co-worker's minds, rather than call or write months later. Baird (2007) offers some important guidelines to follow when asking for a recommendation:

- **Don't Assume.** Before requesting the letter, ask your supervisor if he or she is comfortable writing a supportive letter for you. This may seem silly, especially if you have a good relationship. Many supervisors, if they do not feel they can write a positive letter, will simply suggest that you ask someone else. However, you don't want to take the chance that the letter will be less positive than you expected, especially if it is sent directly to a school or employer.

- **Know What You Need.** Make clear what your future goals are. A letter for graduate school may look somewhat different than a letter for employment. There may also be some jobs for which your supervisor feels you are better qualified than others.

- **Plan Ahead.** Give plenty of notice and let the reference know of the time frame; provide whatever forms and envelopes are needed.

Quick Tip: Preaddressed and stamped envelopes show courtesy and consideration.

Some interns find these conversations disappointing (which does not mean they are not important). Many interns develop powerful feelings of closeness, respect, and affection for their supervisors. In some cases, the feelings are mutual. But remember that while this is a new and unique experience for you, it may not be for your supervisor. It is more typical that the supervisor has enjoyed working with you and watching you grow and will be sorry to see you leave but does not feel personally close to you and has no desire to develop a friendship after the internship is over. If this is the case, try not to let it change the value of your experience or even your feelings about your supervisor.

FOR HELPING AND SERVICE PROFESSIONALS: SAYING GOODBYE TO CLIENTS

Just when I'm really starting to feel comfortable with the kids, it's almost time to end.

STUDENT REFLECTION

Termination is the word used to describe the ending of a therapeutic relationship. To some of you, this is a familiar term. You may have studied it in class, and if your agency does one-to-one counseling, you may have heard the term there as well. For others, the term may be new, somewhat strange, and harsh sounding, especially given its recent use in the movies and on television. Not all of you are doing counseling or psychotherapy in your internship, but if you are doing any sort of direct work with clients, you need to think carefully about ending your work with them and saying goodbye, regardless of the label you apply to the process. If you are not, then this section may not have relevance to your internship, and you can skip ahead to the last section in this chapter, "Bidding Farewell to the Placement Site," on page 323.

As an intern, you are saying goodbye under unusual circumstances. Usually, when a relationship between a patient or client and a helping professional ends, it is because the work is done or the client decides to end it. However, you are dealing with what is called *forced terminations* (Baird, 2007; Gould, 1978). It is the calendar that dictates the ending. The therapeutic work is not necessarily over, and the clients are not necessarily ready to stop or start over with someone else. Ending relationships with clients is bound to evoke emotional reactions for them and for you. Dealing with those feelings and ending well is an important part of the internship. For some of you, it will be relatively easy. For others, it may be one of the biggest challenges you face.

Part of the reason interns in the helping professions have such varied experiences in coming to closure with clients is that there are such great variations in the internship sites. Some sites do a lot of one-to-one counseling. Others, such as adolescent shelters, do some but in the context of daily living and recreation in the *milieu*. Other sites do none at all. There is also great variation in the characteristics of the clients. Some client populations are much more emotionally vulnerable and may take your departure that much harder. Some clients are more autonomous than others. Clients who drop into a senior services center, for example, or a town recreation department are probably functioning pretty well on their own. They appreciate you, but they really don't *need* you. Finally, some clients are simply more aware of your leaving than others. Interns working with clients who have Alzheimer's disease often report having to tell each client over and over that they are leaving. Many of the clients did not remember the interns from day to day, although they seemed glad to see them each day. Leaving was difficult for the interns but probably not for the clients. In any case, there are four important issues for you to think about when bringing closure with clients:

- Deciding when and how to tell them
- Addressing the unfinished business
- Dealing with feelings, both yours and theirs
- Planning for future needs

Timing and Style

There are many different theories and opinions about when and how to tell clients you are leaving. Because internship sites and client populations vary so greatly, you must look to your site, your site supervisor, and your campus supervisor or instructor to guide you in this area; no prescriptions we could make would apply to every intern or even to most of them. Some agencies recommend that you tell clients right away that you will be leaving at the end of the term or the year; in fact, some agencies tell the clients for you. Some clients are used to seeing interns come and go; as soon as you say you are an intern, they know you will not be there for very long. Other agencies recommend that you begin discussion of termination one week, two weeks, or more in advance. Some recommend that you not mention it until right before you go.

There are also a variety of methods of discussing termination, from individual conferences, to group meetings, to letters, or any combination of these approaches. If you do not know how your agency handles termination, you need to take the initiative to ask. Depending on your needs and attitudes about ending, you may feel that the agency's policy is too casual or forces you to focus on something that does not seem like a big deal. Remember that the clients are the primary focus; the decision needs to be made based on what will work best for them, not for you. Still, this is a good opportunity for discussion with your supervisor if you disagree with the agency's policy or if it does not match a theory you have studied. Both of you might learn something, and at the very least, you will be more likely to accept and follow through with the agency's procedures.

Unfinished Business

As you approach the end of your relationship with clients, it is a good time to reflect, certainly in private and perhaps with clients, on their progress. If they had specific goals, this is the time to review them. Even if they did not, your memory and your journal are good sources for remembering what your clients were like when you first arrived as well as what your relationship with them was like. Both clients and workers can become so caught up in the problems and challenges of the present that they don't think much about the change that has already occurred. Add to that the difficult feelings that termination can bring, and you can see why it is important to take time to reflect on the positive.

It is equally important, though, to be clear with yourself and with clients about the work that remains. All of this is especially important if the client is going to continue at the agency, which is usually the case. You may want to use some of your supervision time to review each of your clients with your supervisor or you may want to discuss them as a group. You will want to talk about how you feel about each of these terminations as well as how you think each client is going to react. Baird (2007) discusses a Termination Scale developed by Fair and Bressler that has fifty-five items covering emotional responses and planning. You may want to make use of this resource (Fair and Bressler, 1992, cited in Baird, 2007).

Overarching Feelings

In many cases, the termination process can be an emotional one for everyone involved. For both you and the clients, your leaving may recall echoes of past experiences of separation and loss. In addition, many interns report feeling nervous and apprehensive about discussing termination with clients because they are afraid of what the clients' reactions might be. Given the variety of client personalities and the separation experiences they have had, you can expect an equal variety of reactions to the news that you are leaving. Of course, termination can be a learning opportunity as well. Clients can learn that goodbyes, while often painful, do not have to be traumatic, angry, or hysterical. And the responsibility is yours to model for them positive and empowering ways of dealing with their feelings.

The Client's Experience

Clients may be especially vulnerable to feelings of separation and loss. For many clients—especially those who have been subjected to abuse, neglect, or parents with substance abuse problems—separations have been arbitrary and unpredictable, such as when a parent abandons a child or an alcoholic parent goes on a drinking binge and returns days later. For some people, separation is associated with anger and hysteria, as in the loud and even violent arguments that can lead to the disruption or ending of a marriage or other intimate relationship. You may also find that clients have had particularly traumatic separations, as in the death of a parent, sibling, or close friend. Finally, many of your clients will have worked with several agencies, workers, or several residential settings and have had many termination experiences. If some or all of these experiences have been painful, this termination may bring up those feelings again (Baird, 2007), and you are right to be concerned.

> *Why do you have to go? is probably the most difficult [question]. Many of these children have had people who come and go in their lives, and I don't want them to look at me as one of those people.*

STUDENT REFLECTION

Clients also may feel especially vulnerable to feelings of abandonment. After all, you are leaving because of your school calendar, not because they no longer need to work with you; in fact, they may wonder how much you really cared about them in the first place (Baird, 2007; Stanziani, 1993). Some clients may feel angry and believe their trust has been betrayed, even if you have been clear from the beginning that you will be leaving at the end of the semester. It is important to remember that whatever reaction you receive may be only part of the picture (Penn, 1990). Termination, like the end of any significant relationship, usually creates feelings of anger, sadness, and appreciation. Your client may be expressing only one of those feelings, and indeed that may be the only feeling he or she can access, but it is important for you to remain, open to the other feelings as well and perhaps to help your client be open to them too. It is also important to separate what clients choose to show you from what their deeper feelings may be.

Sometimes these feelings are expressed indirectly, not unlike your own feelings about ending the internship (Penn, 1990). Clinically, this is referred to as *resistance*. Some clients may be genuinely indifferent to your leaving. But others will feign indifference or be indifferent because their real feelings are too hard to accept. Another indirect way of dealing with termination issues is to demand that termination come immediately. In these cases, once you tell clients you will be leaving, they may stop coming to the agency or, in residential settings, ask for another worker to be assigned right away. Other clients may begin to exhibit older, more problematic behaviors, as if to say, "You can't leave me; I still need you." Finally, clients may begin to devalue the work you have done together, since it is easier to say goodbye to something or someone that, after all, wasn't especially important anyway.

The Intern's Experience

For most interns, saying goodbye to clients is an emotional experience as well—sometimes in unexpected ways. You may find yourself feeling guilty that you haven't been able to do more for some or all of your clients, and again, this feeling is often unrelated to how much you have actually accomplished (Baird, 2007). In other cases, you may worry that you are the only one who can work successfully with a particular client or group of clients. It may be that you are the first person to really "get through" to a client. That is a wonderful feeling, but it places extra weight on the termination process. As one intern said, "*They do not receive the same help from others in certain areas. . . . It just makes me sick inside.*" Remember that although the client may regress some when you leave, chances are that at least some of the progress you have made will remain. It can, however, be hard to remember or believe that in the moment. If a client starts to regress, you may begin to question the value of your work or your effectiveness in general (Gould, 1978).

The Future

The last issue to think about in terminating with clients is the future, both for them and for your relationship with them. Just as this is a time to review goals and accomplishments, it

Focus on PRACTICE

Knowing the Pitfalls

All helping professionals are vulnerable to client reactions, but as an intern, especially if this is your first intensive experience in the helping field, you may be particularly vulnerable. The closeness that can happen between worker and client is a heady, heart-warming experience, and it is especially affecting the first time (Baird, 2007). Both of us recall with special vividness our first few clients and the special difficulties that we faced in saying goodbye. There are pitfalls to watch out for as you take in and process clients' reactions to your leaving, and we've identified several common ones for you.

- **Personalizing Reactions**   It may be easy for some of you to take quite personally the client's reactions, as if they were statements about *you*. Of course, they have something to do with you, but remember that you may also be hearing expressions of their current struggles with themselves and echoes of old wounds. Work with your site supervisor to prevent such personalizing.

- **Confusing Needs and Issues**   It is important to separate the clients' needs and issues from those of your own. If, for example, you struggle with a need to be liked, then you may be especially vulnerable to a client who turns on you in anger or devalues the experience. Work with your site supervisor to prevent such confusion.

- **Mixing Feelings and Decisions**   It is important to deal with the feelings and issues separately from the decisions you make about how to respond to your clients. You may need to be liked, but liking you in that moment may not be the best thing for your client. If a client has never been able to express disappointment with anyone, then the ability to do that with you is far more important than your need to be reassured. Work with your site supervisor to ensure that you keep both separate.

- **Seeking Affirmation from Clients**   Some interns unconsciously try to get clients to make the termination easier for them. They seek reassurance that they have done a good job or that whatever problems remain are not their fault (Baird, 2007). In these times, it is especially challenging to keep the client's needs in the foreground. You may very well need some support and some focused attention to help you process your feelings and needs. That is what your campus instructor, site supervisor, and peers are there for; don't turn to your clients for that support. Instead, discuss your need to do so in supervision.

- **Promising to Return**   Another pitfall is promising to come back and visit. Clients may ask you to come back or to maintain contact in some other way. They may be avoiding the difficulties of separation, they may not understand the boundaries of a professional relationship, or they may just be doing what they would do with anyone they have come to care about. Even if they don't ask, you may find yourself wanting to reassure them that you will visit or keep in touch. Use extreme caution before making any such promises. Consult with your supervisor and your faculty instructor. Above all, if you do make commitments, make them with care; the worst thing you can do to some clients is promise to keep in touch and then fail to do so.

is also a time to set or reestablish goals for the future. You also need to think about who will be dealing with your clients after you leave.

TRANSITIONING CLIENTS

In many agencies, intern supervisors plan the transition for clients that will occur when you leave, and you should be part of that planning. If your supervisor is not raising this issue, we suggest that you do. This transition, often referred to as *transferring clients*, requires time and attention if it is to be done well (Baird, 2007; Rogers et al., 2000; Faiver, Eisengart, & Colonna, 1994). The nature of the transfer will depend a great deal on your specific setting. You may actually have some clients that have been assigned to you. For example, there may be individuals who come to a clinic for counseling, a group you have been assigned to lead or co-lead, children and families whom you visit at their homes, or community groups that you meet with on a regular basis. In residential settings, it is not unusual for each resident to have a primary staff member, and sometimes, interns move into this capacity after they have been there for a few weeks. In other settings, there will not be clients who have been officially assigned to you, but you may have developed especially close relationships with certain individuals, and it may be clear to you that you are the one to whom they turn most often. In any case, the issue of who is going to take over is one that should be discussed with your supervisor, with your co-workers, and, of course, with your clients. You should also be sure to reserve some time to work with the person who will be taking over, and you should begin this process well before your last day.

POST-TERMINATION RELATIONSHIPS

What sort of relationship will you have with your students, patients, residents, or clients after you leave? What sort should you have? In many cases, the answer to both questions is "none." At most, you might keep in touch by mail, which is not initiated by you. For others, either because the client asks for more or because you want more, the issue becomes less clear. As we discussed in an earlier chapter, sexual relationships with clients while you are an intern are unacceptable and are a clear violation of the ethical standards of virtually every professional organization in the helping professions. Social relationships with clients are also discouraged, although in some social service settings, they are unavoidable and even beneficial. Now, however, you are leaving, and there may be clients with whom you want to pursue a friendship or a romance. Some of you, particularly those working with adolescents, may find this absurd and out of the question. But others of you are working with clients your own age and, trust us, the issue does come up.

This issue is complicated once again by the wide variety of placements and clients that interns deal with. Post-termination relationships have been the subject of a good deal of discussion in the counseling field. For example, in a survey of attitudes about friendship and sexual relationships with former clients, of the counselors surveyed, 70% said they thought a friendship was acceptable, while 30% thought a sexual relationship could be acceptable. However, they also advised a lengthy waiting period, averaging 25 months for friendship and 62 months for romance (Salisbury & Kiner, 1996). So, if you were thinking that maybe you could just pick right up with a different kind of relationship with one of your clients, you may want to think again. Herlihey and Corey (1992) have provided a thoughtful review of these and many other issues about relationships with clients.

The major concern about post-termination relationships is the unfair power differential between you and your former patient, student, resident, or client (Salisbury & Kiner, 1996). Through working with clients, helping professionals often have knowledge of their most sensitive issues and areas of vulnerability. Those being helped, on the other hand, usually have no such knowledge of the professionals' lives, and this puts them in a vulnerable position in a friendship or romance.

A word to the wise . . .   Not all of you are counselors, nor are you necessarily working with clients who have come to you or the agency because of psychological or emotional issues. If you are working in a recreation center, doing community organizing, or providing health education for college students, the power differential referred to may be quite different. So, unfortunately, there are no hard-and-fast rules to follow—just some important issues to think about. If you are considering pursuing any sort of relationship with someone you've worked with in a helping capacity, you should talk that over very carefully with your campus and site supervisors. Whatever you decide, remember that even though you are no longer working with them, the primary concern in your decision making still must be their welfare, not yours.

BIDDING FAREWELL TO THE PLACEMENT SITE

> *I think to myself this is the last time I am going to be doing this and seeing these people. It's been a busy and scary week.*
>
>   STUDENT REFLECTION

The last week of an internship is hard in many ways. You have had your final evaluation meeting with your supervisor; you have had meetings about the status of the projects you have been working on; if you have clients, you have prepared them for your departure; and, when applicable, you have notified human resources accordingly. It becomes more difficult as you listen to the future of the work you have been doing—work that you will not necessarily see materialize. All this is leading up to the final day—the last day of your field experience.

Rituals and Remembrances

Think about the pace of life at your site. Most agencies in the helping professions are understaffed and extremely busy, just as some areas of the private sector are. In residential settings, large and small, crises erupt all the time. And it's not the last day for anyone else; it is *your* last day! For that matter, it may not even be the end of their day or week. So, it is understandable that they might forget about your departure.

How You Don't Want Your Last Day to Go

Your last day, a day you have been looking forward to with an incredible mixture of feelings for weeks, is approaching fast. You begin to imagine what it will be like when everyone says goodbye to you and what you will say as you leave. As you move through the last day, though, no one says a thing to you about leaving. Your supervisor doesn't seem to be around, and for your co-workers, it seems to be business as usual. Conversations focus on the work, and during quiet times, it's the usual office small talk.

You go about your business, feeling a little confused but certain that someone will say something soon. Maybe they're being sneaky, and there is a little party planned for you later. No such luck. As you are leaving for the day, you say goodbye to people, pretty much as you always do. One of your co-workers looks up from what she's doing and says, *"See you next week."* *"Well, no,"* you respond. *"Today is my last day."* *"Oh my gosh,"* says your colleague. *"I totally forgot! Well, hey, it's been great. Good luck to you."* You start to reply, but the phone rings, a client calls, or a crisis erupts, and your colleague swirls away, back into the normal pace of life at the site.

When driving home, you feel differently than you thought you would. Yes, there is joy and some relief but also some hurt and a vague sense of emptiness. You try not to focus on that feeling, but it gnaws at you. What's wrong with those people anyway? Maybe you didn't mean as much to them as you thought. Maybe they're just rude and not quite the people or professionals you thought they were. You feel as if you deserve better from them after all you have done for the organization.

The point here is that you may need to be proactive in assuring you get that sense of closure you want. This is another time when a ritual may be in order. Some agencies have fairly elaborate rituals when someone leaves. There might be time set aside at a staff meeting or there might be a party or a farewell lunch. Other sites don't have any of these rituals in place. You won't know what your site typically does unless you raise the issue. For example, talk with your supervisor about the best way to mark the end of your internship. Ask what the norms are and be clear about what you need. It may be that there is just no time for a group activity, but at least you will know that, and you can schedule individual fifteen-minute appointments with some of your co-workers. In our experience, most supervisors are glad to help you make something happen, but they might not think of it on their own or may want to know what you prefer.

Even if there is going to be a formal goodbye celebration, there may be individual co-workers to whom you have grown especially close. They may take the initiative to have a final conversation with you or they may not for any number of reasons. Here again, you need to take the initiative and schedule some time with them. This is yet another opportunity to practice your feedback skills; it is a time to let them know specifically what you have learned from them, what you appreciate about working with them, what memories you are left with from the experience, and how you feel about leaving or about them as supervisors and mentors. They may, in turn, do the same for you. Remember, though, the main point is for you to say what you want and need to say, and if you do, the conversation is a success. You don't want to go away with that nagging feeling that you wish you had said such and such to so-and-so. Whatever you get back from them is an extra bonus.

A SELF-EVALUATION:
THE LEARNING CONTRACT AND BEYOND

It is time for one last look at your learning contract as well as what lies beyond it for you. Take some time to be proud of having met your learning goals and objectives as well as to reflect on growth and accomplishments that may not have been in the contract. You should also, however, take some time to consider the next steps for you in your personal, professional, and civic development.

Your formation as a person and a professional is never complete. Now, however, you will most likely be more in charge of the process. Some skills and knowledge, and even values, may be required or encouraged in your next setting, whether that is another internship, a job, or a more advanced degree program. We hope, though, that you have *internalized* a commitment to your own growth and that you will identify and pursue new goals regardless of the extent of the external demands that you do so. Remember to include a commitment to attitudes, skills, and knowledge as well.

Remember, too, that the professional arena is only part of the picture. We have tried to help you look at yourself not just as a professional but as a civic professional with a clear sense of the human and community context of the work you do and the public relevance of your profession. We have also encouraged you to use the internship as a vehicle for your civic development, and we hope that you continue to identify and pursue the values, skills, and knowledge that make you an engaged and effective citizen.

You have also learned a great deal about yourself. We have tried to emphasize that sort of learning in this book, and interns often tell us that this is the most powerful learning of all. You know more about how you respond to challenges and why. And you know that self-understanding is a process, not an accomplishment. You have more tools to pursue your self-understanding, and the practice you have had, if you continue it, will make those tools more and more second nature to you.

Preparing a Portfolio

Many interns we have worked with have found value in creating a portfolio of their internship experience. It is a reflective tool but also a powerful asset as you seek employment or further education. There are a number of pieces you can consider including: reflections on your work, sample projects, case studies (being mindful of confidentiality), performance reviews, summaries of projects, and so on. Whatever time you spend on this endeavor will be time well spent, but if you are going to invest serious energy in this work, we suggest you follow a process for portfolio development (King & Sweitzer, 2002):

- **Consulting with Peers**   Work in small groups to exchange ideas about your portfolio and its contents; begin to sketch it out.
- **Constructing the Vision**   Here is where you begin to think about the contents, although final decisions do not come until later. Write a table of contents and an introductory statement. Make a list of potential contents. Think about how you would like to package the portfolio.
- **Creating the Vision**   Two steps here. In the first, you brainstorm all the possible organizing lenses or metaphors you might use to create and organize your portfolio. Examples include humanitarian student, emerging professional, ethical decision maker, community member, responsible civic professional, or caring advocate. In the second step, you begin collecting the documents you need to substantiate your self-identified strengths.
- **Communicating the Vision**   Here is where you choose the actual format and design of the portfolio.

LOOKING AHEAD AND MOVING ON

Your internship is just about over now, and your future is beginning. Probably, there were times when you thought this moment and this day would never come or times when you were amazed and unnerved at how fast it was approaching. New challenges await you, of course. Some of you may be headed for another internship, some for a job in the field, and still others for the next level of studies and more field-based learning.

Planning for your future includes a reassessment of the role of the internship in your career plans, the role you see yourself taking as a civic professional, and concrete steps to move your life forward, such as preparing your résumé, developing your portfolio, compiling your professional networking list, and adjusting your outgoing message on your voice mail to reflect the professional you have become!

You have learned valuable skills. If you continue to use them, they will grow sharper and more integrated over time. Even if your life and career change directions now, you can still take a lot of what you have learned with you. You have learned what to expect from an internship. The challenges will have a different shape and pace because they will be happening in a different place. You have grown and changed so much that it may feel like these challenges are happening to a different person. And, of course, you will handle the issues in a new way. However, the concerns themselves should seem familiar to you if you can get enough distance and perspective from your everyday activities. Perhaps you will continue to keep a journal and pause for regular reflection on what you have written.

A final word to the wise . . .   Keep in mind that the developmental stages of an internship are not phases you go through just once. In your new position or in another field placement, you are going to go through them again. The concerns about expectations and acceptance and the challenges of keeping yourself moving and growing, of confronting problems, and perhaps of ending well will all visit you again and again. When we tell our students this piece of news, some of them roll their eyes and hold their heads. When they remind each other of how they aptly handled the challenges they faced and how much they learned, they realize that their learning will travel with them into the future.

Think About It

Am I Ready to Move On?
You know well the concrete steps that must be taken to move into the future. If that future is the beginning of your career, then here's a practical to-do list . . . in case you need a reminder!

- Update your résumé to reflect the internship.
- Polish your interviewing skills.
- Fine-tune your networking list.
- Develop your portfolio.
- Contact your references.
- Change your outgoing voice mail messages.
- Visit the career services office on campus.
- Join your professional organization and its Listservs™.
- Start building your library of professional resources.
- Begin the search for your first career position.
- Create your personal web site.
- Think about your next level of learning. . . .

A Fond Farewell . . .

It may seem odd for us to say this, since we have never met you, but we wish you well in your life's work and in your development as a civic professional. We are always glad to hear from former students and welcome hearing from you about how this book may or may not have been useful to your journey in experiential learning. And if there are topics we didn't discuss and you think we should, then let us know and we'll give attention to them. We leave you with these thoughts from one of our students:

> *It's like moving out of your house. You're leaving something that has had such an impact on your life and starting something unknown and exciting. You are leaving a foundation of growth and a place filled with memories. You're going into unknown territory, and that is exciting.*

STUDENT REFLECTION

For Contemplation

PERSONAL REFLECTION:
SELECT THOSE INQUIRIES THAT ARE MOST MEANINGFUL TO YOU.

1. In what ways has your civic development been affected by this internship? How have you become a civic professional as a result of your internship?

2. What is on your agenda right now? What projects need to be finished at your internship? How about other classes? What else do you have to take care of in the next few weeks?

3. Think back to times in your life when you faced endings. What was it like for you then? What feelings did you have? How did you handle the ending?

4. Can you make any general statements about how you tend to handle goodbye? Are you satisfied with your patterns and tendencies in this area?

5. What will be most difficult in ending your internship? What can you do to ensure that you end the way you want to?

6. If you have clients, patients, residents, customers, or students, how will you approach termination with them? When will you tell them you are leaving? How can you come to some closure with them? What future goals do you have for them? How do you feel about saying goodbye?

7. Think for a moment about your supervisors. What are some of the things you have learned about them? What is it that you most appreciate about them? What do you wish had been different in your work with them? Is there anything you want to say to your supervisors before you leave that you know you aren't able to say for whatever reasons?

8. Are there other individuals at your placement that you want to be sure to say good-bye to? How will you do this?

9. What plans do you have to acknowledge or celebrate the end of your internship?

10. What is next for you? Another internship? A job? Another degree? All of the above? As you approach those experiences, take time to review your areas of growth in the personal, professional, and civic arenas and to set some new goals for yourself.

SEMINAR SPRINGBOARDS

1. Those of you who are working with clients may want to take some class time to focus on just how you will let them know you are leaving. This is another opportunity to role-play.

2. Consider Sullivan's concept of the civic professional in class and what that term means to interns in different types of placements and/or different academic majors.

3. Take some time in class to focus on the issue of offering feedback to supervisors. Role-play these situations in triads, with an intern, a "supervisor," and an observer. The observer should use the principles of effective feedback to note strengths and weaknesses.

4. Consider passing along to future interns the lessons you learned from your internship. We ask our students to write their "words of wisdom" on index cards, which we share with interns who follow them. A colleague of ours asks her interns to write Dear Intern letters and then distributes them in booklet form (Kellner, 2007). Regardless of the way you share the lessons you learned, doing so is a gift to future interns.

For Further Exploration

Baird, B. N. (2007). *The internship, practicum, and field placement handbook: A guide for the helping professions* (5th ed.). Upper Saddle River, NJ: Prentice Hall.

This book contains a comprehensive, thoughtful discussion of a variety of issues associated with the ending of an internship.

Brill, N. I., & Levine, J. (2004). *Working with people: The helping process* (8th ed.). Boston: Allyn & Bacon.

Especially helpful in thinking about termination with clients.

Diaz, A. (2002). *The Harvard college guide to careers in public service*. Cambridge, MA: Career Services Publications.

Discussion of a wide range of careers and opportunities.

Gould, R. P. (1978). Students' experience with the termination phase of individual treatment. *Smith College Studies in Social Work, 48*(3), 235–269.

An excellent discussion of the unique nature and demands of "forced terminations."

References

Baird, B. N. (2007). *The internship, practicum, and field placement handbook: A guide for the helping professions* (5th ed.). Upper Saddle River, NJ: Prentice Hall.

Birkenmaier, J., & Berg-Weger, M. (2007). *The practical companion for social work: Integrating class and field work* (2nd ed.). Needham Heights, MA: Allyn & Bacon.

Brill, N. I., & Levine, J. (2004). *Working with people: The helping process* (8th ed.). Boston: Allyn & Bacon.

Diaz, A. (2002). *The Harvard college guide to careers in public service*. Cambridge, MA: Career Services Publications.

Fair, S. M., & Bressler, J. M. (1992). Therapist-initiated termination of psychotherapy. *The Clinical Supervisor, 10*(1), 171–189.

Faiver, C., Eisengart, S., & Colonna, R. (1994). *The counselor intern's handbook*. Belmont, CA: Brooks/Cole.

Garthwait, C. L. (2005). *The social work practicum: A guide and workbook for students.* (3rd ed.). Needham Heights, MA: Allyn & Bacon.

Gould, R. P. (1978). Students' experience with the termination phase of individual treatment. *Smith College Studies in Social Work, 48*(3), 235–269.

Haas, S. (1990). *Hearing voices: Reflections of a psychology intern*. New York: Penguin.

Herlihey, B., & Corey, G. (1992). *Dual relationships in counseling*. Alexandria, VA: American Counseling Association.

Kegan, R. (1982). *The evolving self: Problem and process in human development*. Cambridge, MA: Harvard University Press.

Kellner, L. A. (2007). "Dear Intern": Students' recommendations for transformative internships. *Human Services*, 27(1), 4–17.

King, M. A., & Sweitzer, H. F. (2002, October). Reflection as keystone: Growth and learning in service. Annual conference of the National Organization for Human Service Education, Providence, RI.

Kiser, P. M. (2008). *Getting the most from your human service internship: Learning from experience* (2nd ed.). Belmont, CA: Wadsworth.

Munson, C. E. (1993). *Clinical social work supervision* (2nd ed.). New York: Haworth Press.

Penn, L. S. (1990). When the therapist must leave: Forced termination of psychodynamic psychotherapy. *Professional Psychology: Research and Practice, 21*, 379–384.

Rogers, G., Collins, D., Barlow, C. A., & Grinnell, R. M. (2000). *Guide to the social work practicum*. Itasca, IL: F. E. Peacock.

Salisbury, W. A., & Kiner, R. T. (1996). Post termination friendship between counselors and clients. *Journal of Counseling and Development, 74*(5), 495–500.

Schafer, R. (1973). The termination of brief psychoanalytic psychotherapy. *International Journal of Psychoanalytic Psychotherapy, 11*, 135–148.

Shulman, L. (1983). *Teaching the helping skills: A field instructor's guide*. Itasca, IL: F. E. Peacock.

Stanton, T., & Ali, K. (1994). *The experienced hand: A student manual for making the most of an internship* (2nd ed.). New York: Caroll Press.

Stanziani, P. (1993). *Practicum handbook: A guide to finding, obtaining, and getting the most out of an internship in the mental health field*. Cambridge, MA: Inky Publications.

Suelzle, M., & Borzak, L. (1981). Stages of fieldwork. In L. Borzak (Ed.), *Field study: A sourcebook for experiential learning* (pp. 136–150). Beverly Hills, CA: Sage Publications.

Watzlawick, P., Weaklund, J. H., & Fisch, F. (1974). *Change: Principles of problem formation and problem resolution*. New York: Norton.

Wilson, S. J. (1981). *Field instruction: Techniques for supervisors*. New York: Free Press.

⟡ Index

Note: Page numbers followed by *n* refer
to footnotes.

abortion, 67
abstract conceptualization (AC), 10
abstract random style, 117
abstract sequential style, 117
academic integrity, 39, 275
acceptance
    by clients, 173–74
    of clients, 173
    by co-workers, 124–26
    mainstream standards and, 68
accomplishment, loss of, 220
accomplishments, sense of, 258
acknowledgement, 258
action
    balance with reflection, 193
    goals and, 237–38
active experimentation (AE), 10
active listening, 36
Adams, M., 86
advice, as support, 26
advising and evaluating, 37
affirmation, from support system, 27
agencyspeak, 17–18
aggravated negligence, 289
Albert, G., 223
alcohol, 68
all-channel network, 142
American Civil Liberties Union, 290
American Counselor Association
        (ACA), 284
American Psychological Association
        (APA), 284
analyzing and interpreting, 37
Anticipation stage. *See also* learning
        contract
    concerns and responses, 50, 94–95
    relationships in, 52
    roles and purposes, 94–95
    tasks at hand and concerns, 52, 93–96
    what ifs, 51, 92

anxiety. *See also* anticipation stage
    commonness of, 12
    treated as an intern, 29
apprenticeship level of learning, 105
assessment
    community and site risks, 183–84
    dealing with differences, 63–64
    guides for, 104
    learning contract and, 99, 103–4, 106
attitudes and values, 28–29, 179
audiotaping, 35, 118
authority
    issues, 180
    structure, 141–42
autonomy, 286
    definition of, 67
    power vs., 149–50, 165
    shame and doubt vs., 79–80
Axelson, J. A., 186

Baird, B. N., 7, 15, 47, 59, 270, 329
balanced way of living, 262
Barker, J., 61
Battistoni, R., 18
Beaumont, E., 18, 107
Behavior-Person-Environment (BPE)
        principle, 231*n*
Belenky, M. F., 85
belief, power of, 229
Bell, L. A., 86
beneficence, 286
Benner, P., 223
Berger, R. L., 156
Berg-Weger, M., 47, 59, 181, 186
Birkenmaier, J., 47, 59, 181, 186
Blanchard, K., 129
Blumenfeld, W. J., 86
Bolman, L., 133
Borzak, L., 223
Boylan, J. C., 59
brainstorm the strategies, in
        metamodel, 296
breach of duty, 289

Brill, N., 85, 244, 301, 329
Burn Out (BO), 207

Caine, Bruce Woody, 96
Callanan, P., 85, 178, 186, 301
campus resources, 39
career choice, questioning, 221
CAS Professional Standards for Internship
    Programs, (CAS), 285
CAS Professional Standards for Service-
    Learning Programs, (CAS), 285
Castenada, R., 86
chain pattern, 142
challenges and problems
    Five Cs, and, 30–31
    handling of, 197
    predictable, 197–99
    support for, 27
    unpredictable, 200
change
    change process essentials, 194
    Eight Steps to Creating Change
        metamodel, 231–38
    experiencing, 193
Chiaferi, R., 59–60
Chisholm, L.A., 59–60
civic development
    Competence stage and, 267–68
    definition of, 6
    goals, 152–54, 211
    issues with, 211
    learning, 6
    professionalism, 6–7
civic professionals
    community context and, 159–61
    Competence stage, 266–69
    definition of, 7, 159
    overview, 6
    public relevance to, 7
civil law, 290
clients, in human services work
    acceptance of, 173, 176
    acceptance by, 173–74
    assumptions and stereotypes,
        170–72, 184
    authority issues, 180
    boundary issues, 180–81
    career considerations, 81
    common grounds and empathy
        with, 176
    common pitfalls with, 321
    community context, 160

countertransference, 177
credentialing, 180
cultural profiles, 175
definition of, 4
difficult clients, 177, 210
distinction between needing and
        wanting, 81–82
diversity, 170, 177–78
emotional reactions to, 206–7
feelings of separation and loss, 319–20
future of, 320–22
good-byes to, 317–18
manipulation by, 83, 175
organizational chart and, 142
overidentification, 175
for placement, 81
post-termination relationships, 322
reactions to, 175, 210
resistance from, 174
risk assessment, 40, 182–84
safety concerns with, 161, 181–82
self-disclosure with, 178
success with, 172
transference, 175
transitioning of, 322
unfinished business with, 318–19
values, awareness and, 178–79
Clinchy, M., 85
coaching, 30–31
Cochrane, S. F., 59–60
Codes of Ethics for the Helping
        Professions, 285
Cohen, D. S., 244
Colby, A., 18, 107
Coles, R., 18
colleague. *See* coworkers
Collison, G., 18, 47
Colona, R., 59
comfort, 27
common law, 290
communication
    community context, 163
    conflict resolution, 145
    flow of, 142
    patterns, 142, 145
    response modes, 37
community context
    assets and needs, 162
    basic information, 162
    cultural symbols, 164
    human and social resources, 164
    identifying, 160–61

information sources on, 166
overview, 158–59
politics and decision making, 164
risk assessment and, 183–84
structures of, 163
community, risk assessment, 40
companionship, 27
Compassion Fatigue (CF), 207
competence
children and, 81
definition, 251
internship issues, 80–81
issues, and ethics, 39, 275
Competence stage
challenges during, 249–50, 256–57
civic professionals, 266–69
Confrontation stage and, 54
personal integrity, 255
personal relationships, 259–60
professional awareness, 254–55, 265
success and fulfillment, 257–61
supervisory relationships, 253–54
textbook knowledge, going beyond, 265–66
transformation and empowerment, 251–52
transition issues, 261–65
work quality, 255–56
concrete experience (CE), 10
concrete style, 117
conferences, 313–15
confidentiality, of journals, 32
conflict resolution, 145
Confrontation stage
belief, will, and effort, 228–31
discussion of problems, 54
Eight Steps to Creating Change metamodel, 231–38
expected roles, 54
freezing the moment and stagnation, 56
goals and purposes, 54
interpersonal issues, 54
intrapersonal factors, 54–55
overview, 228
rebuffed intern example case, 239–42
confrontations, 44
connected knowing, for helping professionals, 71
consider the causes, in metamodel, 236
consider the consequences, in metamodel, 296
consider the laws, in metamodel, 296

constitutional law, 290
consult the checklist, in metamodel, 296
consult the codes, in metamodel, 296
consult with colleagues, in metamodel, 296
contextual meaning, of skills, 196
continuity, 30–31
Corey, G., 47, 85–86, 129, 175, 178, 186, 243, 301
Corey, M. S., 47, 85–86, 129, 175, 186, 243, 301
Corrigan, R., 107
Cottone, R. R., 301
Council for Standards in Human Service Education, 104
Council for the Advancement of Standards in Higher Education (CAS), 104, 285
countertransference, 177
Cowan, M. A., 156
co-workers
acceptance by, 124–26
adjustment patterns, 126–27
cliques, 126, 147–48
Comptence stage, 253
concerns about, 204–5
definition of, 4, 110
ethical code guidelines, 284
expectations of, 123–24
identification with the organization, 127
identification with, 127
misunderstandings with, avoiding, 117–18
create the change, in metamodel, 238
credentialing, 180
criminal law, 290
crisis in confidence, 219–21
crisis in management, 263
crisis response, 298–300
the crunch, 262–64
Crutcher, R. A., 107
Culmination stage
clients, saying good-bye to, 317–23
closure of, 57
closure with supervisors, 312–17
competence issues, 310
concerns and responses, 57
expressing feelings, 311
feedback to the supervisor, 315–17
finishing the work, 311–12
learning contract, 325–26
looking forward, 324–27
opportunities, seizing, 308–9

Culmination stage (*continued*)
  overview, 57, 306–7
  placement closure, 323–25
  planning for the future, 311–12
  portfolio preparation, 325–26
  premature disengagement, 309
  relationships, saying good bye to, 307–11
  self-reflection, 313–14
  unfinished business, 57, 310
cultural identity
  attitudes toward other groups, 75–77
  culture, macroculture, and
    microculture, 74
  degree of identification, 75
  dominant vs. subordinate, 77
  notion of privilege, 77
  subgroup membership, 77–78
  subgroups, 74–75
culture, definition, 74
culture, organizational, 150

damages, 290
DART notes, 34
data gathering, from concrete
  experience, 35
Day, 286
Deal, T., 133
decide diligently, in metamodel, 297
decision-making
  agency structure and, 142–43
  politics and, 164–65
determine the goals, in metamodel,
  237–38, 296
developmental stage model. *See also*
  specific stages
  apprenticeship model, 8
  predictable stages, 12
  personal growth and, 54
  recognition of stages, 54–65
Dewey, J., 11
Dhooper, S. S., 6, 129
Diaz, A., 329
difference versus deviance, 63–64
dilemmas, ethical, 294–97
Diller, J. V., 86
discussions, 43–44
Disillusionment stage
  anticipation and actuality, 198
  assessing progress, 194–97, 209
  attitudes and values, 197
  boundary issues, 209–10
  career choice, questioning, 221

changes and, 216
civic development issues, 211–12
clients, emotional reactions to, 206–7,
  210
crisis in confidence, 219–21
crisis of growth, 53
difficulties versus problems, 53, 197–99
discussing disappointments, 199, 218
emotions of, 52, 217–18
expectations, 198
growth, 192
life situations and, 206–7
people issues, 201
placement site issues, 54n, 210–11
reactions and responses during, 199, 207
self-understanding issues, 212–14
seminar classes and, 218
sense of loss, 220
skills and, 196
supervisors and, 201, 218–19
What's wrong with my internship, 53
work issues, 200–1
dissonance, examining, 35
diversion, 27
diversity, 29
Doane, A., 86
Dolgoff, R., 302
Dreyfus, H. L., 223
Dreyfus, S. E., 223
drugs, 68
dysfunctional patterns, 70

effectiveness principle, 94
effort, power of, 231
Egan, G., 156
Eight Principles of Good Practice for
  All Experiential Learning
  Activities, 285
Eight Steps to Creating Change
  metamodel, 231–38
Eisengart, S., 59
Eisenstat, S., 288n
Elbaum, B., 18, 47
Ellis, A., 244
empowerment
  discussions versus confrontations, 43–44
  humor, role of, 44
  mindfulness, 44
  overview of, 42–45
  perspectives and, 43
  positive expectations, 42
Erikson, E., 78

Erlich, T., 18, 107
Ethical Standards of Human Service
    Professionals (NOHS/CSHSE), 285
ethics
  academic integrity, 39
  CAS Professional Standards for
    Internship Programs, 285
  CAS Professional Standards for Service-
    Learning Programs, 285
  codes and values, 280, 285–86
  *Codes of Ethics for the Helping
    Professions*, 285
  confidentiality and privacy, 281
  Council for the Advancement of
    Standards in Higher Education, 285
  decision-making models, 295–97
  definition of, 283–84, 289–90
  dilemmas, 293–97
  *Eight Principles of Good Practice for All
    Experiential Learning Activities*, 285
  Ethical Standards of Human Service
    Professionals, 285
  information disclosure, 281–82
  integrity issues, 275–76
  internship issues, 273–74
  intervention issues, 277
  legal issues, 288–93
  practice issues, 275
  principles of, 282–83, 285–87
  professional conduct, 7–8, 278–79
  professional crisis, managing, 298–300
  relationship issues, 280–81
  role issues, 274–75
  rules and standards, 283–285, 291–92
  vicarious liability, 292
euthanasia, 67
evaluation. *See also* assessment
  concerns about, 203–4
  criteria for, 143–44
  methods of, 143
  performance vs. outcomes, 144
  supervisor closure and, 313–17
  when to withhold advice, 37
expand your thinking, in metamodel,
  234–36
expectations
  of co-workers, 123–24
  Disillusionment stage and, 198
  overview, 93
  power of positive, 42–43
  reality gap, 198–99
  of supervisors, 112–14

experiential education. *See also* reflection
  active learning, 10–11
  four phase cycle of, 10
  learning strenghts, 24
  learning styles, 23
  philosophy of, 11
expressive management style, 115
Eyler, J, 18, 30–31, 47

Faiver, C., 59
family
  patterns, rules, and roles, 67, 73–74
  unresolved issues and, 82–83
Federico, R. C., 156
feedback, 36. *See also* evaluation
fidelity, 287
field instructors. *See* supervisors
field site. *See* placement site
fieldspeak, 17
Fisch, R., 245
Five Cs, reflective techniques, 30–31
flexibility, 28
focus, loss of, 220
focus your attention, in metamodel, 237
forced terminations, 317
for-profit agencies, funding, 138
Frameworks for Assessing Learning and
    Development Outcomes, 104
freedom of speech, 68

Garthwait, C. L. 107, 244
Garvin, D. S., 223
generalization, 140–141
Gibbs, L., 244
Giles, D. E., 10, 18, 30–31, 47
goals
  activities, 104–5
  Controntation stage and, 54
  goals-focused contract, 99
  resources for, 103–4
  self-determined, 260–61
Goldberger, N. R., 85
Goldenberg, G., 86
Goldenberg, I., 86
Goldstein, M. B., 301–2
Gordon, G. R., 59–60, 129, 156, 186
Gould, R. P., 329
Grambrill, E., 244
Grant, R., 60
Green, J. W., 86
Griffin, P., 86
Griffin, M., 59–60

Grobman, L. M., 59, 244
Grossman, B., 60
growth, 11
grunt work, 96

Haavind, S., 18, 47
Hackman, H. W., 86
Hage, H. H., 59–60, 129, 156, 186
Hahn, T. N., 44
Hanley, M. M., 59–60
Harper, R. A., 244
Harrington, D., 302
Health and Human Services
    Privacy Rule, 282
Hersey, P., 129
Hesser, G., 19
Homan, M. S., 167, 244
honesty, 67
Howard, J., 107
human experience, internship as, 8
human resources, 145
human service educators, 169
human services work. *See also* clients, in
    human services work
  community context, 159
  helping and service professionals, 78, 83
  legal guidelines, 290
  psychosocial identity issues, 78
  unfinished business and, 83
humor, role of, 44
hygiene, 68

identify the strategies, in metamodel, 238
imposter syndrome, 94
impulses, 79–80
information disclosure, 281–82
Inkster, R. P., 18–19, 60, 104, 107
instructors. *See* supervisors
instrumental management style, 115
Integrated Processing Model, 34
integrity issues, 275–76
intentional torts, 289
interdisciplinary integration and problem
    solving model, 231n
interns
  common issues, 39, 41
  concerns about role of, 29
  definition of, 4
  ethical obligations, 292–93
  how one learns, 11
  peers and, 15, 205
  rights of, 38, 293
  self-understanding issues, 213

internships. *See also* specific stages
  campus liability, 182
  definition of, 3, 5
  Disillusionment stage, 215–16
  five categories of, 32
  human side of, 192–93
  informed commitment, 96
  as a learning experience, 8
  predictable stages, 12
  purpose of, 5
  theoretical knowledge and, 72
intervention issues, 277
Issues and Ethics in the Helping
    Professions, 178

Jackson, L., 223
Jackson, R., 19, 270
Jennings, M., 61
Johnson, D. E., 129
Johnson, D. W., 47, 243
journals
  assessment tools and, 106
  critical incident, 34
  DART notes, 34
  double entry, 33–34
  Integrated Processing Model,
    34–35
  key phrase, 33
  portfolios, 35
  processing techniques, 34
  as a reflection tool, 31–32
  SOAP notes, 34
  Three Column Processing, 34
  unstructured, 32–33
  Web-based technology and, 32
  what to include in, 33
justice, 286

Kabat-Zinn, J., 47
Kahn, M., 14
Kegan, R., 63, 85
Kenyon, P., 302
Kerson, T., 61
King, M. A., 60
Kiser, P. M., 34–35, 47, 60–61
knowledge
  identifying relevant data, 35
  Integrative Processing Model and, 35
  internship expectations, 38
  procedural knowledge, 71n
  separate and connected knowing,
    30–31, 70–71
  theoretical knowledge, 72

theory and practice integration, 195
  understanding people, 71
Kolb, D., 8, 10, 63
Kolb's learning cycle theory, 10–12,
  234–36
Kotter, J. P., 244
Kretzmann, J. P., 167

Lacoursiere, R., 12, 60
Lamb, D., 61
language, cultural framework of, 151
lateral control, 144
law, and ethical issues, 283, 288–93
learning contract
  activities, 104–5, 107
  assessment, 99, 106–7
  attitude and value goals, 100
  civic development goals, 100–1
  development of, 98, 195–96
  disillusionment stage and, 195
  fundamentals of, 99
  goals statement, 99–101
  goals-focused vs. objectives-focused
    approach, 99–100
  involvment with, 98
  knowledge goals, 100, 195
  learning objectives statement,
    99–103
  levels of learning, 105
  list of verbs, for writing103–4
  outcome statements, 103
  overview, 97–98
  personal development goals,
    100–1
  placement or internship agreement,
    97n
  professional development goals,
    100–1
  skill goals, 100
  timing of, 98
  writing tools, 102–3
learning strenghts, 24
learning styles, 23, 69–70
learning, articulation of, 35
Levine, J., 60, 85, 244, 301, 329
Levine, K., 231
liability insurance, 39
life context
  academics, 23
  anxieties about, 24
  employment, 24
  family, 25
  Five Cs, 30–31

friends, 25
  intimate partners, 25
  other commitments, 25
  roomates, 24
  self-awareness and, 214
  yourself, 26
listening, active, 26
Loewenberg, F., 302
Lum, D., 86

MacCarthy, B., 60
Malley, P. B., 59
malpractice, 289–90
management styles, 148
mastery level of learning, 105
McBride, R. B., 59–60, 129, 156, 186
McClam, T., 243
McEnroe, T., 302
McKenzie, R. H., 19
McKnight, J. L., 167
meaning, loss of, 220
meaning-making system, 62–63, 213
Meara, N. M., 286
mentoring relationship, 113, 253–54
Michelsen, R., 61
Michigan Journal of Community Service
    Learning, 104
Milnes, J. A., 60
Mind Styles theory, 117
mindfulness, 44
minority groups, 77
moral behavior, six fundamental principles
    of, 296–97
moral, definition of, 283
morale, 50
multi-tasking, 45

name the problem, in metamodel,
    233–34, 296
narrow the focus, in metamodel, 296
National Society for Experiential
    Education (NSEE), 11, 14, 104
needs assessment, 78–79
negligence, 289
negligence torts, 289
nonmaleficence, 287
NSEE Foundations Document
    Committee, 19

O'Brien, P., 107
objective-based approach, 103
objectives-focused contract, 99
observability, 96

on-site resources, 39
open-mindedness, 28
operational philosophy, 211
ordinary negligence, 289
organizational chart, 139–140
organizational meaning, of tasks, 96
organizational politics, 148–49
organizational structure, 139
organizational theory, 133–34
organizations, professional, 104
orientation level of learning, 105
overidentification, 175

the paranoid "they", 76–77
Parilla, P., 19
pattern recognition, of families, 73–74
Payne, R. K., 86
Pedersen, P., 186
people, understanding, 71. *See also* clients
perfectionism, 80
personal development
  definition of, 5–6
  overview, 5–6
personal information. *See* self-disclosure
personal power. *See* empowerment
personal safety, 40–42, 181–84
Peters, M. L., 86
placement site
  civic development, 152–54
  cliques, 147–48
  closure, 323–25
  culture of, 150–51
  definition of, 4
  Disillusionment stage issues, 211
  division of responsiblilites and tasks, 139
  expectations and requirements, 38
  external environment, 151–52
  formal roles, 139–40
  funding, 137–39
  goals and objectives, 135
  informal roles, 147
  lateral coordination and control, 144
  mission statements, 134–35
  norms, 145–47
  organizational politics, 148–49
  organizational values, 210–11
  organizations, 133
  overview, 37–38
  philosophy of, 211
  policies and procedures, 143

resources, 39
risk assessment, 40, 183–84
staff development, 148
structure of, 139
systems theory, 133
value statements, 135–37
work coordination and control, 141
planning
  development, 35
  finishing, 80
portfolio development, 34, 325–26
power
  authority vs., 149–50
  of communities, 165
Practical Reasoning, 34, 159
practice issues, 275
praise, as support, 27
prejudice, 75
principle ethics, 286
privacy rule, 282
private nonprofit agencies, funding, 138
privilege, notion of, 77
privileged communication, 281
problem-solving models, 232
procedural knowledge, 71n
process concerns, 111
process recording, 34
processing techniques, for journals, 34
professional conduct issues, 278–82
professional development, definition, 5–6
professional goals, 104
professional issues, 289–300
professional myopia, 64
professional organizations, 104
psychosocial identity issues
  autonomy vs. shame and doubt, 79–80
  children and, 79–81
  competence, 80–81
  industry vs. inferiority, 80–81
  initiative vs. guilt, 80
  trust vs. mistrust, 78–79
public agencies, funding, 138
purpose, loss of, 220

Queralt, M., 156
questioning and probing, 37

Reaching Out, Johnson, D., 37
reaction patterns, 68–69
receptiveness, 28
record keeping, 30

reflection. *See also* self-understanding.
  definition of, 17
  Five Cs, of, 30–31
  journals and, 3
  purpose of, 30
reflective observation (RO), 10
Reframing Organizations, 133
Reilly, E. P., 59
relationships. *See also* support systems
  in Anticipation stage, 93–94
  with clients, 173–76, 179, 209
  co-workers reactions, 125, 204
  developing of, 93–94
  end of, 309–10, 316–17
  intimate partners as life-context, 25
  as learning opportunities, 9
  mentoring, 253–54
  professional issues, 277, 280
  responsible, 259
  self-disclosure and, 178
  self-understanding and, 212–13
  seminars and, 13
  with supervisor, 111–13, 115–16, 202,
    253
  between theory and practice, 7
  triadic, 122
religion, 67
resolution of dilemmas, 295–97
resource issues, 298–99
responsibilities, Competence stage and,
    261–62
responsible relationships, 98
rewards, intrinsic, 259
Richardson, N. F., 302
risk assessment. *See* personal safety
role issues, 274–75
Rompf, E. L., 6, 129
Ronnestadt, M. H., 61, 223
Rosenthal, R., 223
Ross, R. G., 18–19, 60, 104, 107
Rothenberg, P. S., 86
Rothman, J. C., 103, 302
Royse, D., 6, 129
Rushton, S. P., 61
Russo, F. X., 19
Russo, J. R., 126–27, 129, 156, 169

say it out loud, in metamodel, 233
Schmidt, 286
Schneider, C. G., 107
Schon, D. A., 19
Schutz, W., 12, 60, 129, 186

self-actualization, 259
self-assessment guides, 104
self-defined success, 260
self-determination, 259
self-disclosure, 67, 178. *See also*
    self-understanding
self-knowledge, 65
self-understanding. *See also* anxiety;
    reflection
  Barn Raising, 14–15
  The Beauty Context, 14
  career considerations, 81–82
  crises situations, 300
  Disillusionment stage, 214
  The Distinguished House Tour, 14–15
  The Free for All, 14
  issues, 213
  meaning-making system, 63–64
  motivations for placement, 81–82
  projection and, 65–66
  relationships and, 65
  value awareness, 66–68
seminars
  overview, 13–18
  theoretical analysis and problem
    solving, 72
  types of, 14
Senge, P. M., 156
sequential style, 117
Service-Learning Field, 104
Serving and Surviving as a sexuality, 67
Shafir, R., 45, 47, 85
Shearer, P., 60
Shepard, H. A., 245
shifting, with tasks, 45
Shulman, L., 186
Siporin, M., 61
site supervisors. *See* supervisors
site. *See* placement site
skills
  acquisition of, 196
  attitudes and values, 197
  communication, 36–37
  contextual meaning, 196
  Disillusionment stage and, 196
  expansion of, 196
  Integrative Processing Model, 34–35
  reflection and, 29–30,
  Three Column Processing, 34
Skovholt, T. M., 61, 223
SOAP notes, 34
social contract, in the private sector, 212

sociopolitical environment, 152
specialization, 140–41
stages of an internship. *See* developmental
    stage model; specific stages
standards of care, 291
standards of ethics, 283
Steinman, S. O., 302
Stephens, J., 18, 107
Steps to Creating Change metamodel,
    233–23
stereotypes, 75–76, 184
strategies, 104
strict liability torts, 289
subgroups
    dominant vs. subordinate, 77–78
    identification with, 75
    membership in, 74
    minorities, 77–78
    prejudices and stereotypes, 75–76, 184
    unfinished business, 83
success, in Anticipation stage, 94
Sue, D. W., 186
Sullivan, W. M., 7, 19
supervision and ethics, 275
supervision issues, 39
supervisors
    active and conscious involvement of
        intern, 98
    anxieties about, 111–12, 201
    approaches to, 118
    closure with, 312–17
    Comptence stage, 253–54
    conference sytles, 119
    definition of, 4
    direction and support, 113, 116
    Disillusionment stage and, 201–2,
        218–19
    evaluation process, 118, 120–22, 113–44
    expectations for, 112–13, 117
    feedback and evaluation, 122–23, 203,
        313, 315
    idealizing of, 201
    informed participants, 112
    journal confidentiality, 32
    legal issues, 289
    letters of recommendation, 316
    maturity and, 116
    meeting structure, 119
    mentoring relationship, 113, 253–54
    Mind Styles theory, 117
    misunderstandings, avoiding, 117–18

observation in real-time, 118
peer-reports, 119–20
personal and professional viewpoint
    of, 114
process concerns, 111
reactions to, 121
relationship management with,
    112, 122
role models, 113
roles of, 122
self-reports, 119
source of wisdom, 266
sponsor, advocate, or broker roles, 113
style of supervisor, 115, 117
supervision styles, reactions to, 202
supervisor-supervisee relationship, 114
supervisory malpractice, 291–92
task concerns, 111
vicarious liability, 292
supervisory malpractice, 291–92
support systems
    advice, as personal consultants, 26
    affirmation, as comrades, 27
    challenge, as personal coaches 27
    comfort, as chicken soup people, 27
    companionship, as buddies, 27
    Confrontation stage and, 54
    diversion, as playmates, 27
    groups, 15
    listening, as sounding boards, 26
    praise, as fans, 27
    types of, 26–28
Sweitzer, H. F., 60, 86, 244

tape-recorded journals, 32
Tarule, J. M., 85
Tarvydas, V. M., 301
task accomplishment, 50–51, 96
task concerns, 111
task environment, 151–52
task forces, 144
teacher, definition, 113
teams, 141
termination, 317
The Internship as Partnership, 104
The Zen of Listening: Mindful
    Communication in the Age of
    Distraction, 45
theories
    Integrative Processing Model, 35
    modes of responding, 37

theoretical orientation, 115
three-column processing, 34, 234
time management
    crisis in management, 263
    the crunch, 262–63
    8-stepping it, 264
    living a balanced life, 262
    organizing for, 264
time, 68
Tinker, R., 18, 47
torts, 289
transference, 175
Trapido-Lurie, B., 61
Tuckman, 12

unfinished business, 82–83
Universal Precautions, 42
unresolved issues, 82–83

values
    clients and, 179
    definition of, 283
    managing differences and, 178
veracity, 287
verbatims, in pastoral counseling, 34
vertical coordination and control, 141
vicarious liability, 292
Vicarious Traumatization (VT), 207–8
videotaping, 35, 118
virtue ethics, 286
volunteers, 9

Wadsworth Group, 302
Wagner, J., 194
Watzlawick, P., 245
Weaklund, J. H., 245
Web-based technology, 32
Weinstein, G., 34, 70, 85, 244
Wentz, E. A., 61
wheel pattern, 142
will, power of, 229–31
Willis, G., 19
Wilson, S. J., 129
women's development, 70
Women's Ways of Knowing, 70
Woodside, M., 243
Work
    Competence stage and, 257–58
    Culmination stage, 312
    grunt work, 96
    issues with, 200
    needs, balancing of, 200–1
    people issues, 201
    public relevance of, 267
    theoretical orientation, 115
    values and, 67
Working With People, 76

Yarris, E., 61

Zlotkowski, Edward, 104
Zubizarreta, J., 47
Zuniga, X., 86